LIBRARY-LRC
TEXAS HEART INSTITUTE

Coronary Imaging and Physiology

Myeong-Ki Hong
Editor

Coronary Imaging and Physiology

Editor
Myeong-Ki Hong
Division of Cardiology
Severance Cardiovascular Hospital
Seoul
South Korea

ISBN 978-981-10-2786-4   ISBN 978-981-10-2787-1 (eBook)
https://doi.org/10.1007/978-981-10-2787-1

Library of Congress Control Number: 2017957982

© Springer Nature Singapore Pte Ltd. 2018
This work is subject to copyright. All rights are reserved by the Publisher, whether the whole or part of the material is concerned, specifically the rights of translation, reprinting, reuse of illustrations, recitation, broadcasting, reproduction on microfilms or in any other physical way, and transmission or information storage and retrieval, electronic adaptation, computer software, or by similar or dissimilar methodology now known or hereafter developed.
The use of general descriptive names, registered names, trademarks, service marks, etc. in this publication does not imply, even in the absence of a specific statement, that such names are exempt from the relevant protective laws and regulations and therefore free for general use.
The publisher, the authors and the editors are safe to assume that the advice and information in this book are believed to be true and accurate at the date of publication. Neither the publisher nor the authors or the editors give a warranty, express or implied, with respect to the material contained herein or for any errors or omissions that may have been made. The publisher remains neutral with regard to jurisdictional claims in published maps and institutional affiliations.

Printed on acid-free paper

This Springer imprint is published by Springer Nature
The registered company is Springer Nature Singapore Pte Ltd.
The registered company address is: 152 Beach Road, #21-01/04 Gateway East, Singapore 189721, Singapore

Preface

Interventional cardiology is a very exciting and fast-developing area in modern medicine. Since percutaneous coronary angioplasty was introduced in 1977, the inventions of novel devices, such as the balloon catheter, bare metal stent, and drug-eluting stent, have steadily improved clinical outcomes of percutaneous coronary intervention. These advances were undoubtedly based on insights derived from intracoronary imaging or physiologic evaluations.

Intravascular ultrasound is the "gold standard" among intravascular imaging modalities and provides various information about lesional characteristics and interventional therapy. Optical coherence tomography enables visualization of intravascular morphologies clearly based on high resolution. The assessment of fractional flow reserve, as known, guides whether the stenotic lesion needs revascularization. Because these examinations have their own advantages and disadvantages, it is important to know their characteristics and applications. The comprehensive understanding of intravascular imaging and physiology eventually might help to treat patients with coronary artery diseases in daily practice.

It is my honor to provide a state-of-the-art update on the most relevant topics of *coronary imaging and physiology* written by an expert group of Imaging and Physiology on Patients with Cardiovascular Disease (IPOP) in Korea. I appreciate the authors' dedication to this work despite their busy practices. I hope that this book helps clinicians to provide the optimal treatment for patients with coronary artery diseases.

Seoul, South Korea   Myeong-Ki Hong, MD

Introduction: Coronary Anatomy and Circulation

Coronary Anatomy

The coronary artery is the first branch of the aorta and is divided into the left and right coronary arteries. The left main coronary artery is derived from the left coronary cusp and is divided into the left anterior descending artery (LAD) and left circumflex artery (LCX). The LAD is located in the anterior interventricular groove and supplies the anterior wall, septum, and apex. The branches of the LAD are septal perforating arteries and diagonal branches. The septal perforating arteries supply most of the septum, and the diagonal branches supply the lateral wall of the left ventricle. The LCX passes through the atrioventricular groove and supplies the left atrium, as well as most of the lateral and posterior walls of the left ventricle. The branches of the LCX are obtuse marginal branches, and approximately 30–40% of the sinoatrial nodal branch is derived from the LCX [1, 2].

The right coronary artery (RCA) is derived from the right coronary cusp; it runs along the right atriventricular groove and continues to the posterior interventricular sulcus. The RCA supplies the right atrium, right ventricle, sinoatrial node, and atrioventricular node via several branches (conus, right ventricular wall, sinoatrial nodal, atrioventricular nodal branch). At the distal portion of the RCA (i.e., the crux), it divides into two branches: the posterolateral and posterior descending arteries, which supply the inferior portion of the interventricular septum and apex. In more than 80% of cases, the RCA has posterior descending and posterolateral branches. The others are left-dominant systems, in which the LCX gives rise to posterolateral and posterior descending branches, or codominant systems, in which both arteries provide an equal supply (Fig. 1).

The incidence of coronary anomaly is approximately 1%. Common anomalies are separate origin of the LAD and LCX (0.4%), high takeoff (0.25%), single coronary artery (atresia), origin from opposite coronary sinus, and anomalous termination (fistula) [1, 2]. Myocardial bridge is a specific congenital condition in which the epicardial coronary artery travels the intramuscular course, usually in the middle portion of the LAD. Approximately 5–80% of autopsy, 25% of CT scan, and 0.15–25% of cases were detected during coronary angiography as systolic compression of the coronary artery. The myocardial bridge is usually benign, but sometimes it causes chest pain, acute coronary syndrome, left ventricular dysfunction, and arrhythmias [3, 4].

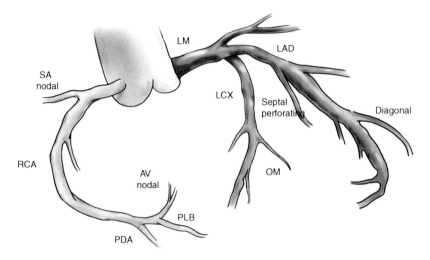

Fig. 1 Anatomy of coronary artery. *LM* left main, *LAD* left anterior descending artery, *LCX* left circumflex artery, *OM* obtuse marginal, *RCA* right coronary artery, *SA nodal* sinoatrial nodal, *AV* atrioventricular, *PDA* posterior descending, *PLB* posterolateral branch

Coronary Circulation

Coronary blood flow is a phasic pattern; main arterial flow in the coronary artery occurs in diastole. During systole, contraction of the myocardium compresses the coronary microvessels, impedes arterial blood flow, and increases venous outflow. Coronary flow is determined by myocardial demand and blood supply. Major determinants of myocardial blood flow are heart rate, myocardial contractility, and myocardial wall stress (preload, afterload). Because coronary blood flow passes from the epicardium to the endocardium, the subendocardial area is susceptible myocardial ischemia. The pressure difference between the epicardial coronary artery and the left ventricle is important to maintain myocardial perfusion. The "potential" for coronary flow to the subendocardium is the difference between diastolic aortic and left ventricular pressures multiplied by the diastolic period. A low aortic pressure or a brief diastolic period (tachycardia) may compromise subendocardial blood flow [5, 6].

The epicardial coronary artery is the conduit to transfer blood to the arteriole, capillary, and myocardium and consists of less than 10% of coronary resistance unless severe stenosis develops. The precapillary arteriole (100–500 μm) connect epicardial conduit to myocardial capillaries; it covers less than 30% of coronary resistance. In a normal state, it gives little contribution to resistance. Distal precapillary arteriolar vessels (<100 μm) are mainly responsible for resistance and flow.

Regulation of Coronary Blood Flow

Coronary blood flow is reasonably constant despite changes in coronary artery pressure to keep myocardial perfusion, although blood pressure changed within certain range, usually between 40–150 mmHg. Below the

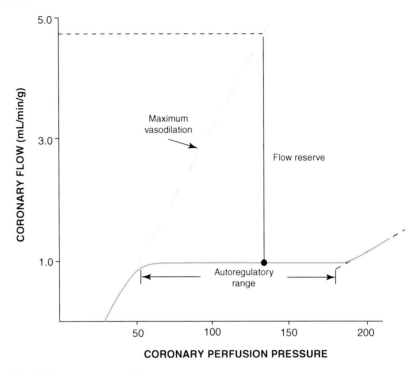

Fig. 2 Coronary autoregulation

autoregulatory range (approximately 60 mmHg), flow is strongly pressure-dependent. Vasodilator reserve is the increase in flow between the prevailing flow and a specified "maximum" vasodilator stimulus. Below the autoregulatory range, vasodilator reserve is exhausted. In a normal coronary artery, blood flow of maximally dilated coronary increases fourfold to sixfold of resting state [7–9] (Fig. 2).

Endothelial-Dependent Regulation

Endothelium-dependent regulation is mediated by nitic oxide (NO). NO is made by NO synthase in endothelial cell. It diffuses into smooth muscle in media, which in turn vasodilate by decreasing intracellular Ca^{++}. Shear stress and paracrine mediators (endothelial-dependent hyperpolarizing factor, endothelin) can influence endothelial function via NO. In a normal coronary artery, acetylcholine dilates coronary artery via increasing NO; however, in case of endothelial denudation, acetylcholine causes vasoconstriction due to decreased NO production [5, 10].

Myogenic Regulation

Myogenic regulation is controlled by coronary smooth muscle, which can change coronary vessel diameter in response to pressure. In normal conditions, smooth muscle of the coronary artery maintains vessel diameter below maximal vasodilation level. According to Laplace law, to decrease wall tension, resistance is inversely related with pressure (Laplace law). If coronary artery pressure increase, it influences smooth tone and results in vasoconstriction via increasing resistance to decrease wall stress. Myogenic regulation is primarily observed in the arteriole (<100 μm) [11].

Metabolic Regulation

Adenosine mainly dilates small coronary arterioles by binding A2 receptor on vascular smooth muscle. It increases cAMP followed by increasing intracellular Ca^{++} mainly small arteriole. Endothelin and hypoxia cause vasoconstriction [5].

Neural Regulation

Increased sympathetic tone stimulates beta-2 receptor followed by coronary vasodilation; however, alpha-1 stimulation leads to vasoconstriction. Although flow-mediated vasodilation is the main mechanism after sympathetic activation in a normal artery, alpha-1-mediated vasoconstriction is developed in case of impaired NO-mediated vasodilation. For cholinergic nervous system, acetylcholine dilates the coronary artery via NO-mediated vasodilation [12, 13].

Extravascular Compression

During systole, coronary blood flow is limited due to the effect of increased resistance as a consequence of coronary artery compression and higher left ventricular pressure than coronary pressure due to myocardial contraction.

Reference Values of Normal Coronary Flow Measurements in Clinical Setting

The characterization of normal coronary blood flow dynamics could provide crucial guidelines for the physiologic assessment of diseased coronary artery. Spectral flow velocity parameters, including average peak velocity (APV), average diastolic peak velocity (ADPV), average systolic peak velocity (ASPV), and diastolic-to-systolic velocity ratio (DSVR), were measured using Doppler wire at baseline and intracoronary adenosine-induced maximal hyperemic state. Coronary flow reserve (CFR) was calculated from the ratio of hyperemia to baseline APV [14–16] (Figs. 3 and 4).

Summary of characteristics of normal coronary flow patterns are as follows:

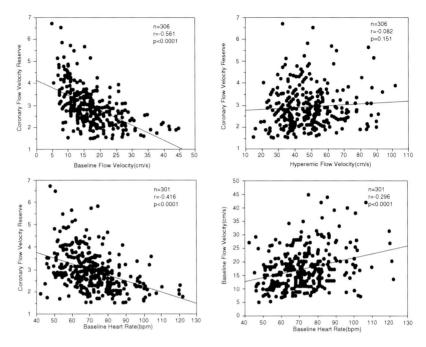

Fig. 3 Correlations between coronary flow velocity and coronary flow reserve (**a**), and between baseline heart rate and coronary flow reserve or baseline coronary flow (**b**). (**a**) CFR had no significant correlation with hyperemic flow velocity, and showed significant inverse correlation with baseline flow velocity. (**b**) Baseline heart rate significantly correlated with baseline flow velocity

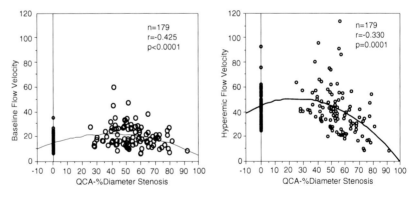

Fig. 4 Correlations between coronary flow and stenosis severity on quantitative coronary angiography at baseline and hyperemia

1. Intracoronary flow velocity was relatively well preserved from proximal to distal segment, especially in the left coronary system (tapered branching model).
2. CFR was preserved from the proximal to distal segments.
3. There was a significant difference in CFR between left and right arteries. CFR of the right artery is significantly higher.

4. CFR has a wide range of individual variation: from 1.6 to 6.7. Incidence of low CFR (<2.0) was 13%.
5. CFR was adversely affected by the level of baseline flow and heart rate at the time of measurement rather than the level of hyperemic flow.
6. In physiologic evaluation of diseased coronary arteries in a real clinical setting, significant regional differences of coronary flow patterns and factors affecting flow pattern, especially baseline hemodynamic status of patients, should be considered.

Coronary Blood Flow Under Coronary Stenosis

When a given coronary artery stenosis is present, pressure drop across a stenosis is influenced by viscous loss and post-stenosis flow separation. According to Poiseuille's Law, ΔP (pressure difference) is inversely related to radius and positively related to stenosis length and flow, so viscous losses are related to stenosis diameter and length. For separation losses, pressure gradient is related to flow, and the relationship is nonlinear.

Under normal coronary autoregulation, coronary blood flow is maintained despite presence of stenosis. In the relationship between pressure drop across a stenosis and coronary blood flow, the pressure drop is that which might be seen across an 80–85% diameter stenosis of a coronary artery. If the aortic pressure is 100 mmHg and the flow is 1.0 ml/min/g myocardium, then the pressure distal to the stenosis will be below the lower limit of autoregulation (approximately 60 mmHg). The patient will probably experience angina, even though flow is greater than an initial resting value of approximately 0.5 ml/min/g myocardium. The pressure-flow relationship curve showed that the pressure drop across the lesion was more prominent as the degree of stenosis was more severe. Because of the nonlinear resistance characteristics of stenoses, the critical narrowing is approximately 80–85% at resting flows but approximately 45% during hyperemia. To prevent myocardial ischemia, the coronary microvasculature dilates to decrease pressure difference across a lesion [14–17].

References

1. Villa AD, Sammut E, Nair A, Rajani R, Bonamini R, Chiribiri A. Coronary artery anomalies overview: the normal and the abnormal. World J Radiol. 2016;8(6):537–55.
2. Yi B, Sun JS, Yang HM, Kang DK. Coronary artery anomalies: assessment with electrocardiography-gated multidetector-row CT at a single center in Korea. J Korean Soc Radiol. 2015;72(4):207–16.
3. Kramer JR, Kitazume H, Prudfit WL. Clinical significance of isolated coronary bridges: benign and frequent condition involving the left anterior descending artery. Am Heart J. 1982;103:283–88.
4. Lee MS, Chen CH. Myocardial bridging: an up-to-date review. J Invasive Cardiol. 2015;27:521–8.
5. Libby P, Bonow RO, Mann DL, Zipes DP. Bruanwald's heart disease: a textbook of cardiovascular medicine. 8th ed. Philadelphia: W.B. Saunders Company; 2008. p. 1167–78.

6. King SB III, Yeung AC. Interventional cardiology. New York: McGraw Hill Medical; 2007. p. 31–62.
7. Kanatsuka H, Lamping KG, Eastham CL, Marcus ML. Heterogeneous changes in epimyocardial microvascular size during graded coronary stenosis. Evidence of the microvascular site for autoregulation. Circ Res. 1990;66:389–96.
8. Kuo L, Chilian WM, Davis MJ. Coronary arteriolar myogenic response is independent of endothelium. Circ Res. 1990;66:860–6.
9. Kuo L, Chilian WM, Davis MJ. Interaction of pressure- and flow-induced responses in porcine coronary resistance vessels. Am J Physiol. 1991;261:H1706–15.
10. Rajendran P, Rengarajan T, Thangavel J, Nishigaki Y, Sakthisekaran D, Sethi G, Nishigaki I. The vascular endothelium and human diseases. Int J Biol Sci. 2013;9:1057–69.
11. Miller FJ Jr, Dellsperger KC, Gutterman DD. Myogenic constriction of human coronary arterioles. Am J Physiol. 1997;273:H257–64.
12. Feigl EO. Coronary physiology. Physiol Rev. 1983;63(1):1–205.
13. Heusch G, Baumgart D, Camici P, Chilian W, Gregorini L, Hess O, Indolfi C, Rimoldi O. Alpha-adrenergic coronary vasoconstriction and myocardial ischemia in humans. Circulation. 2000;101(6):689–94.
14. Tahk SJ, Li YZ, Koh JH, Yoon MH, Choi SY, Cho YH, Lian ZX, Shin JH, Kim HS, Choi BI. Assessment of coronary artery stenosis with intracoronary Doppler guide wire and modified continuity equation method; a comparison with dipyridamole stress Thallium-201 SPECT. Korean Circ J. 1999;29(2):161–73.
15. Tahk SJ, Kim W, Shen JS, Shin JH, Kim HS, Choi BI. Regional differences of coronary blood flow dynamics in angiographically normal coronary artery. Korean Circ J. 1996;26(5):968–77.
16. Tahk SJ. Clinical application of intracoronary flow measurements in coronary artery disease. In: Angioplasty summit: An update on coronary intervention; 2007. p. 56–74.
17. Klocke FJ. Measurements of coronary blood flow and degree of stenosis: current clinical implications and continuing uncertainties. J Am Coll Cardiol. 1983;1(1):31–41.

Suwon, South Korea   Seung-Jea Tahk, MD, PhD

Contents

Part I IVUS

1. **Physical Principles and Equipment: IVUS** 3
   Taek-Geun Kwon, Young Jun Cho, and Jang-Ho Bae

2. **IVUS Artifacts and Image Control** 9
   Hyung-Bok Park, Yun-Hyeong Cho, and Deok-Kyu Cho

3. **Quantitative Measurements of Native Lesion** 19
   Mayank Goyal, Hoyoun Won, and Sang Wook Kim

4. **Qualitative Assessment of Native Lesion** 27
   Young Joon Hong

5. **Clinical Evidence of Intravascular Ultrasound-Guided Percutaneous Coronary Intervention** 37
   Sung-Jin Hong, Yangsoo Jang, and Byeong-Keuk Kim

6. **Pre-Percutaneous Coronary Intervention Lesion Assessment** 49
   Sung Yun Lee

7. **IVUS: Post-Evaluation After Stenting** 63
   Yun-Kyeong Cho and Seung-Ho Hur

8. **Long-Term Complications and Bioresorbable Vascular Scaffolds Evaluation** 75
   Kyeong Ho Yun

9. **Near-Infrared Spectroscopy** 85
   Byoung-joo Choi

Part II OCT

10. **Physical Principles and Equipment of Intravascular Optical Coherence Tomography** 97
    Jinyong Ha

11. **Image Acquisition Techniques** 107
    Ki-Seok Kim

12	**Interpretation of Optical Coherence Tomography: Quantitative Measurement**........................... So-Yeon Choi	115
13	**Qualitative Assessments of Optical Coherence Tomography** .. Ae-Young Her and Yong Hoon Kim	125
14	**Clinical Evidence of Optical Coherence Tomography-Guided Percutaneous Coronary Intervention** .... Seung-Yul Lee, Yangsoo Jang, and Myeong-Ki Hong	133
15	**Pre-interventional Lesion Assessment**................... Hyuck-Jun Yoon	143
16	**Immediate Post-Stent Evaluation with Optical Coherence Tomography**... Seung-Yul Lee, Yangsoo Jang, and Myeong-Ki Hong	155
17	**Late Stent Evaluation (Neoatherosclerosis)** .... Jung-Hee Lee, Yangsoo Jang, and Jung-Sun Kim	165
18	**Bioresorbable Vascular Scaffold Evaluation by Optical Coherence Tomography** .... Soo-Joong Kim	177
19	**Novel Application of OCT in Clinical Practice** .... Sunwon Kim and Jin Won Kim	189

Part III Physiology

20	**Concept of Invasive Coronary Physiology: Focus on FFR**..................................... Bon-Kwon Koo and Joo Myung Lee	203
21	**Setup for Fractional Flow Reserve and Hyperemia** .... Ho-Jun Jang and Sung Gyun Ahn	213
22	**Validation of Fractional Flow Reserve** .... Sung Eun Kim and Jung-Won Suh	223
23	**Practical Learning in Coronary Pressure Measurement** ..... Jin-Sin Koh and Chang-Wook Nam	233
24	**Other Physiologic Indices for Epicardial Stenosis**........... Hong-Seok Lim and Hyoung-Mo Yang	241
25	**Comparison Between Anatomic and Physiologic Indices** ..... Eun-Seok Shin	249
26	**Fractional Flow Reserve in Intermediate or Ambiguous Lesion**.................................. Bong-Ki Lee	259
27	**FFR in Complex Lesions** .... Hyun-Jong Lee and Joon-Hyung Doh	269

28	**FFR in Acute Coronary Syndrome and Noncoronary Disease**	279
	Jang Hoon Lee and Dong-Hyun Choi	
29	**Fractional Flow Reserve in Specific Lesion Subsets**	293
	Hyun-Hee Choi and Sang Yeub Lee	
30	**Invasive Assessment for Microcirculation**	303
	Kyungil Park and Myeong-Ho Yoon	
31	**Non-invasive Assessment of Myocardial Ischemia**	311
	Jin-Ho Choi, Ki-Hyun Jeon, and Hyung-Yoon Kim	

Index 327

Part I
IVUS

Physical Principles and Equipment: IVUS

Taek-Geun Kwon, Young Jun Cho, and Jang-Ho Bae

1.1 Physical Principles

Images made from ultrasound are based on the transmission and reception of sound waves reflected from tissue. Ultrasound transducers use a piezoelectric crystal (usually a ceramic) to generate and receive ultrasound waves. The piezoelectric crystal material has the property of expanding its crystal size through electrical current. When an alternating electric current is applied, the crystal alternately compressed and expands, generating an ultrasound wave [1]. The frequency of sound wave depends on the nature and thickness of the piezoelectric material. When reflected ultrasound waves return to the transducer, an electric current is generated and converted into the image.

The ultrasound beam remains parallel for a short distance (near field) and then diverges (far field), like the light from a flashlight. The beam shape and size depend on transducer frequency, distance from the transducer, and aperture size and shape (Fig. 1.1). The beam shape affects measurement accuracy and contributes to imaging artifacts. The image quality is better in the near field because the beam is narrower and more parallel and has a greater resolution. In addition, the characteristic backscatter from a given tissue is more accurate. The length of the near field (L) depends on the diameter (D) of the transducer and wavelength (λ): $L = D^2/4\lambda$ [2]. Therefore, transducers with lower frequencies are used for examination of large vessels to extend the near field. Image resolution depends on the wavelength and penetrating power of ultrasound wave transmitted by the transducer. Shorter wavelengths penetrate a shorter distance than longer wavelengths. The wavelength for any transducer frequency can be calculated as λ (mm) = $1.54/f$, where 1.54 is the propagation velocity of soundwave in the heart and f is frequency. A 40 MHz transducer has higher image resolution and lower penetrating depth than a 20 MHz transducer (Fig. 1.2).

Image quality can be partially described by spatial resolution and contrast resolution. The spatial resolution is the capacity to differentiate two objects within the ultrasound image and has two principle directions: axial (parallel to the beam) and lateral (perpendicular to both the beam and the catheter). The axial resolution is the ability that the ultrasound technique has to separate the spatial position of two consecutive scatters through its corresponding echoes. The axial resolution (d_r) depends on ultrasound speed (c) and pulse duration (d_t) and is calculated as $d_r = cd_t = cT = c/f = 1.54/f$, where d_t is

T.-G. Kwon · J.-H. Bae (✉)
Department of Cardiology, Konyang University Hospital, Daejeon City, South Korea
e-mail: janghobae@yahoo.co.kr

Y.J. Cho
Department of Radiology, Konyang University Hospital, Daejeon City, South Korea

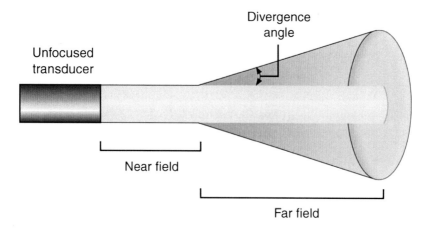

Fig. 1.1 Schematic diagram of beam geometry for an unfocused transducer. The length of the near field and the divergence angle in the far field depend on transducer frequency and aperture

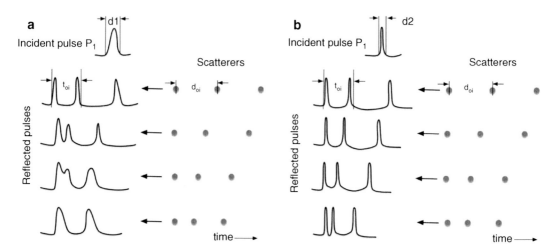

Fig. 1.2 An ultrasound pulse P1 that has width d1 frontally affects a linear scatterer array at a distance d_{oi}. (**a**) Each one of the echoes forms a "train" of pulses temporally distanced according to the equation $t_{oi} = 2|Ri|/c$, Ri being the ith relative emitter/scatterer distance and c the pulse propagation speed. There exists a critical distance width d_t at which the pulses that arrive at the receiver are superposed, therefore not being able to discriminate. The resolution can be improved by diminishing the pulse width d_t, which is equivalent to increasing the frequency of the emitted pulse (**b**)

the pulse width, T is the period of ultrasound wave, and f is the ultrasound frequency. The lateral resolution is the capacity to discern two objects located in the tangential direction and depends on the beam width. The lateral resolution is calculated as $d_\theta = 1.22\lambda/D$, where D is the aperture diameter. For a typical transducer of 40 MHz with aperture diameter 0.6 mm, the axial resolution is approximately ≈ 39 μm, the lateral resolution is $d_\theta \approx 0.8°$, and the focal length is $L = 2.3$ mm. Contrast resolution is the distribution of the gray scale of the reflected signal and is often referred to as dynamic range. The greater the dynamic range, the broader the range of reflected signal (form weakest to strongest) that can be detected, displayed, and differentiated. An image with low dynamic range appears black and white with only a few in-between gray-scale levels. High-dynamic-range images have more shades of gray and can discriminate more different tissue types and more structural elements.

The interaction of ultrasound waves with the tissues of the body can be described in terms of reflection, scattering, refraction, and attenuation (Fig. 1.3). When ultrasound waves encounter a boundary between two tissues such as fat and muscle, the beam will be partially reflected and partially transmitted. The amount of ultrasound reflected depends on the relative change in acoustic impedance between the two tissues and the angle of reflection. For example, imaging of highly calcified structures is associated with acoustic shadowing because of nearly complete reflection of the beam at the soft tissue/calcium interface. Scattering of the ultrasound signal occurs with small structures, such as red blood cells, because the radius of the cell (about 4 μm) is smaller than the wavelength of the ultrasound signal. The extent of scattering depends on particle size (red blood cells), number of particles (hematocrit), ultrasound transducer frequency, and compressibility of blood cells and plasma. Scattering results in a pattern of speckles. The intensity of blood speckle increases exponentially with the frequency of the transducer. Blood stasis resulted from the catheter crossing a tight stenosis increases blood speckle because of red cell clumping or rouleaux formation. Actually, static blood can be more echodense than plaque. Ultrasound waves can be refracted as they pass through a medium with a different acoustic impedance, which can result in ultrasound artifacts including double-image artifact. Attenuation is the loss of signal strength as ultrasound interacts with tissue. As ultrasound penetrates into tissues, signal strength is progressively attenuated (reduced) due to absorption of the ultrasound energy by conversion to heat, as well as by reflection and scattering. Therefore, only a small percentage of the emitted signal returns to the transducer. The received signal is converted to electrical energy and sent to an external signal processing system for amplification, filtering, scan conversion, user-controlled modification, and, finally, graphic presentation [3].

1.2 Equipment for IVUS Examination

The IVUS acquisition system consists of a catheter, a pullback device, and a scanning console (Fig. 1.4).

1.2.1 IVUS Catheter

Currently, IVUS catheters are 150 cm long and have a tip size of 3.2–3.5 French (outer diameter, 1.2–1.5 mm) that can go through 5–6-French guiding catheter [4]. The catheter is visible in angiographic images and is advanced along with a guidewire. The guidewire rail is positioned next to the catheter plastic sheath or within its center.

1.2.2 IVUS Transducer

The current intracoronary ultrasound imaging frequency range of 20–45 MHz provides 70–200 μm axial resolution, 200–400 μm lateral resolution, and 5–10 mm penetration [5, 6].

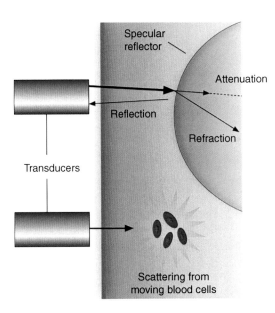

Fig. 1.3 Diagram of the interaction between ultrasound and body tissues

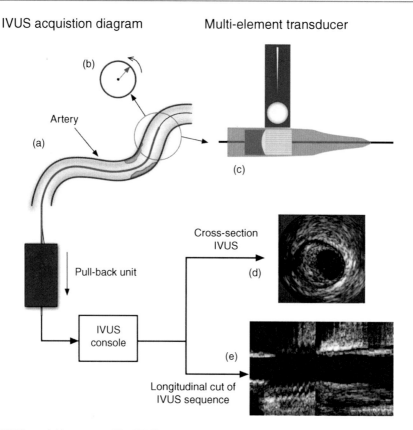

Fig. 1.4 The IVUS acquisition system. The IVUS catheter is manually placed within the artery (**a**) and extracted by a pullback unit at a constant linear velocity and rotated at a constant angular velocity (**b**). Multi-element phased array IVUS transducer (**c**) and its beam shape. The information is transformed by the IVUS console as a unique cross-sectional artery *gray*-level image (**d**) or a longitudinal image sequence (**e**)

There are two different types of IVUS transducers: the mechanically rotating transducer and the electronically switched multi-element array system (Fig. 1.5) [7].

1.2.2.1 Mechanical Systems

A single rotating transducer is driven by a flexible drive cable at 1800 rpm (30 revolutions per second) to sweep a beam almost perpendicular to the catheter. At approximately 1° increments, the transducer sends and receives ultrasound signals. The time delay and amplitude of these pulses provide 256 individual radial scans for each image. The 40 MHz iCross or Atlantis SR Pro catheters (Boston Scientific, Santa Clara, California), the Revolution 45 MHz catheter (Volcano Corp., Rancho Cordova, California), and the 40 MHz LipiScan IVUS (InfraReDx, Burlington, Massachusetts) are commercially available as 6F compatible systems which offer a more uniform pullback and greater resolution. Newer 40 MHz OPTICROSS IVUS catheter (Boston Scientific, Santa Clara, California) features a low profile delivery system allows 5 F guide catheter compatibility and a shorter distal marker to transducer (15 mm).

1.2.2.2 Electronic Systems

Electronic systems, also known as phased array system, use an annular array of small crystals instead of a single rotating transducer. The array can be programmed so that one set of elements transmits while a second set receives simultaneously. The coordinated beam generated by groups of elements is known as a synthetic aperture array. 5F compatible Eagle Eye Catheter (Volcano Corp.) is commercially available.

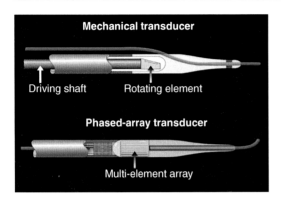

Fig. 1.5 Mechanical transducer catheters use rotating ultrasound sources, while phased array catheters use sequentially flashing ultrasound sources (Courtesy of Boston Scientific)

1.2.3 Catheter Pullback Device

The catheter is manually advanced to the distal end of the lesion of interest in coronary artery and is then pulled back, manually or with an automatic pullback system, at a speed of 0.5–1 mm/s.

Table 1.1 Comparison between mechanical and electronic system

Mechanical transducer		Phased array transducer
Single, rotating	Transducer	Multiple, sequentially firing
6F	Catheter	5F
Rigid, more difficult to pass through tortuous vessels	Flexibility	Flexible, easy to pass through tortuous vessels
Nonuniform rotational distortion	Artifact	Ring-down artifact
Relatively low	Cost	Relatively high

1.2.4 IVUS Scanning Consoles

A scanning console carries a computer that is used for post-processing and storage of recorded IVUS data. A cable from the end of the pullback device is connected with a computer for data processing. During the catheterization procedure, the clinician uses a trackball keyboard and functional buttons to enter the patient information, determine the percentage of stenosis, and apply image processing and possibly tissue characterization techniques to better understand and evaluate atherosclerotic plaques (Table 1.1).

References

1. Foster FS, Pavlin CJ, Harasiewicz KA, Christopher DA, Turnbull DH. Advances in ultrasound biomicroscopy. Ultrasound Med Biol. 2000;26:1–27.
2. Mintz GS, Nissen SE, Anderson WD, et al. American College of Cardiology Clinical Expert Consensus Document on standards for acquisition, measurement and reporting of intravascular ultrasound studies (IVUS). A report of the American College of Cardiology Task Force on clinical expert consensus documents. J Am Coll Cardiol. 2001;37:1478–92.
3. Schoenhagen P, Nissen S. Understanding coronary artery disease: tomographic imaging with intravascular ultrasound. Heart. 2002;88:91–6.
4. Katouzian A, Angelini ED, Carlier SG, Suri JS, Navab N, Laine AF. A state-of-the-art review on segmentation algorithms in intravascular ultrasound (IVUS) images. IEEE Trans Inf Technol Biomed. 2012;16:823–34.
5. Elliott MR, Thrush AJ. Measurement of resolution in intravascular ultrasound images. Physiol Meas. 1996;17:259–65.
6. Brezinski ME, Tearney GJ, Weissman NJ, et al. Assessing atherosclerotic plaque morphology: comparison of optical coherence tomography and high frequency intravascular ultrasound. Heart. 1997;77:397–403.
7. McDaniel MC, Eshtehardi P, Sawaya FJ, Douglas JS Jr, Samady H. Contemporary clinical applications of coronary intravascular ultrasound. JACC Cardiovasc Interv. 2011;4:1155–67.

IVUS Artifacts and Image Control

Hyung-Bok Park, Yun-Hyeong Cho, and Deok-Kyu Cho

2.1 Artifacts

2.1.1 Post-acoustic Shadowing (Severe Calcified Lesions, Metal Stent Struts, and Guidewires)

Since IVUS is a type of ultrasound, when scanning severe calcified lesions, post-acoustic shadowing frequently occurs due to poor penetration of the ultrasound beam into calcium. Due to shadowing, underlying plaque beyond the calcium cannot be evaluated or measured (Fig. 2.1a). In addition, calcified lesion can cause other types of artifacts such as reverberations and side lobes. A large coalesce of calcium seen on IVUS is often revealed to actually be many small calcifications upon histopathologic study [1]. Metal stent struts can also cause a typical sunburst pattern of post-acoustic shadows compromising the plaque evaluation underneath (Fig. 2.1b). Guidewires, at times, cause significant artifacts, especially during bifurcation lesion intervention. The dual guidewires make dual postshadows obscuring significant lesions (Fig. 2.1c). A long monorail catheter could be used as a preventive method. However, when dealing with highly movable vessels that have calcified lesions or even depending on the composition of guidewire tips, these artifacts can be worsened [1].

2.1.2 Ring-Down Artifacts

A luminous ring of false images surrounding the transducer or the catheter of IVUS, which presents as several layers around the catheter that compromise evaluation of the area adjacent to the catheter (Fig. 2.2). Often it is called near-field artifacts when using other medical ultrasound devices. Digital subtraction of a reference mask can suppress ring-down artifact; however, it also limits the ability to distinguish extremely near tissue from the surface of the catheter [2].

2.1.3 Nonuniform Rotational Distortion (NURD)

NURD is the unique motion artifact that can only be observed in IVUS system. It results from hindered constant rotational as well as fullback velocity of the transducer due to nonuniform friction of the coronary artery lumen [1, 3]. It can be

H.-B. Park • Y.-H. Cho • D.-K. Cho (✉)
Division of Cardiology, Cardiovascular Center, Myongji Hospital, Seonam University College of Medicine, Goyang, South Korea
e-mail: chodk123@paran.com

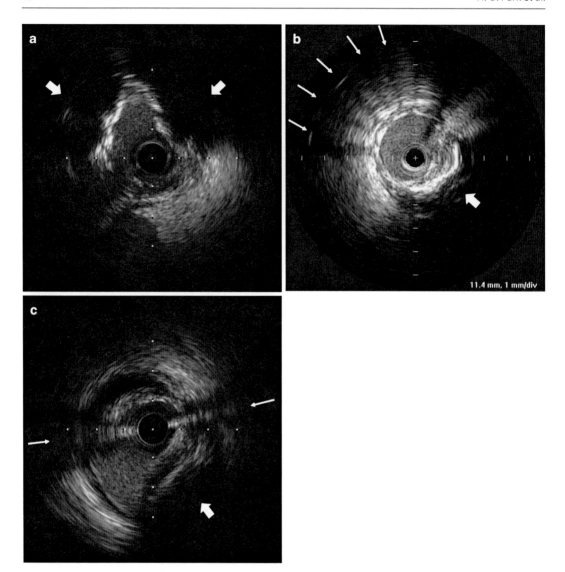

Fig. 2.1 Post-acoustic shadows due to severe calcified lesions, metal stent struts, and guidewires. (**a**) Heavy calcified lesions mask the plaque underneath between the 9 and 3 o'clock positions (*arrowheads*). (**b**) Multiple post-shadows (*arrows*) by stent struts combined with calcification (2–7 o'clock, *arrowhead*) can be observed. (**c**) Dual postshadows are present due to dual guidewires (*arrows*) with calcified plaque (*arrowhead*)

present especially in bending or tortuous vessels, sometimes being influenced by small guided catheter lumens, the kinking of coronary wires in the same catheter, instability of catheter engagement, and the hub or drive machine itself. One of the other problems is the transducer can move up to 5 mm longitudinally according to systolic and diastolic movement of the heart. This movement can also cause significant motion artifacts.

NURD is critically limited for the quantitative analysis of IVUS use (Fig. 2.3).

2.1.4 Side Lobes

Outside of main and high energy ultrasound beams, there are low energy beams called side lobes [1]. If there is a strong echo reflector such as calcium or

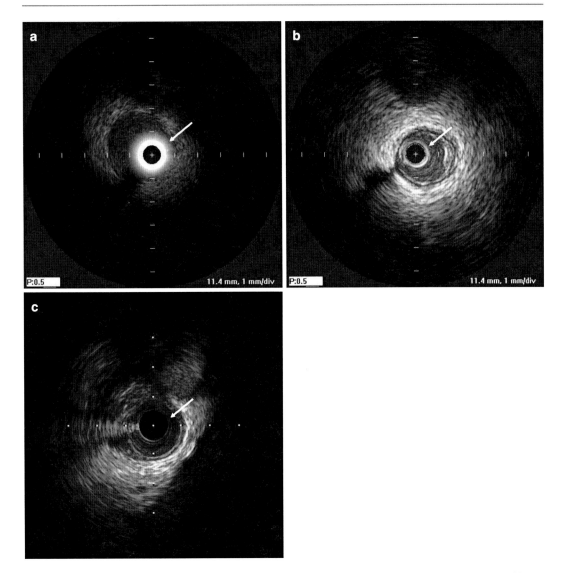

Fig. 2.2 Ring-down or near-field artifacts. (**a**) An extremely displayed ring-down artifact can be noted due to ultrasound element defect (*arrow*). (**b**) Presented here is ring-down artifact of around 10 mm diameter with several bright layers at the center of IVUS image (*arrow*). (**c**) Reduced ring-down artifact is noted (*arrow*)

stent strut, where the side lobe beams should pass, it will reflect these low energy echoes back to the transducer. Hence, false images of circumferential sweep will present adjacent to the calcium or stents. These false images mimic dissection flaps and may compromise precise evaluation of true lumen border (Fig. 2.4). One tip to overcome side lobe artifact is to reduce gain setting.

2.1.5 Reverberations

Reverberations are the production of repetitive false echo images due to reflections between two interfaces with a high acoustic impedance mismatch [2]. When the ultrasound beam bumps into strong reflectors such as calcium, metal stents, guide wires, and guiding catheters, it may

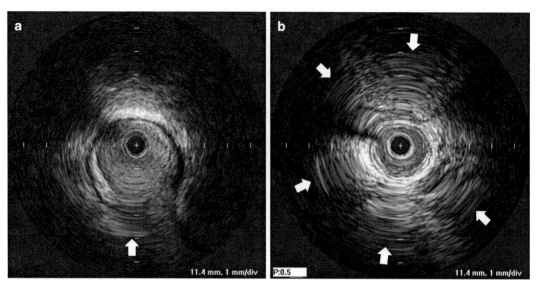

Fig. 2.3 Nonuniform rotational distortion (NURD). (**a**) NURD artifact is present between 5 and 7 o'clock with distortion of the underlying plaque (*arrowhead*). (**b**) Multiple NURD artifacts occurred in a small-sized vessel (less than 2.5 mm diameter, *arrowheads*)

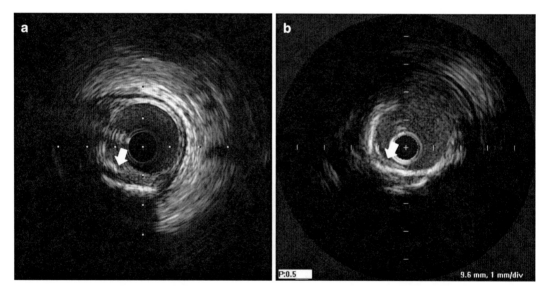

Fig. 2.4 Side lobes. (**a**) False images of circumferential lines are showing between 7 and 8 o'clock (*arrowheads*). (**b**) Slices of circumferential lines are present between 6 and 9 o'clock positions being confused with the dissection flap of the intima (*arrowheads*)

be repeatedly reflected back and forth before returning to the transducer. These repeated reflections are displayed as multiple equidistantly spaced circumferential layers on IVUS (Fig. 2.5).

2.1.6 Ghosts

Stent "ghost" artifacts are false reflected images that present on the opposite side of where the stent metal truly is [2]. They frequently appear after

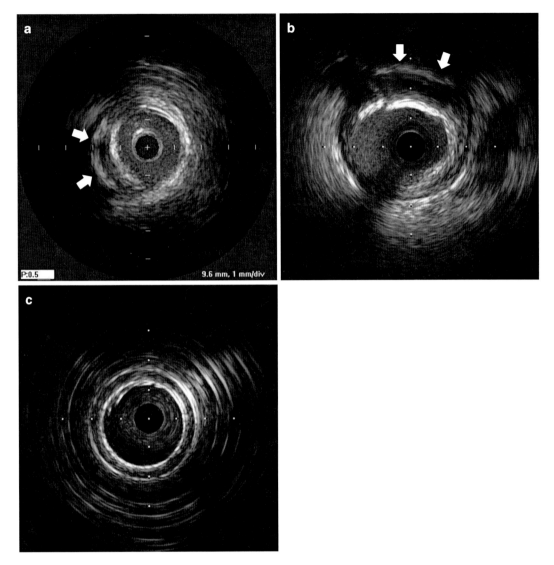

Fig. 2.5 Reverberations. Repetitive false echo images (*arrowheads*) are showing outside the calcified lesion between 7 and 10 o'clock (**a**) and 11 and 1 o'clock (**b**). (**c**) Multiple equidistantly spaced circumferential layers are present due to catheter-derived reverberations

stenting implantation and make it difficult to distinguish the true stent apposition (Fig. 2.6). Ghost artifact can be decreased by reducing overall gain.

2.1.7 Blood Speckle Artifact

High intensity of blood speckles in the coronary lumen make it difficult to distinguish between lumen and plaque (Fig. 2.7). The speckles are the result of decreased velocity of blood due to severe luminal stenosis or at times from using higher than conventional transducer frequencies such as when using 40 MHz [1–3]. Saline or contrast dye flushing would immediately resolve this artifact. Adjusting the time gain control is another option.

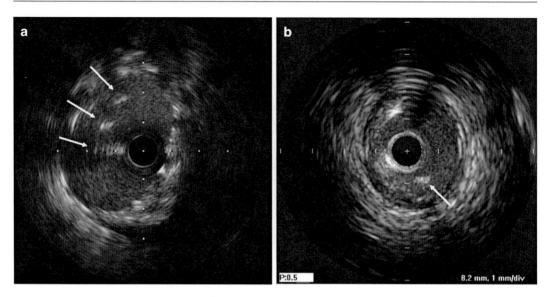

Fig. 2.6 Ghosts (**a**) Circular false images are shown opposite of implanted stent between 9 and 11 o'clock (*arrows*). (**b**) Ghost artifact caused by dense calcium is present (*arrow*)

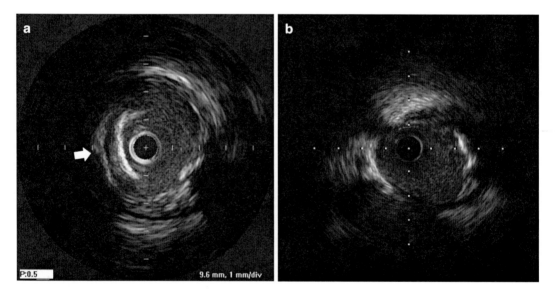

Fig. 2.7 Blood speckle artifact. (**a**) Highly accumulated blood speckles in lumen make it difficult to distinguish between lumen and plaque (9–3 o'clock and 4–6 o'clock). Reverberation artifact is also noted between 7 and 11 o'clock (*arrowhead*). (**b**) Dense blood speckles are shown between 2 and 7 o'clock

2.1.8 Air Bubble Artifact

Air bubble artifact is simply caused by improper catheter saline flushing [2]. Therefore, remaining air bubbles reduce the resolution of IVUS images (Fig. 2.8a). Complete flushing of air bubbles in the catheter with saline will be enough to resolve this problem (Fig. 2.8b). However, it is important to prevent an air embolism from occurring, and thus it is best to remove the IVUS catheter and reintroduce it to the coronary upon successful completion of flushing to remove the air bubbles.

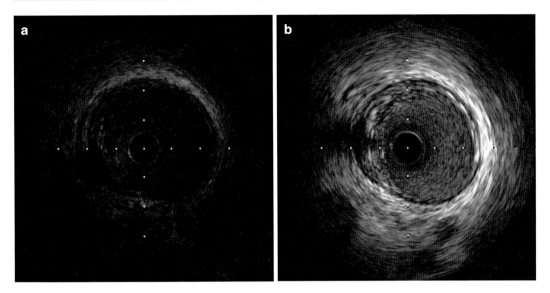

Fig. 2.8 Air bubble artifact. (**a**) Air bubble artifacts are present due to improper catheter saline flushing. (**b**) Upon completion of proper air bubble flushing, an improved image was achieved

Fig. 2.9 Sawtooth artifact. "Sawtooth" appearance is present on the longitudinal reconstructed image due to a rapid swinging coronary artery (*arrows*)

2.1.9 Sawtooth Artifact

This artifact appears when the transducer is excessively swung during the pullback period. Due to this, a "sawtooth" appearance shows on the longitudinal reconstructed image (Fig. 2.9). It frequently happens in tortuous vessels, rapid beating hearts, or rapid moving coronary arteries such as right coronary arteries [2].

2.2 Image Control

2.2.1 Gain Control (Overall Gain and Time Gain Compensation)

Overall gain control is used to increase or decrease the overall brightness of the image through amplification of the return signal without changing the transmitted pulse (Fig. 2.10).

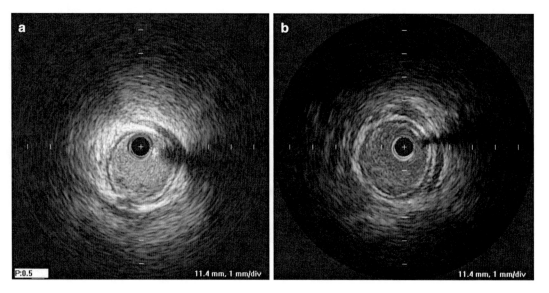

Fig. 2.10 Overall gain control. (**a**) A gain setting being set too high caused amplified noise signals in the lumen. (**b**) In this image, a properly adjusted gain setting can be seen

Therefore, overall gain increase does not improve tissue penetration or resolution power. On the contrary, if overall gain is set too high, it also amplifies noise signals resulting in the reduction of the image resolution [1].

The time gain compensation (TGC) is a way to overcome ultrasound attenuation by increasing signal gain over time it is emitted from the transducer. As emitted ultrasound passes through tissue, wave amplitude becomes attenuated. Hence, TGC is a compensation technique of increasing return signal amplification while penetrating deeper [1]. This correction makes equally echogenic structures appear the same visually regardless of depth. In large vessels such as left main, increasing the far-field attenuation would be helpful. On the other hand, reducing the near-field intensity can compromise assessment of in-stent neointimal tissue, small branches, or CTO lesions.

2.2.2 Rejection

Rejection is a way of increasing image contrast by filtering-out low amplitude noise signals [1]. However, if reject level is set too high, low echogenic structures such as hematoma or small dissection flaps might be missed.

2.2.3 Compression (Dynamic Range)

Compression (also known as dynamic range) is controlling the range of echo intensities by narrowing or broadening the range of the gray scale [1]. If the compression level is set too high, fewer shades of gray will be displayed resulting in higher contrast images with more black and white [2].

2.2.4 Zoom, Depth, and Scale

According to vessel size or lesion morphology, zoom, depth, or scale adjustment should be tactfully chosen. Zoom is only magnifying the image, not providing further structural detail. Therefore, adjusting depth might be useful to gain greater detail within the structure (Fig. 2.11).

2.3 Summary

Artifacts in IVUS imaging are an infrequently inevitable phenomenon due to the inherent limitations of ultrasound modality itself. Therefore, it should be important to distinguish artifacts from

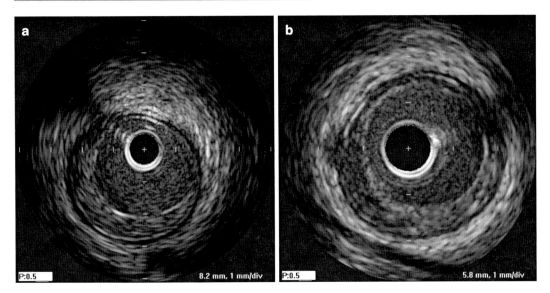

Fig. 2.11 Depth or scale control. Depth or scale was adjusted from 8 mm diameter (**a**) to 6 mm (**b**). It is noted that image resolution has not changed despite magnifying the image

true image findings at each specific clinical scenario and gain image control skills to reduce artifacts as much as possible.

References

1. Mintz GS, Nissen SE, Anderson WD, Bailey SR, Erbel R, Fitzgerald PJ, Pinto FJ, Rosenfield K, Siegel RJ, Tuzcu EM, Yock PG. American College of Cardiology Clinical Expert Consensus Document on Standards for Acquisition, Measurement and Reporting of Intravascular Ultrasound Studies (IVUS). A report of the American College of Cardiology Task Force on Clinical Expert Consensus Documents. J Am Coll Cardiol. 2001;37:1478–92.
2. Mintz GS. Intracoronary ultrasound; 2005.
3. Bangalore S, Bhatt DL. Coronary intravascular ultrasound. Circulation. 2013;127:e868–74.

Quantitative Measurements of Native Lesion

Mayank Goyal, Hoyoun Won, and Sang Wook Kim

Intravascular ultrasound (IVUS) imaging provides a detailed information about the coronary artery including the vessel wall, plaque, and lumen with resolution up to 150 microns. IVUS use is based on the obvious principle, which is angiography cannot accurately represent the sizes (dimensions) or the composition of coronary plaque and luminal narrowings. These information are necessary to assess the lesion morphology as well as percutaneous coronary intervention. The quantitative parameters are summarized in Table 3.1.

3.1 Vessel Wall Identification

Due to reflection of ultrasound waves at tissue interface, generally, there are two such interfaces in the normal coronary artery [1]. It is important to recognize the leading edge of boundaries when measuring the IVUS image quantitatively [2]. Outside the lumen, the second layer is the media, and the third and outer layer consists of the adventitia and periadventitial tissues [3–8].

Interpretation of IVUS image begins with recognition of these two interfaces: blood/intimal (lumen) and medial/adventitial interface. A relative echo-translucency of media compared with intima and adventitia gives rise to a three-layered appearance of coronary wall (bright, dark, bright) in muscular arteries such as the coronary arteries. As intimal layer reflects ultrasound more strongly than media, there is a spillover effect, called blooming, resulting in slight overestimation of intimal layer and correspondingly underestimation of medial layer. However, the medial-adventitial border is accurately identified as a step-up in echoreflectivity occurring at this border with no blooming effect. In diseased arteries, media may not appear as a distinct layer around the vessel. The adventitia and periadventitial structures are similar in echodensity so that a distinct outer-adventitial border cannot be defined (Fig. 3.1).

3.2 Lumen Measurements

The most important parameter of quantitative analysis is the measurement of coronary lumen. Lumen measurements are performed using the interface between the lumen and the leading edge of the intima. The leading edge of the innermost echogenic layer should be used as the lumen boundary. The intimal leading edge can be easily identified because the intima has thickened enough to be resolved as a separate layer and has

---

M. Goyal
Rajiv Gandhi Super Speciality Hospital,
New Delhi, India

H. Won • S.W. Kim (✉)
Cardiovascular-Arrhythmia Center, Chung-Ang University Hospital, Seoul, South Korea
e-mail: swivus@gmail.com

© Springer Nature Singapore Pte Ltd. 2018
M.-K. Hong (ed.), *Coronary Imaging and Physiology*,
https://doi.org/10.1007/978-981-10-2787-1_3

Table 3.1 Quantitative measurements of IVUS

EEM and lumen measurement	
Lumen cross-sectional area	The area bounded by the luminal border
Minimum lumen diameter	The shortest diameter through the center point of the lumen
Maximum lumen diameter	The longest diameter through the center point of the lumen
Lumen eccentricity	[(Maximum lumen diameter - minimum lumen diameter)/maximum lumen diameter]
Lumen area stenosis	(Reference lumen CSA - minimum lumen CSA)/reference lumen CSA
External elastic membrane cross-sectional area	EEM is an interface at the border between the media and the adventitia. Synonyms: vessel area, total vessel area
Vessel volume	Vessel area measurements can be added to calculate volumes (Simpson's rule)
Lumen volume	Lumen area measurements can be added to calculate volumes (Simpson's rule)
Plaque measurement	
Plaque + media CSA (atheroma area)	EEM CSA-lumen CSA
Maximum plaque + media(or atheroma) thickness	The largest distance from the intimal leading edge to the EEM along any line passing through the center of the lumen
Minimum plaque + media(or atheroma) thickness	The shortest distance from the intimal leading edge to the EEM along any line passing through the luminal center of mass
Plaque + media (or atheroma) eccentricity	(Maximum plaque plus media thickness - minimum plaque plus media thickness)/maximum plaque plus media thickness
Plaque (or atheroma) burden	Plaque + media CSA/EEM CSA
Plaque volume	Plaque + media CSA measurements can be added to calculate volumes (Simpson's rule)
Calcium measurement	
Superficial/deep calcium	The leading edge of the acoustic shadowing appears within the most shallow/the deepest 50% of the plaque plus media thickness
Arc	Measured in degrees by using an electronic protractor centered on the lumen
Semiquantitation	Absent or subtracting one, two, three, or four quadrants

CSA cross-sectional area, *EEM* external elastic membrane

sufficiently different acoustic impedance from the lumen in normal segments. The vessel wall has a single-layer appearance because the intima cannot be resolved as a discrete layer with a thin, inner echolucent band corresponding to the intima and media, particularly in younger normal subjects (e.g., posttransplantation). The thickness of this layer will be <160 μm and will be a negligible error to the measurement.

3.3 EEM Measurements

The third and outer layer consists of the adventitia and periadventitial tissues. But external elastic membrane area is the outer layer of the vessel in IVUS measurement because the border of adventitia and periadventitial tissue is not distinct. A discrete interface at the border between the media and the adventitia is usually present within IVUS images due to the relative echotranslucency of media. The term of this measurement is *EEM CSA*, rather than alternative terms such as "vessel area". The measurement of EEM border should be avoided at the sites where large side branches originate or with extensive calcification and acoustic shadowing. If acoustic shadowing involves a relatively small arc (<90°), planimetry of the circumference can be performed by estimation from the closest identifiable EEM borders, although the accuracy and reproducibility will be reduced. If calcification is more extensive than 90° of arc, EEM measurements should not be reported. Normal arteries are

3 Quantitative Measurements of Native Lesion

Direct Measurements:
Lesion length = 12 mm
Reference Lumen Area = 6.84 mm²
Reference EEM Area = 9.90 mm²
Lesion Lumen Area = 1.69 mm²
Lesion EEM Area = 6.49 mm²
Maximum Plaque thickness = 1.3 mm
Minimum Plaque thickness = 0.15 mm
Lesion Maximum Lumen Diameter = 1.58 mm
Lesion Minimum Lumen Diameter = 1.38 mm
Arc of Calcium = 40 degree

Derived Measurements:
Lesion Plaque Area = 4.79 mm²
Lesion Plaque Burden = 73.9 %
Remodeling Index = 0.7
Lesion Lumen Eccentricity Index = 0.83
Plaque Eccentricity Index = 0.12

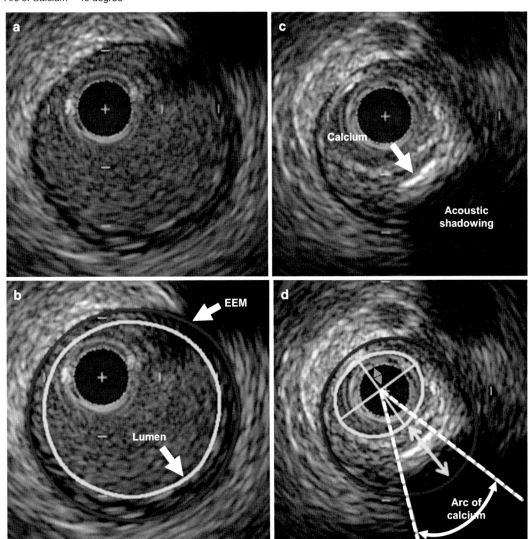

Fig. 3.1 Quantitative measurements of IVUS. Panels **a** and **b** show the reference segment. Panels **c** and **d** represent the target lesion. The EEM and lumen areas are traced (**b**). The minimum and maximum lumen diameters are shown inside the lumen. In panel **d**, the minimum and maximum plaque plus media thickness is also assessed using *double-headed arrows* (*blue* for minimum and *yellow* for maximum). The EEM and lumen areas are demonstrated, and the arc of calcification (*dotted line*) is shown. Panel **e** represents the measurement of lesion length. EEM external elastic membrane

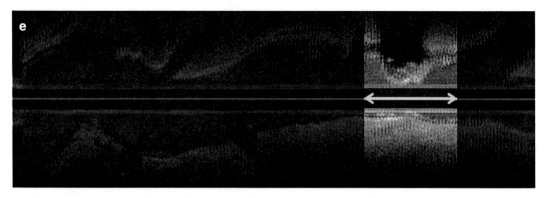

Fig. 3.1 (continued)

circular, but a noncircular configuration is often due to the atherosclerotic remodeling.

3.4 Plaque Measurements

IVUS revealed a much larger plaque burden than what would be predicted by angiography. The plaque plus media area is used to calculate a surrogate for the true atheroma area, because the leading edge of the internal elastic membrane (the media) is not well delineated and IVUS measurements cannot determine true histologically atheroma area (the area bounded by the internal elastic membrane) [1]. The media represent only a very small percent of the atheroma CSA, and it does not conclude a major limitation of IVUS. Thus, the term "plaque plus media" can be used and that the measurements can be performed.

3.5 Reference Segment Measurements

Usually reference segment is defined as the most normal-looking (site with largest lumen with minimal plaque burden) area within 10 mm from the lesion site (maximum stenosis) with no intervening major side branches. Even normal-looking reference segment on angiography has shown to be on an average having 35–50% plaque burden on IVUS [9, 10]. Once the reference segments are selected, quantitative assessment should be similar to the lesion.

3.6 Calcium Measurements

Intravascular ultrasound is a sensitive method to detect the coronary calcium in vivo [11, 12]. Calcific deposits represent as bright echoes that block the penetration of ultrasound signal; it produces a phenomenon known as "acoustic shadowing" behind the calcium. IVUS can delineate only the leading edge and cannot identify its thickness. The ultrasound signal is oscillated between the transducer and calcium and causes concentric arcs (reverberations or multiple reflections) in the image at reproducible distances. Calcium deposits can be described qualitatively according to the location (Fig. 3.2).

3.7 Remodeling

Figure 3.3 represents the IVUS measurement of vascular remodeling. The superiority of IVUS is to measure the vascular remodeling about coronary artery disease compared to other imaging device [13–19]. Vascular remodeling was originally described from necropsy specimens by Glagov et al. [20]. The EEM area is expanded or shrinked during the development of atherosclerosis. If EEM area increases during atheroma development, the process is termed "positive remodeling." If the EEM area decreases, the pro-

a. Superficial calcium

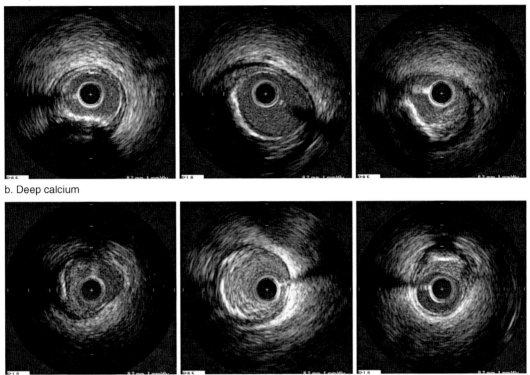

b. Deep calcium

Fig. 3.2 IVUS measurements of calcium. Panel **a** represents the superficial calcium, and panel **b** shows the deep calcium

cess is termed "negative" or "constrictive remodeling." In positive remodeling, the EEM area increase may *overcompensate* for the increasing plaque area and a net increase in lumen size. An index that describes the magnitude and direction of remodeling is expressed as: lesion EEM CSA/reference EEM CSA. If the lesion EEM area is greater than the reference EEM area, positive remodeling has occurred, and the index will be ≥1.05. If the lesion EEM area is smaller than the reference EEM area, negative remodeling has occurred, and the index will be <0.95. A number of dichotomous definitions of remodeling have been proposed [13–19]. The reference segment(s) used in studies of remodeling should be measured without any major intervening side branches. However, the interpretation of remodeling should be careful in the situation which both reference and lesion sites have undergone changes in EEM area due to the atherosclerotic disease process.

3.8 Length Measurements

IVUS measurements of the length can be performed only using the motorized transducer pullback. The length of the lesion, stenosis, calcium, or any other longitudinal feature can be determined.

3.9 Summary

IVUS is a reliable and established imaging modality to quantify the coronary lesion with high sensitivity and specificity. It has an important role in assessing the target lesion and planning intervention strategy. Although IVUS cannot replace the noninvasive or invasive functional assessment, it has a key role to evaluate the lesion morphology including the vessel size, luminal narrowing, plaque composition, vascular remodeling, and lesion length.

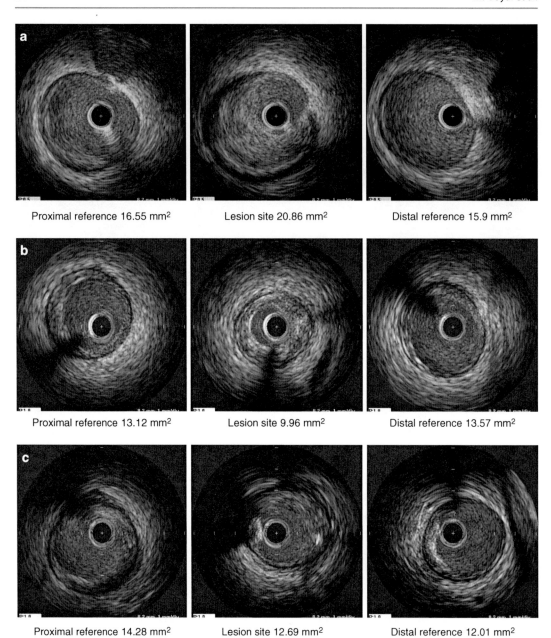

Fig. 3.3 IVUS represents the vascular remodeling. Panel a shows the positive remodeling (remodeling index 1.3), panel b shows the negative remodeling (remodeling index 0.7), and panel c shows the intermediate remodeling (remodeling index 1.0)

References

1. Wong M, Edelstein J, Wollman J, Bond MG. Ultrasonic-pathological comparison of the human arterial wall. Verification of intima-media thickness. Arterioscler Thromb. 1993;13:482–6.
2. Mintz GS, Nissen SE, Anderson WD, Bailey SR, Erbel R, Fitzgerald PJ, et al. American College of Cardiology clinical expert consensus document on

standards for acquisition, measurement and reporting of intravascular ultrasound studies (IVUS): a report of the American College of Cardiology Task Force on Clinical Expert Consensus Documents. J Am Coll Cardiol. 2001;37:1478–92.
3. Fitzgerald PJ, St. Goar FG, Connolly AJ, et al. Intravascular ultrasound imaging of coronary arteries. Is three layers the norm? Circulation. 1992;86:154–8.
4. Gussenhoven EJ, Essed CE, Lancee CT, et al. Arterial wall characteristics determined by intravascular ultrasound imaging: an in vitro study. J Am Coll Cardiol. 1989;14:947–52.
5. Lockwood GR, Ryan LK, Gotlieb AI, et al. In vitro high resolution intravascular imaging in muscular and elastic arteries. J Am Coll Cardiol. 1992;20:153–60.
6. Mintz GS, Douek P, Pichard AD, et al. Target lesion calcification in coronary artery disease: an intravascular ultrasound study. J Am Coll Cardiol. 1992;20:1149–55.
7. Nishimura RA, Edwards WD, Warnes CA, et al. Intravascular ultrasound imaging: in vitro validation and pathologic correlation. J Am Coll Cardiol. 1990;16:145–54.
8. Potkin BN, Bartorelli AL, Gessert JM, et al. Coronary artery imaging with intravascular high-frequency ultrasound. Circulation. 1990;81:1575–85.
9. Mintz GS, Painter JA, Pichard AD, et al. Atherosclerosis in angiographically "normal" coronary artery reference segment: an intravascular ultrasound study with clinical correlation. J Am Coll Cardiol. 1995;25:1479–85.
10. St Goar FG, Pinto FJ, Aldervan EL, et al. IVUS imaging of angiographically normal coronary arteries: an in vivo comparison with quantitative angiography. J Am Coll Cardiol. 1991;18:952–8.
11. Metz JA, Yock PG, Fitzgerald PJ. Intravascular ultrasound: basic interpretation. Cardiol Clin. 1997;15:1–15.
12. Tuzcu EM, Berkalp B, De Franco AC, et al. The dilemma of diagnosing coronary calcification: angiography versus intravascular ultrasound. J Am Coll Cardiol. 1996;27:832–8.
13. Hermiller JB, Tenaglia AN, Kisslo KB, et al. In vivo validation of compensatory enlargement of atherosclerotic coronary arteries. Am J Cardiol. 1993;71:665–8.
14. Losordo DW, Rosenfield K, Kaufman J, Pieczek A, Isner JM. Focal compensatory enlargement of human arteries in response to progressive atherosclerosis. In vivo documentation using intravascular ultrasound. Circulation. 1994;89:2570–7.
15. Mintz GS, Kent KM, Pichard AD, Satler LF, Popma JJ, Leon MB. Contribution of inadequate arterial remodeling to the development of focal coronary artery stenoses. An intravascular ultrasound study. Circulation. 1997;95:1791–8.
16. Nishioka T, Luo H, Eigler NL, Berglund H, Kim CJ, Siegel RJ. Contribution of inadequate compensatory enlargement to development of human coronary artery stenosis: an in vivo intravascular ultrasound study. J Am Coll Cardiol. 1996;27:1571–6.
17. Pasterkamp G, Wensing PJ, Post MJ, Hillen B, Mali WP, Borst C. Paradoxical arterial wall shrinkage may contribute to luminal narrowing of human atherosclerotic femoral arteries. Circulation. 1995;91:1444–9.
18. Pasterkamp G, Schoneveld AH, van der Wal AC, et al. Relation of arterial geometry to luminal narrowing and histologic markers for plaque vulnerability: the remodeling paradox. J Am Coll Cardiol. 1998;32:655–62.
19. Schoenhagen P, Ziada KM, Kapadia SR, Crowe TD, Nissen SE, Tuzcu EM. Extent and direction of arterial remodeling in stable versus unstable coronary syndromes: an intravascular ultrasound study. Circulation. 2000;101:598–603.
20. Glagov S, Weisenberg E, Zarins CK, Stankunavicius R, Kolettis GJ. Compensatory enlargement of human atherosclerotic coronary arteries. N Engl J Med. 1987;316:1371–5.

Qualitative Assessment of Native Lesion

Young Joon Hong

Qualitative assessment was performed according to criteria of the American College of Cardiology clinical expert consensus document on IVUS and the Study Group on Intracoronary Imaging of the Working Group of Coronary Circulation and of the Subgroup on IVUS of the Working Group of Echocardiography of the European Society of Cardiology [1, 2].

4.1 Plaque Morphology

IVUS cannot be used to detect and quantify specific histologic contents. The threshold between normal and abnormal is the subject of some debate, but more than 0.3 mm of intimal thickening is probably abnormal and can be used to distinguish from atherosclerosis. The maximal thickness of the intima-media complex or, more appropriately, the percentage of the total vessel area occupied by plaque is the most common quantitative indices used to define the severity of atherosclerotic involvement. Atherosclerotic lesions may be present in segments which are angiographically normal because compensatory total vessel enlargement in the early phases of atherosclerosis tends to keep the lumen constant.

Lumen reduction does not occur, according to these pathology studies, until the plaque occupies more than 40% of the total cross-sectional vessel area. However, atherosclerotic lesions occupying less than 20 and 40% of the total vessel area can still be considered as lesions with a minimal and moderate atherosclerotic burden, respectively (Table 4.1) [1].

Atherosclerotic lesions are heterogeneous and include varying amounts of calcium, dense fibrous tissue, lipid, smooth muscle cells, thrombus, etc. By IVUS, imaging can grossly separate lesions into subtypes according to echodensity and the presence or absence of shadowing and reverberations (Table 4.2) [3].

4.1.1 Soft (Echolucent) Plaque

The term "soft" refers not to the plaque's structural characteristics but rather to the acoustic

Table 4.1 Atherosclerotic burden

Normal intima	Single-layer appearance or three-layer appearance with intimal thickness < 0.3 mm
Minimal atherosclerotic burden	≤20% of VA occupied by plaque
Moderate atherosclerotic burden	>20%, ≤40% of VA occupied by plaque
Large atherosclerotic burden	>40%, ≤60% of VA occupied by plaque
Massive atherosclerotic burden	>60% of VA occupied by plaque

VA total vessel area

Y.J. Hong
Division of Cardiology, Chonnam National University Hospital, Gwangju, South Korea
e-mail: hyj200@hanmail.net

Table 4.2 Intravascular ultrasound plaque characteristics

Homogeneous[a]			Mixed[b]
Soft	Fibrous	Calcific[c]	Soft/fibrous
Low echoreflective	High echoreflective	High echoreflective with shadowing	Soft/calcific
			Fibrocalcific

[a]>80% area constituted by the same plaque components; no calcium or focal calcium deposits (arc of calcium <10°)
[b]The presence of multiple plaque components not matching the 80% criterion of prevalence
[c]Total calcific arc greater than 180°

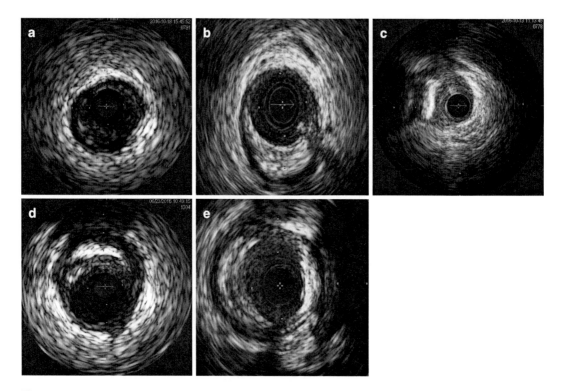

Fig. 4.1 Plaque morphology. (**a**) Soft (echolucent) plaque. Soft plaque is less bright compared with reference adventitia. Soft plaque contains varying amounts of fibrous and fatty tissue. (**b**) Fibrous plaque has an intermediate echogenicity between soft (echolucent) atheromas and highly echogenic calcific plaques. Fibrous plaque is as bright as or brighter than the adventitia without shadowing. (**c**, **d**) Calcified plaque is echodense (hyperechoic) plaque (brighter than the reference adventitia) that shadows using IVUS calcium can be localized and characterized as superficial [closer to tissue-lumen interface (**c**)] and deep [closer to the media-adventitia junction (**d**)]. (**e**) When there is no dominant plaque composition, the plaque was considered "mixed." Mixed plaque is also called as "fibrocalcific," "fibrofatty" plaque

signal that arises from low echogenicity. Soft plaque is less bright compared with reference adventitia. Soft plaque contains varying amounts of fibrous and fatty tissue. Reduced echogenicity may also result from a necrotic zone within the plaque, an intramural hemorrhage, or a thrombus (Fig. 4.1a).

4.1.2 Fibrous Plaque

Fibrous plaque has an intermediate echogenicity between soft (echolucent) atheromas and highly echogenic calcific plaques. Fibrous plaque represents the majority of atherosclerotic lesion. Fibrous plaque is as bright as or brighter than the adventitia without shadowing. Very dense fibrous plaques may produce sufficient attenuation to be misclassified as calcification with acoustic shadowing. Both calcified and fibrotic plaques are hyperechoic (Fig. 4.1b).

4.1.3 Calcified Plaque

Calcium is a powerful reflector of ultrasound. Calcific deposits appear as bright echoes that obstruct the penetration of ultrasound, a phenomenon known as "acoustic shadowing." In practice calcium is echodense (hyperechoic) plaque (brighter than the reference adventitia) that shadows using IVUS calcium can be localized and characterized as superficial (closer to tissue-lumen interface) and deep (closer to the media-adventitia junction) and quantified according to its arc and length. The arc of calcium can be measured (in degrees) by using an electronic protractor centered on the lumen. Because of beam-spread variability at given depths within the transmitted beam, this measurement is usually valid only to ±15°. Semiquantitative grading has also been described, which classifies calcium as absent or subtending 1, 2, 3, or 4 quadrants. The length of the calcific deposit can be measured using motorized transducer pullback (Fig. 4.1c, d).

4.1.4 Mixed Plaque

Plaques frequently contain more than one acoustical subtype. When there is no dominant plaque composition, the plaque was considered "mixed." Mixed plaque is also called as "fibrocalcific," "fibrofatty" plaque (Fig. 4.1e).

4.2 Vulnerable Plaque

No definitive IVUS features define a plaque as vulnerable. However, necropsy studies demonstrated that unstable coronary lesions are usually lipid rich with a thin fibrous cap. Accordingly, hypoechoic plaques without a well-formed fibrous cap are presumed to represent potentially vulnerable atherosclerotic lesions. The important mechanisms leading to the development of acute coronary syndrome (ACS) are rupture of a vulnerable plaque and subsequent thrombus formation. The majority of ACS events are the result of sudden luminal thrombosis, with 55–60% due to plaque rupture, 30–35% caused by plaque erosion, and a small portion resulting from a calcified nodule.

4.2.1 Plaque Rupture

Rupture of vulnerable plaque and/or endothelial erosions with subsequent thrombus formation are considered the main mechanisms implicated in the pathogenesis of ACS. Ruptured plaque contains a cavity that communicated with the lumen with an overlying residual fibrous cap fragment [4] (Figs. 4.2 and 4.5). Rupture sites separated by a length of artery containing smooth lumen contours without cavities are considered to represent different plaque ruptures (multiple plaque ruptures) (Fig. 4.3). Plaque rupture is closely related to obstructive thrombus formation, and the longitudinal morphology of plaque rupture also affects the coronary flow. The presence of thrombi may obscure IVUS detection of plaque rupture.

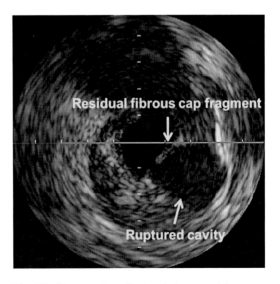

Fig. 4.2 Plaque rupture. Ruptured plaque contains a cavity that communicated with the lumen with an overlying residual fibrous cap fragment

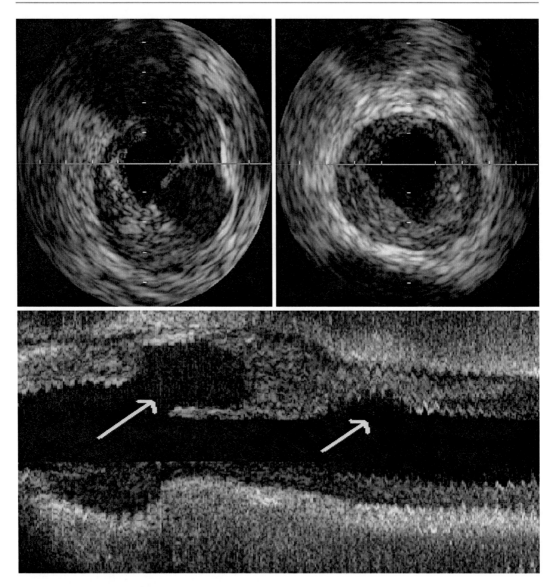

Fig. 4.3 Multiple plaque rupture. Rupture sites separated by a length of artery containing smooth lumen contours without cavities are considered to represent different plaque ruptures (multiple plaque ruptures)

4.2.2 Thrombus

The identification of thrombus is one of the most difficult aspects of IVUS imaging [5, 6] (Figs. 4.4 and 4.5).

Clues to the presence of thrombus include the following [5, 6]:

1. Sparkling or scintillating appearance
2. Lobulated mass projecting into the lumen
3. A distinct interface between the suspected thrombus and underlying plaque
4. Identification of blood speckle within the thrombus indicating microchannels through the thrombus
5. Mobility

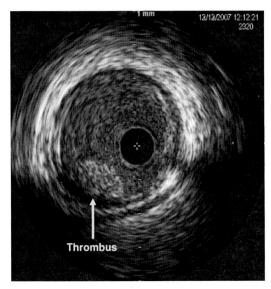

Fig. 4.4 Thrombus. Thrombus shows sparkling or scintillating appearance and lobulated mass projecting into the lumen and a distinct interface between the suspected thrombus and underlying plaque

Because of limited resolution of IVUS, the detection rate of thrombus by IVUS is not high. Injection of contrast or saline may disperse the stagnant flow, clear the lumen, and allow differentiation of stasis from thrombosis. However, none of these features is pathognomic for thrombus, and the diagnosis of thrombus by IVUS should always be considered presumptive.

4.2.3 Attenuated Plaque

Attenuated plaque is defined as hypoechoic plaque with deep ultrasound attenuation without calcification or very dense fibrous plaque [7] (Fig. 4.6). Wu et al. [8] reported that 78% of the AMI patients had attenuated plaques in HORIZONS-AMI trial. Lee et al. [9] reported that attenuated plaque was observed in 39.6% of STEMI and 17.6% of NSTEMI ($p < 0.001$), and the level of C-reactive protein (CRP) was higher; angiographic thrombus and initial Thrombolysis In Myocardial Infarction (TIMI) flow grade <2

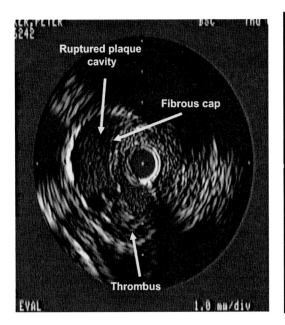

Fig. 4.5 Representative intravascular ultrasound finding in patient with acute myocardial infarction. This figure shows plaque rupture with overlying fibrous cap with surrounding thrombus

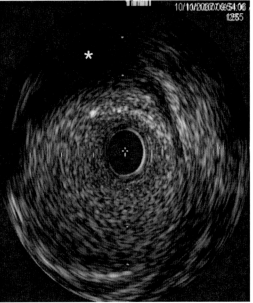

Fig. 4.6 Attenuated plaque. Attenuated plaque is defined as hypoechoic plaque with deep ultrasound attenuation (*asterisk*) without calcification or very dense fibrous plaque

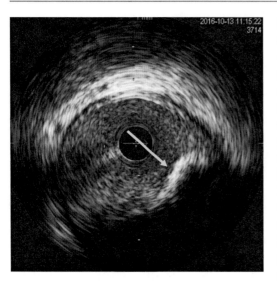

Fig. 4.7 Calcified nodule. Calcified nodule (*arrow*) is an eruptive, dense, calcified mass often having an irregular surface appearance

were more common; IVUS lesion site plaque burden and remodeling index were significantly greater; lesion site luminal dimensions were significantly smaller; and thrombus, positive remodeling and plaque rupture were more common in AMI patients with attenuated plaque compared with those without attenuated plaque.

4.2.4 Calcified Nodule

Calcified nodule is an eruptive, dense, calcified mass often having an irregular surface appearance (Fig. 4.7). Lee et al. [10] reported the IVUS characteristics of calcified nodules, which include (1) a convex shape of the luminal surface, (2) a convex shape of the luminal side of calcium, (3) an irregular luminal surface, and (4) an irregular leading edge of calcium. Although calcified nodule is a marker for atherosclerosis, it is associated with fewer future cardiac events, suggesting quiescence rather than ongoing activity.

4.3 Angiographic Aneurysms

Maehara et al. [4] reported IVUS findings in 77 consecutive patients with an aneurysmal dilatation (defined as a lesion lumen diameter 25% larger than reference) in a native coronary artery diagnosed by angiography. IVUS true aneurysms were defined as having an intact vessel wall and a maximum lumen area 50% larger than proximal reference (Fig. 4.8). IVUS pseudoaneurysms had a loss of vessel wall integrity and damage to adventitia or perivascular tissue. Complex plaques were lesions with ruptured plaque or spontaneous or unhealed dissection. Twenty-one lesions (27%) were classified as true aneurysms, 3 (4%) were classified as pseudoaneurysms, 12 (16%) were complex plaques, and the other 41 (53%) were normal arterial segments adjacent to ≥1 stenosis. Therefore, only one third of angiographically diagnosed aneurysms had the IVUS appearance of a true or pseudoaneurysm. Instead, most angiographically diagnosed aneurysms had the morphology of complex plaques or normal segments with adjacent stenoses.

4.4 Angiographically Ambiguous Lesions

Angiographically ambiguous lesions may include (1) intermediate lesions of uncertain stenotic severity, (2) aneurysmal lesions, (3) ostial stenoses, (4) disease at branching sites, (5) tortuous vessels, (6) left main stem lesions, (7) sites with focal spasm, (8) sites with plaque rupture, (9) dissection after coronary angioplasty, (10) intraluminal filling defects, (11) angiographically hazy lesions, and (12) lesions with local flow disturbances. IVUS is frequently employed to examine lesions with the above characteristics, in some cases providing additional evidence useful in determining whether the stenosis is clinically significant (i.e., difficult to assess left main or borderline stenosis with continued symptoms). However, it must be emphasized that IVUS does not provide physiologic information per se.

4.5 Myocardial Bridge

The muscle overlying the intramyocardial segment of an epicardial coronary artery is termed a myocardial bridge, and the artery coursing within

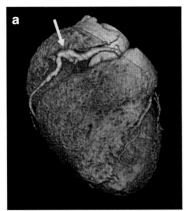
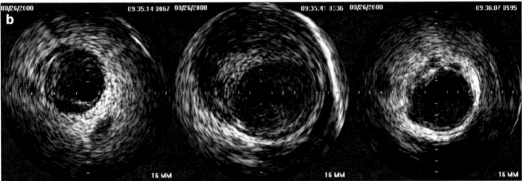

Fig. 4.8 True aneurysm. True aneurysm in 30-year-old female with acute myocardial infarction. (**a**) Computed tomography showed true huge aneurysm (*arrow*) in middle right coronary artery. (**b**) Intravascular ultrasound true aneurysm which was defined as having an intact vessel wall and a maximum lumen area 50% larger than proximal reference

the myocardium is called a tunneled artery. It is characterized by systolic compression of the tunneled segment. The "half-moon phenomenon" is a characteristic IVUS finding. It seems specific for the existence of myocardial bridging inasmuch as it is only found in tunneled segments but not in proximal or distal segments or in other arteries. In the presence of a half-moon phenomenon on IVUS, milking can be provoked by intracoronary provocation tests, even if the bridge was angiographically undetectable. IVUS pullback studies supported the absence of atherosclerosis within tunneled segments, although ≈90% of patients showed plaque formation proximal to the bridge. When deep tunneled segments approach the right ventricular subendocardium, the trabeculated right chamber myocardium and the right ventricular cavity may be visible on IVUS.

4.6 Spontaneous Dissection

Spontaneous coronary artery dissection is a rare cause of acute coronary syndrome (ACS) and sudden death. IVUS showed an entry intimal tear with an intimal flap dividing the true lumen from the false one at the dissection site. Usually, the false lumen has a larger area than the true lumen (Fig. 4.9).

4.7 Chronic Total Occlusion

Fujii et al. [11] published an excellent paper describing IVUS findings in 83 CTOs interrogated with IVUS immediately after antegrade wire crossing and small caliber balloon inflation.

4.8 Summary

IVUS is reliable and established imaging modality to evaluate coronary lesion with high sensitivity and specificity. IVUS is very helpful to define plaque morphology according to echodensity and the presence or absence of shadowing and reverberations and to detect vulnerable plaque and aneurysm and to define angiographically ambiguous lesions and to detect myocardial bridge, spontaneous dissection, and chronic total occlusion.

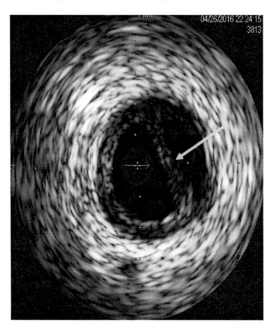

Fig. 4.9 Spontaneous dissection. IVUS showed an entry intimal tear with an intimal flap (*arrow*) dividing the true lumen from the false one at the dissection site

Procedures were performed in four Japanese centers with procedures performed between 2003 and 2005. This study found nearly all lesions contained calcium (96%) although in many, (68%), this was mild. The authors were able to define the proximal cap according to angiographic landmarks and observing abrupt morphology change on IVUS. The proximal cap was a location where calcium was concentrated, particularly in blunt stump CTOs. A calcified arc was demonstrated in the wall opposite the side branch in 74% of this morphology. A smaller proportion had calcification found perpendicular to the side-branch origin, and a small number were found on the ipsilateral aspect as the side branch.

Suzuki et al. [12] described IVUS findings in 79 CTO lesions and found moderately strong correlations between lesion age and indices of calcification assessed by IVUS. Some very recent CTOs were heavily calcified suggesting that the CTO had arisen in a vessel with well-entrenched atheroma.

References

1. Di Mario C, Görge G, Peters R, Kearney P, Pinto F, Hausmann D, et al. Clinical application and image interpretation in intracoronary ultrasound. Study Group on Intracoronary Imaging of the Working Group of Coronary Circulation and of the Subgroup on Intravascular Ultrasound of the Working Group of Echocardiography of the European Society of Cardiology. Eur Heart J. 1998;19:207–29.
2. Mintz GS, Nissen SE, Anderson WD, Bailey SR, Erbel R, Fitzgerald PJ, et al. American College of Cardiology clinical expert consensus document on standards for acquisition, measurement and reporting of intravascular ultrasound studies (IVUS): a report of the American College of Cardiology Task Force on Clinical Expert Consensus Documents. J Am Coll Cardiol. 2001;37:1478–92.
3. Tobis JM, Mallery J, Mahon D, Lehmann K, Zalesky P, Griffith J, et al. Intravascular ultrasound imaging of human coronary arteries in vivo. Analysis of tissue characterizations with comparison to in vitro histological specimems. Circulation. 1991;83:913–26.
4. Maehara A, Mintz GS, Bui AB, Walter OR, Castagna MT, Canos D, et al. Morphologic and angiographic features of coronary plaque rupture detected by intravascular ultrasound. J Am Coll Cardiol. 2002;40:904–10.
5. Chemarin-Alibelli MJ, Pieraggi MT, Elbaz M, Carrié D, Fourcade J, Puel J, Tobis J. Identification of coronary thrombus after myocardial infarction by intracoronary ultrasound compared with histology of tissues sampled by atherectomy. Am J Cardiol. 1996;77:344–9.
6. Frimerman A, Miller HI, Hallman M, Laniado S, Keren G. Intravascular ultrasound characterization of thrombi of different composition. Am J Cardiol. 1994;73:1053–7.
7. Endo M, Hibi K, Shimizu T, Komura N, Kusama I, Otsuka F, et al. Impact of ultrasound attenuation and

plaque rupture as detected by intravascular ultrasound on the incidence of no-reflow phenomenon after percutaneous coronary intervention in ST-segment elevation myocardial infarction. JACC Cardiovasc Interv. 2010;3:540–9.
8. Wu X, Mintz GS, Xu K, Lansky AJ, Witzenbichler B, Guagliumi G, et al. The relationship between attenuated plaque identified by intravascular ultrasound and no-reflow after stenting in acute myocardial infarction: the HORIZONS-AMI (harmonizing outcomes with revascularization and stents in acute myocardial infarction) trial. JACC Cardiovasc Interv. 2011;4:495–502.
9. Lee SY, Mintz GS, Kim SY, Hong YJ, Kim SW, Okabe T, et al. Attenuated plaque detected by intravascular ultrasound: clinical, angiographic, and morphologic features and post-percutaneous coronary intervention complications in patients with acute coronary syndromes. JACC Cardiovasc Interv. 2009;2:65–72.
10. Lee JB, Mintz GS, Lisauskas JB, Biro SG, Pu J, Sum ST, et al. Histopathologic validation of the intravascular ultrasound diagnosis of calcified coronary artery nodules. Am J Cardiol. 2011;108:1547–51.
11. Fujii K, Ochiai M, Mintz GS, Kan Y, Awano K, Masutani M, et al. Procedural implications of intravascular ultrasound morphologic features of chronic total coronary occlusions. Am J Cardiol. 2006;97:1455–62.
12. Suzuki T, Hosokawa H, Yokoya K, Kojima A, Kinoshita Y, Miyata S, et al. Time-dependent morphologic characteristics in angiographic chronic total coronary occlusions. Am J Cardiol. 2001;88:167–9.

Clinical Evidence of Intravascular Ultrasound-Guided Percutaneous Coronary Intervention

Sung-Jin Hong, Yangsoo Jang, and Byeong-Keuk Kim

Intravascular ultrasound (IVUS) provides anatomic information regarding the coronary artery lumen, wall, and plaques, which can help the accurate evaluation of lesion characteristics with vessel sizing. In addition, after stent implantation, underexpansion, malapposition, or edge dissections can be detected by IVUS. Thus, through further intervention based on these IVUS findings, stent optimization can be achieved, causing the improved clinical outcomes. Current guidelines recommend the use of IVUS to optimize stent implantation for select patients (Class of recommendation IIa, Level of evidence B) [1, 2]. However, recently, many evidences demonstrating the clinical usefulness of IVUS have been accumulated since the prior guidelines were released. In this chapter, clinical evidences of IVUS-guided percutaneous coronary intervention (PCI) will be discussed from observational studies, randomized studies, and meta-analysis.

5.1 Clinical Studies Evaluating Clinical Usefulness of IVUS-Guidance PCI

Several randomized clinical trials were performed to demonstrate clinical usefulness of IVUS-guidance during PCI. Recently conducted randomized controlled trials comparing IVUS-guidance vs. angiography-guidance particularly using the drug-eluting stent (DES) are summarized in Table 5.1 [3–10]. The first two trials by Jakabacin et al. and Cheiffo et al. failed to prove the clinical benefit of IVUS-guidance because of relatively small number of patients, less than 150 patients in each group were included in their studies [3, 4]. Kim et al. reported that IVUS usage for diffuse long lesions was associated with improved clinical outcomes particularly when used by operators' decision. In the per-protocol analysis, IVUS-guidance group significantly had lower 1-year major adverse cardiovascular event (MACE) (4.0% vs. 8.1%, $p = 0.048$), although the strategy of routine IVUS for DES implantation did not improve the MACE rates in the intention-to-treat analysis [5]. Recent randomized trials which showed statistically significant clinical benefit were performed mainly for complex lesions, such as left main lesions [7], chronic

S.-J. Hong · Y. Jang · B.-K. Kim (✉)
Division of Cardiology, Severance Cardiovascular Hospital, Yonsei University College of Medicine, Seoul, South Korea
e-mail: kimbk@yuhs.ac

Table 5.1 Recent randomized studies comparing clinical usefulness between IVUS-guided and angiography-guided PCI

Study	Year	N (IVUS vs. angiography)	Enrolled patients	Follow-up, m	Primary endpoint	Major findings (IVUS vs. angiography)
Jakabacin et al. [3]	2010	105 vs. 105	Complex cases and high clinical risk profile	18	Composite of death, MI, TLR	No significant differences (11% vs. 12%)
Chieffo et al. [4]	2013	142 vs. 142	Complex lesions	24	Post-procedural in-lesion MLD	IVUS group had greater MLD (2.70 mm vs. 2.51 mm, $p = 0.002$)
Kim et al. [5]	2013	269 vs. 274	Long lesions (implanted stent ≥ 28 mm in length)	12	Composite of cardiovascular death, MI, stent thrombosis, or TVR	No significant differences by intention-to-treat analysis; but IVUS group had lower primary endpoint by per-protocol analysis (4.0% vs. 8.1%, $p = 0.048$)
MOZART [6]	2014	41 vs. 42	High risk of contrast-induced acute kidney injury or volume overload	–	Total volume contrast agent used during PCI	IVUS group had lower volume contrast agent (20 ml vs. 65 ml, $p < 0.001$)
Tan et al. [7]	2015	62 vs. 61	Unprotected LM in the elderly (aged 70 or older)	24	Composite of death, non-fatal MI, or TLR	IVUS group had lower primary endpoint (13.1% vs. 29.3%, $p = 0.031$)
CTO-IVUS [8]	2015	201 vs. 201	Chronic total occlusion	12	Cardiac death	No significant differences in primary endpoint; but IVUS group had lower secondary endpoint (the composite of cardiac death, MI, or TVR) (2.6% vs. 7.1%, $p = 0.035$)
Tian et al. [9]	2015	115 vs. 115	Chronic total occlusion	12	Late lumen loss	IVUS group had a lesser late lumen loss (0.28 mm vs. 0.46 mm, $p = 0.025$)
IVUS-XPL [10]	2015	700 vs. 700	Long lesions (implanted stent ≥28 mm in length)	12	Composite of cardiac death, MI, or TLR	IVUS group had lower primary endpoint (2.9% vs. 5.8%, $p = 0.007$)

IVUS intravascular ultrasound, *LM* left main, *MI* myocardial infarction, *MLD* minimal lumen diameter, *PCI* percutaneous coronary intervention, *TLR* target-lesion revascularization, *TVR* target-vessel revascularization

total occlusions (CTO) [8, 9], and diffuse long lesions [10]. The CTO-IVUS (Chronic Total Occlusion InterVention with drUg-eluting Stents) study, the first randomized trial for CTO lesions, demonstrated that IVUS-guided PCI may improve 12-month MACE rates after DES implantation when compared with conventional angiography-guided CTO-PCI [8]. In the IVUS-XPL (Impact of Intravascular Ultrasound Guidance on Outcomes of Xience Prime Stents in Long Lesions) trial, IVUS-guided DES implantation compared with angiography-guided DES implantation resulted in a significantly lower rate of the composite of MACE (a composite of cardiac death, myocardial infarction [MI], or target-lesion revascularization

[TLR]) at 1 year (2.9% vs. 5.8%, hazard ratio [HR] = 0.48, $p = 0.007$) [10]. These differences were primarily due to lower risk of TLR (2.5% vs. 5.0%, HR = 0.51, $p = 0.02$).

According to the ADAPT-DES (The assessment of dual antiplatelet therapy with drug-eluting stents) study, the most recent largest observational study with all-comers ($n = 8583$) [11], IVUS was utilized in 3349 patients (39%), and larger-diameter devices, longer stents, and/or higher inflation pressure were used in the IVUS-guided cases. At 1 year, propensity-adjusted multivariable analysis revealed IVUS-guidance vs. angiography-guidance was associated with a reduced definite/probable stent thrombosis (0.6% vs. 1.0%, $p = 0.003$), MI (2.5% vs. 3.7%, $p = 0.004$), and composite adjudicated major cardiac events (cardiac death, MI, or stent thrombosis) (3.1% vs. 4.7%, $p = 0.002$). The benefits of IVUS were especially evident in patients with acute coronary syndromes and complex lesions [11]. Further recent observational studies evaluating clinical usefulness of IVUS-guided PCI are summarized in Table 5.2 [11–17].

Lastly, meta-analyses comparing the IVUS-guidance and angiography-guidance are presented in Table 5.3 [18–22]. Shin et al. reported the results of meta-analysis with individual patient-level data from 2345 randomized patients. IVUS-guided new-generation DES implantation vs. angiography-guided DES implantation was associated with a favorable outcome, particularly the occurrence of hard clinical endpoint (the composite of cardiac death, MI, or stent thrombosis) for complex lesions [22]. Of note, the primary endpoint of this meta-analysis did not include TLR. Therefore, different from the IVUS-XPL trial showing the benefit of IVUS due primarily to the less frequent TLR events [10], MACEs, even excluding the TLR events in this meta-analysis, were less frequent with IVUS-guidance than angiography-guidance [22].

Table 5.2 Recent observational studies comparing clinical outcomes between IVUS-guided and angiography-guided PCI

Study	Year	N (IVUS vs. angiography)	Enrolled patients	Follow-up, m	Major findings (IVUS vs. angiography)
Witzenbichler et al. [11]	2014	3349 vs. 5234	All comers	12	Definite/probable ST: 0.6% vs. 1.0%, $p = 0.003$ MI: 2.5% vs. 3.7%, $p = 0.004$ Composite of cardiac death, ST, MI; 3.1% vs. 4.7%, $p = 0.002$
Roy et al. [12]	2008	884 vs. 884 by matching	All comers	12	Definite ST: 0.7% vs. 2.0%, $p = 0.014$
Park et al. [13]	2013	463 vs. 463 by matching	Nearly all comers	12	Composite of cardiac death, MI, TLR: 4.3% vs. 2.4, $p = 0.047$
Youn et al. [14]	2011	125 vs. 216	Primary PCI cases	36	Composite of death, MI, TLR, TVR: 12.8% vs. 18.1%, $p = NS$
Kim et al. [15]	2011	487 vs. 487 by matching	Non-left main bifurcation	36	Death or MI: 3.8% vs. 7.8%, $p = 0.03$
Hong et al. [16]	2014	201 vs. 201 by matching	Chronic total occlusion	24	Definite/probable ST: 0% vs. 3.0%, $p = 0.014$ MI: 1.0% vs. 4.0%, $p = 0.058$
de la Torre Hernandez et al. [17]	2014	505 vs. 505 by matching	Left main lesions	36	Composite of cardiac death, MI, TLR: 11.3% vs. 16.4%, $p = 0.04$ Definite/probable ST: 0.6% vs. 2.2%, $p = 0.04$

IVUS intravascular ultrasound, *MI* myocardial infarction, *PCI* percutaneous coronary intervention, *ST* stent thrombosis, *TLR* target-lesion revascularization, *TVR* target-vessel revascularization, *NS* non-significant

Table 5.3 Recent meta-analyses comparing clinical outcomes between IVUS-guided and angiography-guided PCI

Study	Year	N (analyzed studies)	N (IVUS vs. angiography)	Data analysis	Major findings (IVUS vs. angiography)
Jang et al. [18]	2014	3 RCTs and 12 observational studies with DES implantation	11,793 vs. 13,056	Study-level meta-analysis	IVUS had lower MACE (OR = 0.79, $p = 0.001$), all-cause mortality (OR = 0.64, $p < 0.001$), MI (OR = 0.57, $p < 0.001$), TVR (OR = 0.81, $p = 0.01$), and ST (OR = 0.59, $p = 0.002$)
Ahn et al. [19]	2014	3 RCTs and 14 observational studies with DES implantation	12,499 vs. 14,004	Study-level meta-analysis	IVUS had lower TLR (OR = 0.81, $p = 0.046$), death (OR = 0.61, $p < 0.001$), MI (OR = 0.57, $p < 0.001$), and ST (OR = 0.59, $p < 0.001$)
Elgendy et al. [20]	2016	7 RCTs with DES implantation	1593 vs. 1599	Study-level meta-analysis	IVUS group had lower MACE at a mean of 15 months (6.5% vs. 10.3%, $p < 0.0001$), mainly because of reduction in the risk of TLR (4.1% vs. 6.6%, $p = 0.003$)
Steinvil et al. [21]	2016	7 RCTs and 18 observational studies with DES implantation	14,659 vs. 16,624	Study-level meta-analysis	IVUS group had lower MACE (OR = 0.76, $p < 0.001$), death (OR = 0.62, $p < 0.001$), MI (OR = 0.67, $p < 0.001$), ST (OR = 0.58, $p < 0.001$), TLR (OR = 0.77, $P = 0.005$), and TVR (OR = 0.85 $p < 0.001$)
Shin et al. [22]	2016	3 RCTs with new-generation DES implantation	1170 vs. 1175	Individual patient-level meta-analysis	IVUS group had a lower occurrence of hard clinical outcome (composite of cardiac death, MI, or ST) at 1 year (0.4% vs. 1.2%, $p = 0.04$)

DES drug-eluting stent, *IVUS* intravascular ultrasound, *MACE* major adverse cardiovascular event, *MI* myocardial infarction, *OR* odds ratio, *RCT* randomized clinical trial, *ST* stent thrombosis, *MI* myocardial infarction, *TLR* target-lesion revascularization, *TVR* = target-vessel revascularization

5.2 Left Main Lesion

Procedural complication or failure of left main lesion of PCI is critical. Thus, IVUS-guidance PCI for left main lesion is currently recommended as a class IIa or class IIb recommendation [1, 2]. In addition to the stent optimization, particularly for left main lesions, functionally significant lesion can be relatively accurately predicted by IVUS examination for intermediate lesions because of the limited variability of left main coronary artery length, diameter, and the amount of supplied myocardium. Minimal lumen area (MLA) less than 4.5 mm^2 predicted the fractional flow reserve (FFR) less than 0.80 with sensitivity of 77% and specificity of 82% [23]. Other studies also reported the optimal cut-off value of MLA by IVUS for predicting functionally significant left main lesions (FFR less than 0.75) were 5.9 mm^2 and 4.8 mm^2, respectively [24, 25]. IVUS is also essential for the optimization to reduce the restenosis. A previous study showed that the cut-off values of post-stenting MLA that best predicted in-stent restenosis were 5.0 mm^2 in ostial left circumflex, 6.3 mm^2 in ostial left anterior descending, 7.2 mm^2 in polygon of confluence, and 8.2 mm^2 in left main [26].

Recently, a randomized trial for unprotected left main lesions revealed that IVUS-guided group had a lower composite of death, non-fatal MI, or TLR (13.1% vs. 29.3%, $p = 0.031$), although small number of patients were studied in this study [7]. Also, a recent pooled analysis from 4 Spanish registries demonstrated that IVUS-guided DES implantation for unprotected left main showed a lower 3-year composite rate of cardiac death, MI, and TLR compared with the angiography-guided DES implantation (11.3% vs 16.4%, $p = 0.04$), and a more prominent in the subgroup with distal left main lesions (10.0% vs 19.3%, $p = 0.03$) [17].

5.3 Bifurcation Lesion

There were no randomized studies performed particularly for the bifurcation lesions. According to the observational studies, Kim et al. demonstrated that the 3-year cumulative incidence of death or MI was significantly lower in the IVUS-guided PCI group than the angiography-guided PCI group (3.8% vs 7.8%, $p = 0.03$) [15]. Another observational study with bifurcation lesions, the rate of TLR was significantly lower in the IVUS-guided PCI group (6% vs 21%, $p = 0.001$) [27]. In the first study, two-stent technique and final kissing balloon were more frequently used in the IVUS-guidance group [15], whereas in the second study, the number of implanted stents was significantly lower in the IVUS-guidance group [27]. In this regard, although further studies are needed to determine the optimal stent strategies including the stent number particularly for bifurcation lesions, the role of IVUS for the decision of stent strategies may be important to improve clinical outcomes for the complex bifurcation lesions. A previous study evaluated the IVUS parameters predicting the IVUS ≥ 4 mm^2 at 9-month follow-up IVUS for both main vessel and side branch after bifurcation T-stenting with first-generation DES [28]. Inadequate post-procedural minimal stent area (MSA) with increased neointimal hyperplasia may cause the side branch ostium to be the most frequent restenotic site after bifurcation PCI and the optimal cut-off value of post-procedural MSA was 4.83 mm^2 [28].

5.4 Chronic Total Occlusion

The roles of IVUS for CTO intervention could be classified into 3 different uses: (1) wire-crossing for the stumpless CTO lesions, (2) pre-stenting use, and (3) post-stent use. For the stumpless CTO lesions, IVUS-guidance has been reported to lead a higher success rate and to be useful in revealing the entry point of occlusion and in repositioning a guidewire in the event of inadvertent sub-intimal passage [29]. Pre-stenting IVUS could provide the accurate information regarding vessel size and lesion length and cause resultant appropriate stent size and length for stent optimization. CTO vessel often increases in size following the successful CTO PCI. An IVUS follow-up study at 6 month after CTO PCI revealed that distal lumen diameter was increased in two thirds of patients by 0.4 mm ($p < 0.001$) [30]. Post-stent IVUS may detect PCI complications or suboptimal stent expansion and could lead to stent optimization and finally can decide the need for additional stenting or ballooning. However, there had been a lack of evidence regarding the beneficial role of IVUS-guided CTO intervention using current-generation DES for the improved clinical outcomes after stent implantation. Two randomized trials were performed particularly for CTO lesions [8, 9]. In the CTO-IVUS trial, 402 patients with CTOs were randomized to the IVUS-guided group ($n = 201$) or the angiography-guided group ($n = 201$) after successful guidewire crossing [8]. Although IVUS-guided CTO intervention did not significantly reduce cardiac mortality, IVUS-guided CTO intervention improved 12-month MACE rate after new-generation DES implantation when compared with conventional angiography-guided CTO intervention. In this study, IVUS-guidance group had a higher proportion of high-pressure post-stent dilation (51% vs. 41%, $p = 0.045$) with a higher maximum post-stent balloon pressure (14.6 vs. 13.8 atm, $p = 0.040$). Consequently, the post-procedural minimal lumen diameter was significantly larger in the IVUS-guidance vs. angiography-guidance (2.64 vs. 2.56 mm, $p = 0.025$).

In the second randomized trial, Tian et al. reported stent late lumen loss at 1 year between IVUS- vs. angiography-guidance [9]. Late lumen loss was significantly lower in the IVUS-guided group compared with the angiography-guided group (0.28 vs 0.46 mm, $p = 0.025$), although these angiographic findings were not translated into the improvement of clinical outcomes.

5.5 Diffuse Long Lesion

A long lesion inevitably increases the length of implanted stent, and long stent increases the incidence of stent underexpansion. In the IVUS-XPL enrolling 1400 patients requiring ≥28 mm everolimus-eluting stents, adjunct post-stent balloon dilation was more frequently performed in the IVUS-guided stent group (76%) than in the angiography-guided stent group (76% vs 57%, $p < 0.001$) [10]. The mean final balloon size was larger in the IVUS-guided group than in the angiography-guided group. On post-procedural quantitative angiography analysis, minimum lumen diameter was greater and diameter stenosis was smaller in the IVUS-guided stent group than in the angiography-guided stent group [10]. In addition, in the post hoc analysis in that study among the patients within the IVUS-guided stent group, the patients who did not meet the IVUS criteria ($n = 315$, 46%) had a significantly higher incidence of the primary endpoint compared with those meeting the IVUS criteria for stent optimization ($n = 363$, 54%) (4.6% vs 1.5%, $p = 0.02$), when IVUS criteria for stent optimization after PCI was defined as an MLA greater than the lumen cross-sectional area at the distal reference segments [10].

5.6 In-Stent Restenosis

The use of IVUS to guide PCI for the treatment of restenosis is a class IIa recommendation in the current PCI guidelines [1, 2]. IVUS can differentiate whether restenosis is related to intimal hyperplasia or mechanical complications, such as stent fracture or underexpansion. According to the recent IVUS study comparing the mechanisms and patterns of in-stent restenosis among bare metal stents and DES, restenotic first- and second-generation DES were characterized by less neointimal hyperplasia, smaller stent areas, longer stent lengths, and more stent fractures [31].

5.7 Patients with Chronic Kidney Disease

Patients with chronic kidney disease (CKD) comprise a challenging subset of patients because of the increased incidence of contrast-induced acute kidney injury following angiography and PCI. Considerable efforts have been made to reduce contrast volume particularly in patients with CKD. Although most randomized clinical trials measured clinical or angiographic outcomes, the MOZART (Minimizing cOntrast utiliZation With IVUS Guidance in coRonary angioplasty) trials measured the total volume contrast agent used during PCI as the primary endpoint [6]. In this trial, a total of 83 patients with a high risk of contrast-induced acute kidney injury or volume overload were randomized to IVUS-guided PCI or angiography-guided PCI, and IVUS group had a lower total volume of contrast (20 ml vs. 65 ml, $p < 0.001$). Also, recent another study with a total of 31 patients with advanced CKD (creatinine = 4.2 mg/dL) revealed that PCI without contrast using IVUS and physiologic guidance may be performed safely with high procedural success and without complications [32].

5.8 IVUS Predictors for the Better Clinical Outcomes: Stent Optimization

The IVUS predictors of stent failure after DES implantation are underexpansion, dissections, and significant plaque burden (Table 5.4) [33–35]. When a total of 804 patients who underwent both post-intervention IVUS examination after long everolimus-eluting stent (≥28 mm in length) implantation were analyzed from two randomized trials (RESET trial and IVUS-XPL trial), the predictors of MACE were the post-intervention MLA at the target lesion and the ratio of MLA/distal reference segment lumen area [33]. The MLA and MLA-to-distal reference segment lumen area ratio that best predicted patients with MACE from those without it were 5.0 mm^2 and 1.0, respectively. Patients with an MLA < 5.0 mm^2 or a distal reference segment lumen area had a higher risk of MACE than those without MACE (HR = 6.2, $p = 0.003$). Similarly, Song et al. reported that the optimal MSA to predict angiographic restenosis at

9 months were 5.3 mm² for zotarolimus-eluting stents and 5.4 mm² for everolimus-eluting stents [34]. Therefore, the confirmation of sufficient MLA by IVUS is important after DES implantation. Figure 5.1 represents the stent underexpansion detected by IVUS examination despite of angiographically acceptable diameter stenosis, suggesting the need of post-stent adjuvant ballooning. Figure 5.2 represents the achievement of sufficient MLA after post-stent adjuvant ballooning.

Table 5.4 IVUS parameter after newer generation DES implantation predicting angiographic restenosis or MACE

	N	Follow-up endpoint	Stent	IVUS parameter after stenting	Cut-off value	Accuracy
Lee et al. [33]	804	MACE (cardiac death, MI, and TLR)	EES	MLA MLA/distal reference lumen area	5.0 mm² 1.0	Patients with an MLA < 5.0 mm² or a distal reference segment lumen area had a higher risk of MACE (hazard ratio = 6.231, p = 0.003) than those without MACE
Song et al. [34]	229 220	Angiographic in-stent restenosis	EES ZES	MSA	5.4 mm² 5.3 mm²	Sensitivity 60%, specificity 60% Sensitivity 57%, specificity 62%
Kang et al. [35]	433 422 813	Angiographic edge restenosis	E-ZES R-ZES EES	Edge plaque burden	56.3% 57.3% 54.2%	Sensitivity 67%, specificity 86% Sensitivity 80%, specificity 87% Sensitivity 86%, specificity 80%

EES everolimus-eluting stent, *E-ZES* Endeavor zotarolimus-eluting stents, *IVUS* intravascular ultrasound, *MACE* major adverse cardiovascular event, *MLA* minimal lumen area, *MSA* minimal stent area, *R-ZES* Resolute zotarolimus-eluting stents

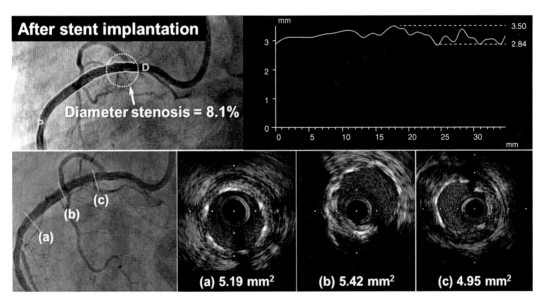

Fig. 5.1 Representative case showing the stent underexpansion by IVUS despite of angiographically acceptable diameter stenosis. After implantation of everolimus-eluting stent (Xience prime 2.75 × 38 mm, Abbott Vascular) for diffuse stenosis of right coronary artery, the residual stenosis by angiography at proximal portion of the stent was 8.1%, which was angiographically acceptable. However, on IVUS evaluation, the MLA was measured 4.95 mm² (**c**), which was smaller than the distal reference lumen area of 5.19 mm² (**a**) and less than 5 mm², suggesting the need of post-stent adjuvant ballooning

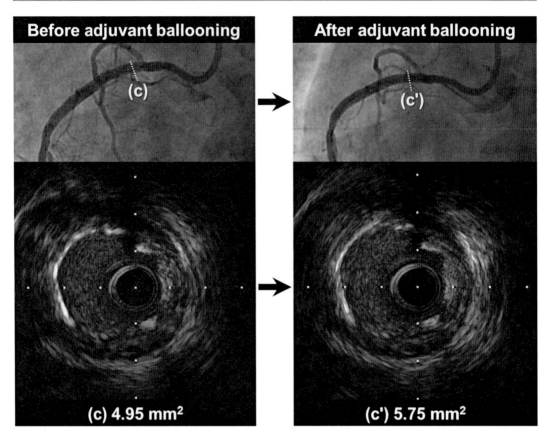

Fig. 5.2 Representative case showing the achievement of sufficient minimal lumen area measured by IVUS after post-stent adjuvant ballooning. After post-stent adjuvant ballooning with 3.0 mm-sized non-compliant balloon catheter based on the findings of IVUS, the minimal lumen was increased from 4.95 mm^2 to 5.75 mm^2. Same patients presented in Fig. 5.1

Kang et al. evaluated IVUS predictors for angiographic edge restenosis after newer generation DES [35]. The predictive cut-off of the reference plaque burden was 56.3% for Endeavor zotarolimus-eluting stents, 57.3% for Resolute zotarolimus-eluting stents, and 54.2% for everolimus-eluting stents. Figure 5.3 presents the representative case showing the need of additional stenting at proximal segment of stent because of edge dissection and residual plaque more than 60%, even though angiographic findings were acceptable.

Although IVUS studies have reported that the late stent malapposition is a predictor of late or very late stent thrombosis, there is no data linking isolated acute stent malapposition without stent underexpansion to early stent thrombosis or restenosis [36].

From the bare-metal stent era, the need for a standard to examine the stent optimization led to the formation of IVUS defined criteria. IVUS criteria for stent optimization used in the recent randomized clinical trials were summarized in Table 5.5 [3, 8–10]. Despite the need for a consensus, several different criteria have been employed in different clinical studies. However, according to the previous studies and the criteria used in recent trials, the achievement of sufficient lumen area by IVUS may be imperative.

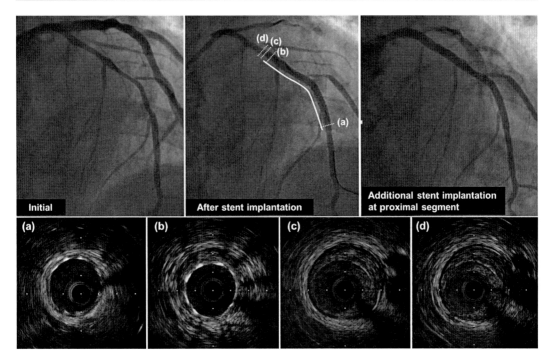

Fig. 5.3 Representative case showing the need of additional stenting at proximal segment of stent because of edge dissection and residual plaque. Although angiographic findings were acceptable, another stent was additionally implanted at the proximal segment based on the IVUS finding, (**c**) the dissection at the proximal edge of the stent with (**d**) residual plaque more than 60%

Table 5.5 IVUS criteria for stent optimization used in recent randomized clinical trials

Study	IVUS criteria for stent optimization
Jakabacin et al. [3]	(1) Good apposition (2) Optimal stent expansion [with MSA of 5 mm^2] or CSA > 90% of distal reference lumen CSA for small vessel (3) No edge dissection (5-mm margins proximal and distal to the stent)
CTO-IVUS [8]	(1) MSA ≥ distal reference lumen area (2) Stent area at CTO segment ≥5 mm^2 as far as vessel area permits (3) Complete stent apposition
Tian et al. [9]	(1) Good apposition (2) Stent MSA >80% of reference vessel area (3) Symmetric index >70% (4) No > Type B dissection
IVUS-XPL [10]	(1) A minimal lumen CSA greater than the lumen CSA at distal reference segments

CSA cross-sectional area, *MSA* minimal stent area

References

1. Levine GN, Bates ER, Blankenship JC, et al. 2011 ACCF/AHA/SCAI guideline for percutaneous coronary intervention. A report of the American College of Cardiology Foundation/American Heart Association task force on practice guidelines and the society for cardiovascular angiography and interventions. J Am Coll Cardiol. 2011;58:e44–122.
2. Windecker S, Kolh P, Alfonso F, et al. 2014 ESC/EACTS guidelines on myocardial revascularization: the task force on myocardial revascularization of the European society of cardiology (ESC) and the European Association for Cardio-Thoracic Surgery (EACTS) developed with the special contribution of the European Association of Percutaneous Cardiovascular Interventions (EAPCI). Eur Heart J. 2014;35:2541–619.
3. Jakabcin J, Spacek R, Bystron M, et al. Long-term health outcome and mortality evaluation after invasive coronary treatment using drug eluting stents with or without the ivus guidance. Randomized control

trial. HOME DES IVUS. Catheter Cardiovasc Interv. 2010;75:578–83.
4. Chieffo A, Latib A, Caussin C, et al. A prospective, randomized trial of intravascular-ultrasound guided compared to angiography guided stent implantation in complex coronary lesions: the AVIO trial. Am Heart J. 2013;165:65–72.
5. Kim JS, Kang TS, Mintz GS, et al. Randomized comparison of clinical outcomes between intravascular ultrasound and angiography-guided drug-eluting stent implantation for long coronary artery stenoses. JACC Cardiovasc Interv. 2013;6:369–76.
6. Mariani J Jr, Guedes C, Soares P, et al. Intravascular ultrasound guidance to minimize the use of iodine contrast in percutaneous coronary intervention: the MOZART (minimizing contrast utilization with ivus guidance in coronary angioplasty) randomized controlled trial. JACC Cardiovasc Interv. 2014;7:1287–93.
7. Tan Q, Wang Q, Liu D, et al. Intravascular ultrasound-guided unprotected left main coronary artery stenting in the elderly. Saudi Med J. 2015;36:549–53.
8. Kim BK, Shin DH, Hong MK, et al. Clinical impact of intravascular ultrasound-guided chronic total occlusion intervention with zotarolimus-eluting versus biolimus-eluting stent implantation: randomized study. Circ Cardiovasc Interv. 2015;8:e002592.
9. Tian NL, Gami SK, Ye F, et al. Angiographic and clinical comparisons of intravascular ultrasound- versus angiography-guided drug-eluting stent implantation for patients with chronic total occlusion lesions: two-year results from a randomised AIR-CTO study. EuroIntervention. 2015;10:1409–17.
10. Hong SJ, Kim BK, Shin DH, et al. Effect of intravascular ultrasound-guided vs angiography-guided everolimus-eluting stent implantation: the ivus-xpl randomized clinical trial. JAMA. 2015;314:2155–63.
11. Witzenbichler B, Maehara A, Weisz G, et al. Relationship between intravascular ultrasound guidance and clinical outcomes after drug-eluting stents: the assessment of dual antiplatelet therapy with drug-eluting stents (ADAPT-DES) study. Circulation. 2014;129:463–70.
12. Roy P, Steinberg DH, Sushinsky SJ, et al. The potential clinical utility of intravascular ultrasound guidance in patients undergoing percutaneous coronary intervention with drug-eluting stents. Eur Heart J. 2008;29:1851–7.
13. Park KW, Kang SH, Yang HM, et al. Impact of intravascular ultrasound guidance in routine percutaneous coronary intervention for conventional lesions: data from the EXCELLENT trial. Int J Cardiol. 2013;167:721–6.
14. Youn YJ, Yoon J, Lee JW, et al. Intravascular ultrasound-guided primary percutaneous coronary intervention with drug-eluting stent implantation in patients with ST-segment elevation myocardial infarction. Clin Cardiol. 2011;34:706–13.
15. Kim JS, Hong MK, Ko YG, Choi D, Yoon JH, Choi SH, Hahn JY, Gwon HC, Jeong MH, Kim HS, Seong IW, Yang JY, Rha SW, Tahk SJ, Seung KB, Park SJ, Jang Y. Impact of intravascular ultrasound guidance on long-term clinical outcomes in patients treated with drug-eluting stent for bifurcation lesions: data from a korean multicenter bifurcation registry. Am Heart J. 2011;161:180–7.
16. Hong SJ, Kim BK, Shin DH, et al. Usefulness of intravascular ultrasound guidance in percutaneous coronary intervention with second-generation drug-eluting stents for chronic total occlusions (from the multicenter Korean-chronic total occlusion registry). Am J Cardiol. 2014;114:534–40.
17. de la Torre Hernandez JM, Baz Alonso JA, Gomez Hospital JA, et al. Clinical impact of intravascular ultrasound guidance in drug-eluting stent implantation for unprotected left main coronary disease: pooled analysis at the patient-level of 4 registries. JACC Cardiovasc Interv. 2014;7:244–54.
18. Jang JS, Song YJ, Kang W, et al. Intravascular ultrasound-guided implantation of drug-eluting stents to improve outcome: a meta-analysis. JACC Cardiovasc Interv. 2014;7:233–43.
19. Ahn JM, Kang SJ, Yoon SH, et al. Meta-analysis of outcomes after intravascular ultrasound-guided versus angiography-guided drug-eluting stent implantation in 26,503 patients enrolled in three randomized trials and 14 observational studies. Am J Cardiol. 2014;113:1338–47.
20. Elgendy IY, Mahmoud AN, Elgendy AY, et al. Outcomes with intravascular ultrasound-guided stent implantation: a meta-analysis of randomized trials in the era of drug-eluting stents. Circ Cardiovas Interv. 2016;9:e003700.
21. Steinvil A, Zhang YJ, Lee SY, et al. Intravascular ultrasound-guided drug-eluting stent implantation: an updated meta-analysis of randomized control trials and observational studies. Int J Cardiol. 2016;216:133–9.
22. Shin DH, Hong SJ, Mintz GS, et al. Effects of intravascular ultrasound-guided versus angiography-guided new-generation drug-eluting stent implantation: meta-analysis with individual patient-level data from 2,345 randomized patients. JACC Cardiovasc Interv. 2016;9:2232.
23. Park SJ, Ahn JM, Kang SJ, et al. Intravascular ultrasound-derived minimal lumen area criteria for functionally significant left main coronary artery stenosis. JACC Cardiovasc Interv. 2014;7:868–74.
24. Jasti V, Ivan E, Yalamanchili V, et al. Correlations between fractional flow reserve and intravascular ultrasound in patients with an ambiguous left main coronary artery stenosis. Circulation. 2004;110:2831–6.
25. Kang SJ, Lee JY, Ahn JM, et al. Intravascular ultrasound-derived predictors for fractional flow reserve in intermediate left main disease. JACC Cardiovasc Interv. 2011;4:1168–74.
26. Kang SJ, Ahn JM, Song H, et al. Comprehensive intravascular ultrasound assessment of stent area and its impact on restenosis and adverse cardiac events in

403 patients with unprotected left main disease. Circ Cardiovasc Interv. 2011;4:562–9.
27. Patel Y, Depta JP, Novak E, et al. Long-term outcomes with use of intravascular ultrasound for the treatment of coronary bifurcation lesions. Am J Cardiol. 2012;109:960–5.
28. Hahn JY, Song YB, Lee SY, et al. Serial intravascular ultrasound analysis of the main and side branches in bifurcation lesions treated with the t-stenting technique. J Am Coll Cardiol. 2009;54:110–7.
29. Park Y, Park HS, Jang GL, et al. Intravascular ultrasound guided recanalization of stumpless chronic total occlusion. Int J Cardiol. 2011;148:174–8.
30. Park JJ, Chae IH, Cho YS, et al. The recanalization of chronic total occlusion leads to lumen area increase in distal reference segments in selected patients: an intravascular ultrasound study. JACC Cardiovasc Interv. 2012;5:827–36.
31. Goto K, Zhao Z, Matsumura M, et al. Mechanisms and patterns of intravascular ultrasound in-stent restenosis among bare metal stents and first- and second-generation drug-eluting stents. Am J Cardiol. 2015;116:1351–7.
32. Ali ZA, Karimi Galougahi K, Nazif T, et al. Imaging- and physiology-guided percutaneous coronary intervention without contrast administration in advanced renal failure: a feasibility, safety, and outcome study. Eur Heart J. 2016;37:3090–5.
33. Lee SY, Shin DH, Kim JS, et al. Intravascular ultrasound predictors of major adverse cardiovascular events after implantation of everolimus-eluting stents for long coronary lesions. Rev Esp Cardiol. 2017;70:88.
34. Song HG, Kang SJ, Ahn JM, et al. Intravascular ultrasound assessment of optimal stent area to prevent in-stent restenosis after zotarolimus-, everolimus-, and sirolimus-eluting stent implantation. Catheter Cardiovasc Interv. 2014;83:873–8.
35. Kang SJ, Cho YR, Park GM, et al. Intravascular ultrasound predictors for edge restenosis after newer generation drug-eluting stent implantation. Am J Cardiol. 2013;111:1408–14.
36. Hassan AK, Bergheanu SC, Stijnen T, et al. Late stent malapposition risk is higher after drug-eluting stent compared with bare-metal stent implantation and associates with late stent thrombosis. Eur Heart J. 2010;31:1172–80.

Pre-Percutaneous Coronary Intervention Lesion Assessment

Sung Yun Lee

6.1 Introduction

Traditionally, quantitative coronary angiography (QCA) was the major imaging modality to assess the severity of CAD for coronary lesion assessment when coronary artery disease is treated with catheter-based coronary interventions. But only provides lumenogram or shadowgram a planar two-dimensional silhouette of the lumen contains only about 25% of the total coronary blood flow and is unsuitable for the precise assessment of atherosclerosis. Intravascular ultrasound (IVUS) provides a unique real-time, tomographic assessment of coronary artery assessment of lesion characteristics, lumen diameters, cross-sectional area, plaque area, and distribution. Generally coronary angiography underestimates the severity and extent of disease, IVUS is golden standard for accurate evaluation for pre-intervention lesion assessment.

Current USA [1] and European guidelines [2] for coronary revascularization recommend IVUS use with class IIa for assessment of angiographically indeterminant left main disease, stent stenosis or failure lesion, stent optimization for selected patients, and evaluation for cardiac allograft vasculopathy (Table 6.1).

Table 6.1 Class IIa recommendation for clinical values of intravascular ultrasound of current coronary revascularization guidelines

2011 ACCF/AHA/SCAI Guideline [1]	Level of evidence
Assessment of angiographically indeterminant left main CAD	B
Reasonable 4–6 weeks and 1 year after cardiac transplantation to exclude donor CAD, detect rapidly progressive cardiac allograft vasculopathy, and provide prognostic information	B
Determine the mechanism of stent restenosis	C
2014 ESC/EACTS guidelines on myocardial revascularization [2]	
Optimize stent implantation in selected patients[a]	B
Assess severity and optimize treatment of unprotected left main lesions	B
Assess mechanism of stent failure	B

[a]In reducing restenosis and adverse events after bare metal stent implantation, better clinical and angiographic results may be obtained under IVUS guidance

6.2 Angiographic Indeterminant Non-Left Main Coronary Artery Stenosis

Fractional flow reserve (FFR; the ratio of distal to proximal pressure at maximum hyperemia) is the standard method for assessing the physiologic significance of a non-left main coronary artery (LMCA) lesion. IVUS has been corrected for vessel size, but IVUS has not been able to factor

S.Y. Lee
Inje University Ilsan Paik Hospital,
Goyang, South Korea
e-mail: im2pci@gmail.com

in the amount of subtended viable myocardium. IVUS minimal lumen area (MLA) in predicting hemodynamic significance in non-LMCA lesions is that the functional effects of a lesion are dependent on additional factors besides dimension. These include lesion location in the coronary tree, lesion length, eccentricity, entrance and exit angles, shear forces, reference vessel dimensions, and the amount of viable myocardium subtended by the lesion [3].

Therefore, in non-LMCA lesions there is only moderate correlation between anatomic dimensions by IVUS and ischemia by physiological assessment. Many studies have attempted to identify invasive IVUS minimum lumen area (MLA) criteria that are equivalent to FFR, reported IVUS MLA cut-off thresholds range from 2.3 to 3.9 mm^2 (Table 6.2) [4–14].

In earlier study IVUS MLA < 4.0 mm^2 correlates with ischemia on single-photon emission computed tomography and also correlates moderately well with an FFR < 0.75. Importantly, low event rates are observed in intermediate lesions when intervention is deferred with an IVUS MLA ≥ 4 mm^2 [15–17]. In the largest study to date, IVUS was compared with FFR in 544 lesions [13]. The optimal cut-off value for predicting an FFR ≤ 0.80 was an MLA = 2.9 mm^2 by IVUS, but the overall accuracy was only 66%. Moreover, of the 240 lesions that had an MLA < 2.9 mm^2, only 47% was hemodynamically significant by FFR. Similarly concerning, 19% of lesions with an MLA > 2.9 mm^2 had an FFR < 0.80, limiting the utility of IVUS for lesion assessment. Kang et al. [7, 8] evaluated 236 angiographically intermediate coronary lesions in which both IVUS and FFR measurements were performed. An IVUS MLA_2.4 mm^2 had the maximum accuracy for predicting FFR < 0.80. However, the overall diagnostic accuracy was 68% with a confidence interval ranging from 1.8 to 2.6 mm^2. FIRST was a multicenter prospective registry of patients who underwent elective coronary angiography and had intermediate coronary stenosis (40–80%) [12]. An IVUS-measured MLA < 3.07 mm^2 had the best sensitivity and specificity (64% and 64.9%, respectively) for correlating with FFR < 0.80.

So, FFR should be considered the standard for assessing the hemodynamic significance of intermediate non-LMCA lesions and better validated than IVUS as a physiologic assessment. An MLA < 4.0 mm^2 has reasonable accuracy in identifying non-significant lesions for which percutaneous coronary intervention (PCI) can be safely deferred [18]. However, an MLA < 4.0 mm^2 does not accurately predict a hemodynamically significant lesion and should not be used in the absence of supporting functional data (such as DEFER, FAME-I, or FAME-II with FFR) to recommend revascularization [3]. An MLA < 3.0 mm^2 is most likely a significant stenosis, but due to its only modest sensitivity and specificity, physiologic testing is desirable before

Table 6.2 Studies correlating intravascular ultrasound to FFR in non-left main intermediate disease

	Takagi et al [4]	Briguori et al [5]	Lee et al [6]	Kang et al [7, 8]	Ben-Dor et al [9, 10]	Koo et al [11]	Waksman et al [12]	VERDICT/ FIRST [13]	Kang et al [14]
No of lesion	51	53	94	236	205	267	304	544	700 LAD
Angiographic DS %	30–70	40–70	30–75	30–75	40–70	30–70	40–80	40–80	30–75
IVUS mean MLA (mm^2)	3.9	3.9	2.3	2.6	3.5	3.0	3.5	3.0	2.5
IVUS MLA cut-off (mm^2)	4.0	4.0	2.0	2.4	3.1	2.8	3.07	3.0	2.5
Year of publication	1999	2001	2010	2011	2011	2011	2013	2013	2013

FFR fractional flow reserve, *LAD* left anterior descending, *DS* diameter stenosis, *IVUS* intravascular ultrasound, *MLA* minimal lumen area, *A* study with Asian populations

Table 6.3 Studies correlating intravascular ultrasound to FFR to identify functional significant LMCA lesion

	N	FFR cut-off	Route of adenosine	IVUS correlation with FFR	Defer	Survival defer (%)	Revascularization	Survival revascularization (%)
Jasti et al [20]	55	0.75	IC	MLA 5.9 mm² MLD 2.8 mm	24	100	20 PCI 11 CABG	100
Park et al [22]	112	0.80	IV	MLA 4.5 mm²	NA	NA	NA	NA
Kang et al. [23]	55	0.80 0.75	IV	MLA 4.8 mm² MLA 4.1 mm²	25	NA	29 PCI 1 CABG	NA

FFR fractional flow reserve, *LMCA* left main coronary artery, *IC* intracoronary, *IV* intravenous, *MLA* minimal lumen area, *MLD* minimal lumen diameter, *PCI* percutaneous coronary intervention, *CABG* coronary artery bypass graft, *NA* not available

proceeding with revascularization. It may be acceptable to defer an intervention in selected situations based on MLA size, IVUS should never be used to justify an intervention.

6.3 Left Main Coronary Artery Lesion

Left main coronary artery (LMCA) lesion has greatest angiographic assessment variability. Small real-world analysis showed that less than half of intermediate LMCA had significant stenosis by IVUS assessment, especially for lesions located at the left main ostium [19].

IVUS evaluation for LMCA stenosis can be valuable when coronary angiography gives equivocal or ambiguous images. Both IVUS and FFR have theoretical and practical limitations for LMCA lesion, proximal LAD and/or LCX disease can impact FFR of LMCA stenosis. With IVUS, distal LMCA lesions can be difficult to accurately image, and often require pullback from both the LCX and LAD. But limited variability in LMCA length, diameter, and amount of supplied myocardium explains the better correlation in LMCA with FFR than non-LMCA stenosis, the most widely used parameter is MLA in LMCA stenosis.

Jasti et al. [20] showed good correlation between FFR and IVUS, with good sensitivities and specificities >0.90. In a study of 55 intermediate LMCA lesions (reference diameter 4.2 mm), an MLA <5.9 mm² and an MLD <2.8 mm correlated well with FFR < 0.75.

A prospective application of these criteria was tested in the LITRO study [21]. LMCA revascularization was performed in 90.5% of patients with an MLA < 6 mm² and was deferred in 96% of patients with an MLA > 6 mm². In a 2-year follow-up period, cardiac death-free survival was 97.7% in the deferred group versus 94.5% in the revascularized group (P = ns), and event-free survival was 87.3% versus 80.6%, respectively (P = ns). At 2-year follow-up, only eight (4.4%) patients in the deferred group required subsequent LMCA revascularization, none of who had an MI. Thus, it is safe to defer LMCA revascularization with MLA > 6 mm². Additionally, the data confirms that MLA < 6.0 mm² is clinically significant, correlates with FFR < 0.75 (Tables 6.3 and 6.4).

More recently study with Asian populations with smaller normal coronary diameters, an MLA cut-off < 4.8 mm² correlates with reduced FFR < 0.8 and <4.1 mm² with FFR < 0.75 [22, 23].

Table 6.4 Challenges treating severely calcified coronary lesions

Respond poorly to angioplasty
Difficult to completely dilate
Prone to dissection during balloon angioplasty or predilatation
Preclude stent delivery to the desired location
Can prevent adequate stent expansion, maybe increased risk of stent thrombosis
May result in stent malapposition
Insufficient drug penetration and subsequent restenosis

6.4 Calcified Lesion

Calcium is under-recognized angiographically. In IVUS study, most of visible calcification by angiogram is correlated with arc of calcium involved, length of calcium involved, and where calcium is located [24]. So visible calcified lesion in angiography means significant calcification nearly encircled the vessel wall and spread along the vessel.

Coronary calcification has been considered a stable coronary lesion. But recent studies, however, it is not really stable because lots of microcalcification and calcified nodule had observed in unstable plaque. Patients with moderate or severe target lesion calcification (TLC) were older, had more renal insufficiency, had lower ejection fractions, and were more likely to have had a STEMI compared with patients with no or mild TLC. On the other hand, lesions with moderate or severe TLC also have other characteristics that are unfavorable, including longer lesion length, more total occlusions, more visible thrombi, and more triple-vessel disease [25].

Calcification may prevent complete expansion of the stent or interfere with stent delivery, resulting in damage either to the structure of the stent or to the polymer in the case of drug-eluting stent (DES). A malapposed, incompletely expanded, or damaged stent increases the risks for stent thrombosis of target lesion. There is general agreement that the greater the arc and length of IVUS-associated lesion calcium the greater the likelihood of underexpansion, but published or agreed criteria for recommending lesion modification prior to stent implantation does not exist. And IVUS has limitation for measure calcium thickness because of acoustic shadow, which may be an important limit to stent expansion ([26], http://www.acc.org/latest-in-cardiology/articles/2016/06/13/10/01/ivus-in-pci-guidance).

On the other hand, and most of the time, iterative IVUS imaging in conjunction with preparation and debulking of the lesion with rotational atherectomy, special balloons such as cutting or scoring and wire-cutting technique and repeated high-pressure adjunctive balloon inflations can be used to correct post-procedure stent underexpansion even in the setting of significant calcification (Fig. 6.1). Nevertheless, it is easier to prevent stent underexpansion than it is to

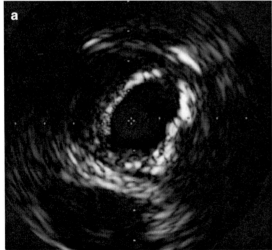
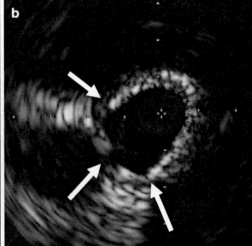

Fig. 6.1 Iterative IVUS imaging for calcified plaque at left anterior descending artery of stable angina patient. (**a**) Pre-intervention intravascular ultrasound showed superficial calcified plaque with 250° of arc. (**b**) Post-intervention intravascular ultrasound revealed luminal gain and few small cracks on superficial calcium (*white arrow*) after AngioScuplt Scout balloon® angioplasty

struggle to correct it such as stent ablation procedure. IVUS studies have shown that localized calcium deposits or the transition from calcified to non-calcified plaque (or to normal vessel wall) are foci for PCI-associated dissections. More extensive dissections occur in segments of arteries that are heavily calcified, and stent implantation into calcified lesions is more often associated with stent fracture.

6.5 Bifurcation Lesion

Coronary bifurcation PCI represents 10–15% of PCI procedures. Bifurcation lesions may show dynamic changes during PCI, with plaque/carina shift or dissection leading to side branch compromise and requiring adjustment to the interventional approach. Therefore, accurate anatomic characterization of bifurcation lesions may improve stent sizing and deployment techniques. The most important role of IVUS is correct measurement of reference vessel size of both main and side branch (SB), if operator decided to use two stent technique with proximal optimization technique.

Also IVUS can detect the distribution of plaques not only in the main branch but also in the ostium of the SB. One study revealed that SB occlusion occurred in 35% of the plaque-containing lesions at the SB ostium after PCI as compared to the 8.2% occlusion rate of plaque-free lesions at the SB ostium [27]. Therefore, wiring the SB to protect it before PCI should be considered if IVUS reveals plaque involvement at the SB ostium, but there do not appear to be reliable IVUS predictors of functional SB compromise after crossover stenting.

In IVUS study [28] regarding complex bifurcation lesions (nearly 90% of the lesions were medina class 1, 1, 1), the number of implanted stents was significantly lower in the IVUS-guided PCI group. Also, the rate of TLR was significantly lower in the IVUS-guided PCI group (6% vs 21%, $P = 0.001$). In this regard, the role of IVUS in decreasing the TLR rate may become more important, a decrease in the number of stents in the IVUS-guided PCI group may contribute to reduce the TLR rate. So, liberal and active use of IVUS in bifurcation PCI is encouraged.

6.6 Vulnerable Plaque

To Identify thrombosis or embolization-prone "vulnerable" plaques before they rupture, catheter-based intravascular imaging modalities are being developed to visualize pathologies in coronary arteries in vivo. Mounting evidences have shown three distinctive histopathological features—the presence of a thin fibrous cap (<65 μm), a lipid-rich necrotic core (>40% of total lesion area), and numerous infiltrating macrophages in the fibrous cap—are key markers of increased vulnerability in atherosclerotic plaques [29].

In the early days of coronary intervention, many coronary angiographic predictors for no-reflow or CK-MB elevation after and during PCI were identified (Table 6.5). After that to visualize these changes, the majority of catheter-based imaging modalities used IVUS with integrated tis-

Table 6.5 Predictors for no-reflow phenomenon or CK-MB elevation after or during PCI

Angiographic characteristics
Accumulated thrombus (>5 mm) proximal to the occlusion
Presence of floating thrombus
Persistent dye stasis distal to the obstruction
Reference lumen diameter of the IRA > 4 mm
Gray scale IVUS
Large plaque burden > 70% of plaque burden
Attenuated plaque
Calcified nodule
Intraluminal mass
Soft plaque, especially lipid pool-like imaging
Positive remodeling
VH-IVUS
Large necrotic core area
VH-thin cap fibrous atheroma
Plaque burden > 40%
necrotic core > 10% of Plaque area
Necrotic core contact lumen at least 3 image slices
Arc of necrotic core > 36° along lumen
Spectroscopy
Max LCBI 4 mm value > 500

sue characterization techniques and OCT to enhance the characterization of vulnerable plaques.

Several studies have evaluated with IVUS to characterize morphologic predictors of plaque vulnerability, the most consistent for this phenomenon, determined by gray scale IVUS, are the presence of a large plaque burden, attenuated plaque, calcified nodule, intraluminal mass (suggestive finding for thrombus), lipid pool-like imaging, and positive vessel remodeling.

Attenuated plaque is defined as the absence of ultrasound signal behind plaque that was either hypoechoic or isoechoic to reference adventitia, but without bright calcium (Fig. 6.2). By definition, echo-attenuated plaque excludes attenuation (or, more correctly, shadowing) behind hyperechoic calcium. The hypothesis that microcalcification and thrombus with underlying advanced atherosclerosis maybe the mechanism of echo attenuation in unstable plaques. Predictors of myonecrosis during stent implantation are a large, grayscale IVUS attenuated plaque especially. When shadowing begins closer to the lumen than to the adventitia [30–32]; a large virtual histology and intravascular ultrasound (VH-IVUS) necrotic core, VH- thin-cap fibroatheroma (TCFA) [33]; a large lipid-rich plaque detected by using near infrared spectroscopy (NIRS) [34–36] (Fig. 6.3); and the presence of plaque rupture [37].

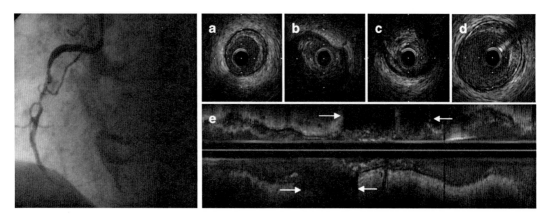

Fig. 6.2 A 75-year-old female presenting with ST elevation myocardial infarction. Pre-intervention intravascular ultrasound showed attenuated plaque (*white arrow*) on both cross-sectional (**b** and **c**) and longitudinal (**e**) intravascular ultrasound image in culprit lesion, but no echo atenuation was found on proximal (**d**) and distal (**a**) refernce segment. After stent deployment, no-reflow phenomenon was developed

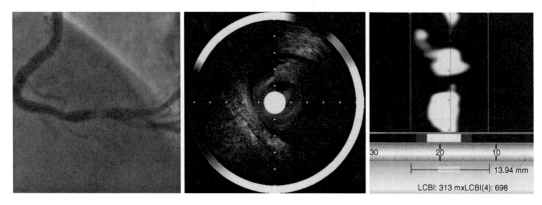

Fig. 6.3 Intravascular ultrasound for distal right coronary artery lesion of acute coronary syndrome patient showed attenuated plaque (unusual echo attenuation without calcification). Near infrared spectroscopy reveled very high maximum LCBI 4 mm value of 698

In patients with acute coronary syndrome, a calcified nodule is observed in 2–7%. IVUS characteristics of a calcified nodule were shown to be: (1) a convex shape of the luminal surface (94.1% of calcified nodules vs. 9.7% of non-nodular calcium); (2) a convex shape of the luminal side of calcium (100% vs. 16.0%); (3) an irregular luminal surface (64.7% vs. 11.6%); and (4) an irregular leading edge of calcium (88.2% vs. 19.0%) [38] (Fig. 6.4). Calcified nodules, especially close to the luminal surface of the plaque, can protrude through and rupture the fibrous cap, leading to thrombus formation and acute coronary syndromes.

However, these studies associated with gray scale IVUS for dangerous plaque are limited by their retrospective or cross-sectional design and small sample size, neither the prognostic utility (risk of future events caused by vulnerable plaques) nor the clinical utility (impact on physician decision making and/or patient outcomes) has been prospectively validated [39], and not informative about the natural history of culprit lesion formation.

More informative color-coded tissue characterization technology has been proved useful tool for TCFA imaging. Of these, VH-IVUS has correlated plaque composition with human coronary atherectomy specimens; however, considering the axial resolution of 200 μm, VH-IVUS is limited in its ability to identify TCFA. To partially overcome this limitation, the VH-TCFA definition was created [40]; VH-TCFA is defined by a focal, necrotic core-containing (10% of the total plaque area) in direct contact with the lumen at least 3 image slices, arc of NC > 36 degree along lumen and in the presence of a percent atheroma volume 40% (Fig. 6.5).

Thrombus aspiration or distal protection device deployment before PCI is recommended if

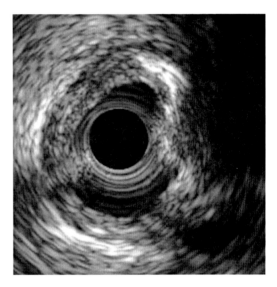

Fig. 6.4 Typical intravascular ultrasound findings of calcified nodule. (1) A convex shape of the luminal surface; (2) a convex shape of the luminal side of calcium; (3) an irregular luminal surface; and (4) an irregular leading edge of calcium

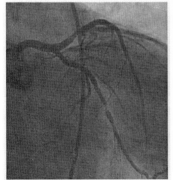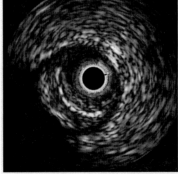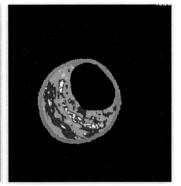

Fig. 6.5 Virtual histology and intravascular ultrasound for left circumflex lesion of acute coronary syndrome patient. (**a**) Gray scale IVUS revealed 76% plaque burden mainly soft plaque with lipid pool-like imaging. (**b**) Virtual histology showed typical findings of VH-thin cap fibroatheroma (necrotic area 49% with lumen contact)

such lesions are found. Furthermore, sometimes interventionists may encounter multiple borderline angiographic lesions without critical narrowing during acute coronary syndrome intervention. Even though inferior to OCT, IVUS as well as tissue characterization may play an important role in locating the culprit lesion where plaque rupture or TCFA has been developed [41].

To date only VH-IVUS has been shown to predict future nonculprit events. In the PROSPECT study, predictors of nonculprit events at 3 years were a VH-TCFA, an IVUS MLA < 4.0 mm^2, and an IVUS plaque burden > 70% [42]. These findings, especially the importance of a large plaque burden [37], were supported by the VIVA (VH-IVUS in Vulnerable Atherosclerosis) and ATHEROREMO-IVUS (European Collaborative Project on Inflammation and Vascular Wall Remodeling in Atherosclerosis–Intravascular Ultrasound) studies [43, 44].

PROSPECT II study is an ongoing overall prospective observational study using multimodality imaging that will examine the natural history of patients with unstable atherosclerotic coronary artery disease with IVUS and Near InfraRed Spectroscopy (NIRS), to identify plaques prone to future rupture and clinical events, plaque Burden (PB) ≥ 70% as the primary threshold defining vulnerable plaques. Currently, we cannot predict which plaques carry a risk of complications high enough to warrant prophylactic therapy, although a randomized substudy within the PROSPECT-II study will attempt to address this issue (https://clinicaltrials.gov/ct2/show/NCT02171065).

6.7 Stent Failure

Recurrence of symptoms or ischemia after PCI is the result of restenosis, incomplete initial revascularization, or disease progression. In both bare metal stents (BMS) and DES, the IVUS predictors of early stent thrombosis or in-stent restenosis (ISR) are underexpanded stent and inflow/outflow track disease (e.g., dissections, significant plaque burden, edge stenosis), but not acute stent malapposition as long as the stent is well expanded [37]. Underexpansion refers to the size of the stent, whereas malapposition refers to the contact of the stent with the vessel wall.

The use of intracoronary imaging has also been advocated in patients with stent failure, including restenosis and stent thrombosis, in order to explicate and correct underlying mechanical factors (Fig. 6.6).

6.8 Restenosis and Neoatherosclerosis

The presence of an underexpanded stent should, if possible, be corrected using repeat aggressive high-pressure noncompliant balloon angioplasty during the repeat procedure.

IVUS criteria of stent underexpansion depend on lesion location or size of reference vessel size.

Restenosis associated with stent underexpansion, repeat aggressive high-pressure balloon dilation should be used to correct underlying, stent-related, predisposing, mechanical problems revascularization and repeat PCI remains the strategy of choice for these patients if technically feasible (Table 6.6).

Recent studies have reported that one-third of patients with in-stent restenosis of bare BMS presented with acute coronary syndrome that is not regarded as clinically benign. Furthermore, both clinical and histologic studies of DES have demonstrated evidence of continuous neointimal growth during long-term follow-up, which is designated as "late catch-up" phenomenon.

In-stent neoatherosclerosis is an important substrate for late stent failure for both BMS and DES, especially in the extended phase. In light of the rapid progression in DES, early detection of neoatherosclerosis may be beneficial to improving long-term outcome of patients with DES implants [45].

Gray scale IVUS cannot discriminate neoatherosclerosis from neointimal hyperplasia. It is difficult for IVUS to determine or classify neointimal tissue because of the signal interference from metal struts, there are several reports attempting discrimination of neointimal tissues by IVUS. A case report described calcified neo-

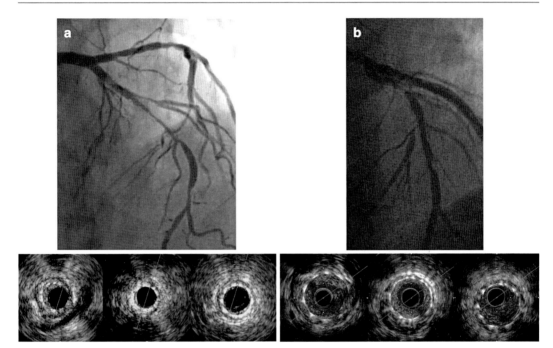

Fig. 6.6 Examples of intravascular ultrasound of in-stent restenosis. (**a**) Pre-intervention intravascular ultrasound measured minimal stent area 2.42 mm² for diffuse in-stent restenosis of PICO elite stent (diameter 3 mm) of mid-LAD, typical example of stent underexpansion. (**b**) Pre-intervention intravascular ultrasound measured minimal stent area 5.83 mm² and neointial hyperplasia area 3.84 mm² (65.9% of stent area) for in-stent restenosis of Endeavor stent (diameter 3 mm) of mid-LAD, typical example of intimal hyperplasia

Table 6.6 IVUS criteria of stent underexpansion and neoatherosclerosis

IVUS minimal stent area criteria of stent underexpansion
Left main above polygon of confluence <8 mm² in LMCA ostium or mid shaft <7 mm² in distal LMCA <6 mm² in LAD ostium <5 mm² in LCX ostium <5.0–5.5 mm² in general
In small vessel disease <80% of the average proximal and distal reference
Lumen <90% of the distal reference lumen area
VH-IVUS and NIRS findings suggest neoatherosclerosis
In-stent necrotic core and dense calcium maxLCBI 4 mm predict OCT-TCNA with a cut-off value of >144

IVUS intravascular ultrasound, *LMCA* left main coronary artery, *LAD* left anterior descending artery, *LCX* left circumflex artery, *VH-IVUS* virtual histology and intravascular ultrasound, *OCT-TCNA* OCT derived thin fibrous cap neoatheromas

intima on gray scale IVUS 8 years after BMS deployment [46], and other reports demonstrated plaque rupture and a flaplike dissection inside a restenotic stent [47, 48].

In addition, VH-IVUS has recently been reported to identify neointimal hyperplasia with unstable morphology that mimics a TCFA as in native arteries.

Using VH-IVUS, tissue characterization of restenotic in-stent neointima after DES ($n = 70$) and BMS ($n = 47$) implantation was assessed in 117 lesions with angiographic in-stent restenosis and intimal hyperplasia (IH) > 50% of the stent area. Both groups had greater percent necrotic core and percent dense calcium at maximal percent IH and maximal percent necrotic core sites, especially in stents that had been implanted for longer periods. VH-IVUS analysis showed that BMS- and DES-treated lesions develop in-stent necrotic core and dense calcium, suggesting the development of in-stent neoatherosclerosis [49] (Fig. 6.7).

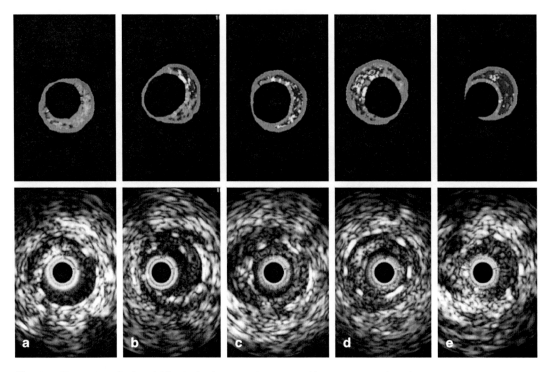

Fig. 6.7 Examples of virtual histologic intravascular ultrasound composition of the neointima at maximal percent intimal hyperplasia sites. Follow-ups of paclitaxel-eluting stent implantation at (**a**) 6 months (necrotic core 10%, dense calcium 2%), (**b**) 9 months (necrotic core 28%, dense calcium 8%), and (**c**) 22 months (necrotic core 39%, dense calcium 20%) and bare metal stent implantation at (**d**) 48 months (necrotic core 40%, dense calcium 25%) and (**e**) 57 months (necrotic core 57%, dense calcium 15%) (Kang et al. Am J Cardiol 2010;106:1561–1565)

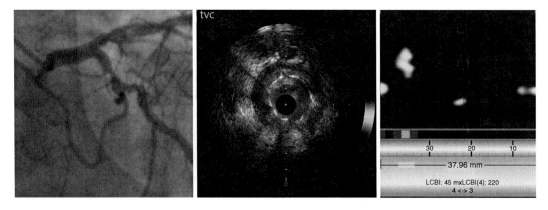

Fig. 6.8 IVUS for LCX ISR lesion showed modest intima hyperplasia and stent fracture. Near infrared spectroscopy revealed maximum LCBI 4 mm value of 220, suggested presence of thin fibrous cap neoatheromas

Recent study evaluated ability of NIRS to detect OCT derived thin fibrous cap neoatheromas [TCNA; thin fibrous cap covering the lipid core (<65 μm)] are prone to rupture and higher risk of late stent failure. In 39 drug-eluting stents with ISR, values of LCBI derived by NIRS were compared with the OCT-derived thickness of the fibrous cap covering neoatherosclerotic lesions. A total of 22 (49%) in-stent neointimas were identified as lipid rich by both NIRS and

OCT. There was good agreement between OCT and NIRS in identifying lipid within in-stent neointima. OCT identified TCNA in 12 stents (23%), the minimal cap thickness of in-stent neoatherosclerotic plaque measured by OCT correlated with the maxLCBI 4mm (maximal LCBI per 4 mm) within the stent ($r = -0.77$, $P < 0.01$). Moreover, maxLCBI 4 mm was able to accurately predict TCNA with a cut-off value of >144. NIRS correlates with OCT identification of lipids in stented vessels and is able to predict the presence of thin fibrous cap neoatheroma [50] (Fig. 6.8).

References

1. Levine GN, Bates ER, Blankenship JC, Bailey SR, Bittl JA, Cercek B, Chambers CE, Ellis SG, Guyton RA, Hollenberg SM, Khot UN, Lange RA, Mauri L, Mehran R, Moussa ID, Mukherjee D, Nallamothu BK, Ting HH, American College of Cardiology Foundation, American Heart Association Task Force on Practice Guidelines, Society for Cardiovascular Angiography and Interventions. 2011 ACCF/AHA/SCAI guideline for percutaneous coronary intervention. A report of the American College of Cardiology Foundation/American Heart Association Task Force on Practice Guidelines and the Society for Cardiovascular Angiography and Interventions. J Am Coll Cardiol. 2011;58:44–122.
2. Authors/Task Force Members, Windecker S, Kolh P, Alfonso F, Collet JP, Cremer J, Falk V, Filippatos G, Hamm C, Head SJ, Jüni P, Kappetein AP, Kastrati A, Knuuti J, Landmesser U, Laufer G, Neumann FJ, Richter DJ, Schauerte P, Sousa Uva M, Stefanini GG, Taggart DP, Torracca L, Valgimigli M, Wijns W, Witkowski A. 2014 ESC/EACTS guidelines on myocardial revascularization: the Task Force on Myocardial Revascularization of the European Society of Cardiology (ESC) and the European Association for Cardio-Thoracic Surgery (EACTS)Developed with the special contribution of the European Association of Percutaneous Cardiovascular Interventions (EAPCI). Eur Heart J. 2014;35:2541–619.
3. Pijls NH, Sels JW. Functional measurement of coronary stenosis. J Am Coll Cardiol. 2012;59:1045–57.
4. Takagi A, Tsurumi Y, Ishii Y, et al. Clinical potential of intravascular ultrasound for physiological assessment of coronary stenosis: relationship between quantitative ultrasound tomography and pressure-derived fractional flow reserve. Circulation. 1999;100:250–5.
5. Briguori C, Anzuini A, Airoldi F, et al. Intravascular ultrasound criteria for the assessment of the functional significance of intermediate coronary artery stenoses and comparison with fractional flow reserve. Am J Cardiol. 2001;87:136–41.
6. Lee CH, Tai BC, Soon CY, et al. New set of intravascular ultrasound-derived anatomic criteria for defining functionally significant stenoses in small coronary arteries (results from intravascular ultrasound diagnostic evaluation of atherosclerosis in Singapore [IDEAS] study). Am J Cardiol. 2010;105:1378–84.
7. Kang SJ, Lee JY, Ahn JM, et al. Validation of intravascular ultrasound-derived parameters with fractional flow reserve for assessment of coronary stenosis severity. Circ Cardiovasc Interv. 2011;4:65–71.
8. Kang SJ, Ahn JM, Song H, et al. Usefulness of minimal luminal coronary area determined by intravascular ultrasound to predict functional significance in stable and unstable angina pectoris. Am J Cardiol. 2012;109:947–53.
9. Ben-Dor I, Torguson R, Gaglia MA Jr, et al. Correlation between fractional flow reserve and intravascular ultrasound lumen area in intermediate coronary artery stenosis. EuroIntervention. 2011;7:225–33.
10. Ben-Dor I, Torguson R, Deksissa T, et al. Intravascular ultrasound lumen area parameters for assessment of physiological ischemia by fractional flow reserve in intermediate coronary artery stenosis. Cardiovasc Revasc Med. 2012;13:177–82.
11. Koo BK, Yang HM, Doh JH, et al. Optimal intravascular ultrasound criteria and their accuracy for defining the functional significance of intermediate coronary stenoses of different locations. J Am Coll Cardiol Intv. 2011;4:803–11.
12. Waksman R, Legutko J, Singh J, et al. FIRST: fractional flow reserve and intravascular ultrasound relationship study. J Am Coll Cardiol. 2013;61:917–23.
13. Stone GW. VERDICT/FIRST: prospective, multicenter study examining the correlation between IVUS and FFR parameters in intermediate lesions. Available at https://www.tctmd.com/slide/verdictfirst-prospective-multicenter-study-examining-correlation-between-ivus-and-ffr. 2013.
14. Kang SJ, Ahn JM, Han S, et al. Sex differences in the visual-functional mismatch between coronary angiography or intravascular ultrasound versus fractional flow reserve. J Am Coll Cardiol Intv. 2013;6:562–8.
15. Abizaid A, Mintz GS, Pichard AD, Kent KM, Satler LF, Walsh CL, Popma JJ, Leon MB. Clinical, intravascular ultrasound, and quantitative angiographic determinants of the coronary flow reserve before and after percutaneous transluminal coronary angioplasty. Am J Cardiol. 1998;82:423–8.
16. Nishioka T, Amanullah AM, Luo H, Berglund H, Kim CJ, et al. Clinical validation of intravascular ultrasound imaging for assessment of coronary stenosis severity: comparison with stress myocardial perfusion imaging. J Am Coll Cardiol. 1999;33:1870–8.
17. Abizaid AS, Mintz GS, Mehran R, Abizaid A, Lansky AJ, et al. Long-term follow-up after percutaneous transluminal coronary angioplasty was not performed based on intravascular ultrasound findings:

importance of lumen dimensions. Circulation. 1999;100:256–61.
18. Lotfi A, Jeremias A, Fearon WF, Feldman MD, Mehran R, Messenger JC, Grines CL, Dean LS, Kern MJ, Klein LW, Society of Cardiovascular Angiography and Interventions. Expert consensus statement on the use of fractional flow reserve, intravascular ultrasound, and optical coherence tomography: a consensus statement of the Society of Cardiovascular Angiography and Interventions. Catheter Cardiovasc Interv. 2014;83(4):509–18.
19. Sano K, Mintz GS, Carlier SG, de Ribamar Costa J Jr, Qian J, Missel E, Shan S, Franklin-Bond T, Boland P, Weisz G, Moussa I, Dangas GD, Mehran R, Lansky AJ, Kreps EM, Collins MB, Stone GW, Leon MB, Moses JW. Assessing intermediate left main coronary lesions using intravascular ultrasound. Am Heart J. 2007;154:983–8.
20. Jasti V, Ivan E, Yalamanchili V, Wongpraparut N, Leesar MA. Correlations between fractional flow reserve and intravascular ultrasound in patients with an ambiguous left main coronary artery stenosis. Circulation. 2004;110:2831–6.
21. de la Torre Hernandez JM, Hernandez Hernandez F, Alfonso F, Rumoroso JR, Lopez-Palop R, et al. Prospective application of pre-defined intravascular ultrasound criteria for assessment of intermediate left main coronary artery lesions results from the multicenter LITRO study. J Am Coll Cardiol. 2011;58:351–8.
22. Park SJ, Ahn JM, Kang SJ, et al. Intravascular ultrasound-derived minimal lumen area criteria for functionally significant left main coronary artery stenosis. JACC Cardiovasc Interv. 2014;7:868–74.
23. Kang SJ, Lee JY, Ahn JM, Song HG, Kim WJ, et al. Intravascular ultrasound-derived predictors for fractional flow reserve in intermediate left main disease. JACC Cardiovasc Interv. 2011;4:1168–74.
24. Mintz GS, Popma JJ, Pichard AD, Kent KM, Satler LF, Chuang YC, Ditrano CJ, Leon MB. Patterns of calcification in coronary artery disease. A statistical analysis of intravascular ultrasound and coronary angiography in 1155 lesions. Circulation. 1995;91:1959–65.
25. Généreux P, Madhavan MV, Mintz GS, Maehara A, Palmerini T, Lasalle L, Xu K, McAndrew T, Kirtane A, Lansky AJ, Brener SJ, Mehran R, Stone GW. Ischemic outcomes after coronary intervention of calcified vessels in acute coronary syndromes. Pooled analysis from the HORIZONS-AMI (harmonizing outcomes with revascularization and stents in acute myocardial infarction) and ACUITY (acute catheterization and urgent intervention triage strategy) TRIALS. J Am Coll Cardiol. 2014;63:1845–54.
26. Mintz GS. Intravascular imaging of coronary calcification and its clinical implications. JACC Cardiovasc Imaging. 2015;8:461–7.
27. Furukawa E, Hibi K, Kosuge M, et al. Intravascular ultrasound predictors of side branch occlusion in bifurcation lesions after percutaneous coronary intervention. Circ J. 2005;69:325–30.
28. Patel Y, Depta JP, Novak E, et al. Long-term outcomes with use of intravascular ultrasound for the treatment of coronary bifurcation lesions. Am J Cardiol. 2012;109:960–5.
29. Virmani R, Burke AP, Farb A, Kolodgie FD. Pathology of the vulnerable plaque. J Am Coll Cardiol. 2006;47:13–8.
30. Lee SY, Mintz GS, Kim SY, et al. Attenuated plaque detected by intravascular ultrasound: clinical, angiographic, and morphologic features and postpercutaneous coronary intervention complications in patients with acute coronary syndromes. J Am Coll Cardiol Intv. 2009;2:65–72.
31. Wu X, Mintz GS, Xu K, et al. The relationship between attenuated plaque identified by intravascular ultrasound and no-reflow after stenting in acute myocardial infarction: the HORIZONS-AMI (harmonizing outcomes with revascularization and stents in acute myocardial infarction) trial. J Am Coll Cardiol Intv. 2011;4:495–502.
32. Shiono Y, Kubo T, Tanaka A, et al. Impact of attenuated plaque as detected by intravascular ultrasound on the occurrence of microvascular obstruction after percutaneous coronary intervention in patients with ST-segment elevation myocardial infarction. J Am Coll Cardiol Intv. 2013;6:847–53.
33. Claessen BE, Maehara A, Fahy M, Xu K, Stone GW, Mintz GS. Plaque composition by intravascular ultrasound and distal embolization after percutaneous coronary intervention. J Am Coll Cardiol Img. 2012;5:S111–8.
34. Goldstein JA, Maini B, Dixon SR, et al. Detection of lipid-core plaques by intracoronary near-infrared spectroscopy identifies high risk of periprocedural myocardial infarction. Circ Cardiovasc Interv. 2011;4:429–37.
35. Raghunathan D, Abdel-Karim AR, Papayannis AC, et al. Relation between the presence and extent of coronary lipid core plaques detected by near-infrared spectroscopy with postpercutaneous coronary intervention myocardial infarction. Am J Cardiol. 2011;107:1613–8.
36. Brilakis ES, Abdel-Karim AR, Papayannis AC, et al. Embolic protection device utilization during stenting of native coronary artery lesions with large lipid core plaques as detected by nearinfrared spectroscopy. Catheter Cardiovasc Interv. 2012;80:1157–62.
37. Mintz GS. Clinical utility of intravascular imaging and physiology in coronary artery disease. J Am Coll Cardiol. 2014;64:207–22.
38. Lee JB, Mintz GS, Lisauskas JB, et al. Histopathologic validation of the intravascular ultrasound diagnosis of calcified coronary artery nodules. Am J Cardiol. 2011;108:1547–51.
39. Alsheikh-Ali AA, Kitsios GD, Balk EM, Lau J, Ip S. The vulnerable atherosclerotic plaque: scope of the literature. Ann Intern Med. 2010;153:387–95.

40. Rodriguez-Granillo GA, Garcia-Garcia HM, McFadden EP, Valgimigli M, Aoki J, de Feyter P, Serruys PW. In vivo intravascular ultrasoundderived thin-cap fibroatheroma detection using ultrasound radiofrequency data analysis. J Am Coll Cardiol. 2005;46:2038–42.
41. Wu X, Maehara A, Mintz GS, et al. Virtual histology intravascular ultrasound analysis of non-culprit attenuated plaques detected by grayscale intravascular ultrasound in patients with acute coronary syndromes. Am J Cardiol. 2010;105:48–53.
42. Stone GW, Maehara A, Lansky AJ, et al. A prospective natural-history study of coronary atherosclerosis. N Engl J Med. 2011;364:226–35.
43. Calvert PA, Obaid DR, O'Sullivan M, et al. Association between IVUS findings and adverse outcomes in patients with coronary artery disease: the VIVA (VH-IVUS in vulnerable atherosclerosis) study. JACC Cardiovasc Imaging. 2011;4:894–901.
44. Cheng JM, Garcia-Garcia HM, de Boer SP, et al. In vivo detection of high-risk coronary plaques by radiofrequency intravascular ultrasound and cardiovascular outcome: results of the ATHEROREMO-IVUS study. Eur Heart J. 2014;35:639–47.
45. Park SJ, Kang SJ, Virmani R, Nakano M, Ueda Y. In-stent neoatherosclerosis: a final common pathway of late stent failure. J Am Coll Cardiol. 2012;59:2051–7.
46. Appleby CE, Bui S, Dzavı'k V. A calcified neointima-"stent" within a stent. J Invasive Cardiol 2009;21:141–143.
47. Fineschi M, Carrera A, Gori T. Atheromatous degeneration of the neointima in a bare metal stent: intravascular ultrasound evidence. J Cardiovasc Med. 2009;10:572–3.
48. Hoole SP, Starovoytov A, Hamburger JN. In-stent restenotic lesions can rupture: a case against plaque sealing. Catheter Cardiovasc Interv. 2010;77:841–2.
49. Kang SJ, Mintz GS, Park DW, Lee SW, Kim YH, Lee CW, Han KH, Kim JJ, Park SW, Park SJ. Tissue characterization of in-stent neointima using intravascular ultrasound radiofrequency data analysis. Am J Cardiol. 2010;106:1561–5.
50. Roleder T, Karimi Galougahi K, Chin CY, Bhatti NK, Brilakis E, Nazif TM, Kirtane AJ, Karmpaliotis D, Wojakowski W, Leon MB, Mintz GS, Maehara A, Stone GW, Ali ZA. Utility of near-infrared spectroscopy for detection of thin-cap neoatherosclerosis. Eur Heart J Cardiovasc Imaging. 2017;18:663. doi:10.1093/ehjci/jew198.

IVUS: Post-Evaluation After Stenting

Yun-Kyeong Cho and Seung-Ho Hur

7.1 Introduction

Although the coronary angiography (CAG) can visualize the improvement of luminal narrowing after stent implantation in coronary atherosclerotic lesions, it only provides indirect vessel information using contrast medium because of a shadow image at stented segments as well as adjacent reference segments. Intravascular ultrasound (IVUS) is capable of generating a cross-sectional anatomy of the vessel wall comparable to corresponding histologic image, resulting in providing more information of atherosclerotic coronary plaque either quantitatively or qualitatively. On the other hand, stent struts appear as focal, bright spots at cross-sectional and longitudinal images owing to a strong echoreflection by ultrasound beam. Thus, it allows detailed information regarding stent strut expansion, intrastent luminal condition, and plaque characteristics at adjacent reference vessel area [1]. The routine use of IVUS in daily practice is still a matter of debate in current drug-eluting stent (DES) era, however, stent optimization by IVUS immediately after stent deployment has reported to improve clinical outcomes, especially during complex percutaneous coronary intervention (PCI) [2, 3]. This chapter reviews important IVUS findings after stent implantation and its clinical relevance.

Y.-K. Cho • S.-H. Hur (✉)
Interventional Cardiology, Cardiovascular Medicine,
Keimyung University Dongsan Hospital,
Daegu, South Korea
e-mail: shur@dsmc.or.kr

7.2 Evaluation of Stent Symmetry and Eccentricity

Symmetry index (SI) defines minimum stent diameter/maximum stent diameter (Fig. 7.1) [4]. Asymmetry index (AI) also can be used to express the stent symmetry: (1 − minimum stent diameter/maximum stent diameter) [5]. Because maximum and minimum stent diameters are the values throughout an entire stented segment, these diameters can derive from different cross section in the stented segment. A stent was characterized as asymmetric when the value of AI was over 0.3 (which corresponds to SI of 0.70 from the MUSIC study). Post-procedural asymmetry of device was associated with unfavorable clinical outcomes [6].

Eccentricity index (EI) was calculated as minimum stent diameter/maximum stent diameter to show the circularity of the cross section. Therefore, the calculation of minimum and maximum stent diameters were derived from the same cross section frame by frame and value was expressed as an average. A stent with EI ≥ 0.7 was defined as concentric while EI < 0.7 was defined as eccentric [7, 8]. The eccentricity of DES had been previously considered as one of the factors for restenosis, because of the higher possibility of the uneven diffusion of the drug into the arterial wall [9]. However, subsequent reports showed that eccentricity of DES did not have any clinical impact because DES powerfully suppressed the neointimal formation [8, 10].

7.3 Measurement of Minimal Stent Area

Minimal stent area (MSA) of bare metal stent (BMS) for long-term patency was considered as 6.4–6.5 mm² [11, 12], and adequate post-interventional MSA of DES was 5.0–5.7 mm² (Fig. 7.2) [13–15]. In left main lesions, optimal

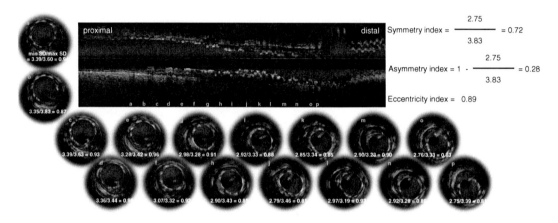

Fig. 7.1 A representative images showing stent symmetry and eccentricity. Minimum and maximum stent diameters with 1 mm interval over the length of the device were shown. Stent (Xience alpine, 3.5 × 15 mm) showed symmetric and concentric expansion

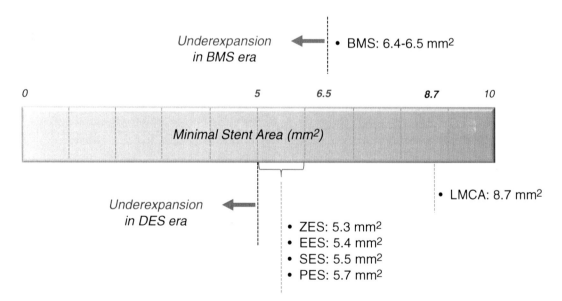

Fig. 7.2 Minimal stent area (MSA) to prevent in-stent restenosis or target vessel revascularization. Best cutoff of bare metal stent (BMS) was 6.4–6.5 mm² and the value of drug-eluting stent (DES) was 5.0–5.7 mm². In case of left main coronary artery (LMCA), 8.7 mm² was suggested

Fig. 7.3 Minimal stent area (MSA) for left main bifurcation lesion. Considering 4 segments of left main bifurcation, the best MSA criteria were 5.0 mm² (ostial left circumflex artery), 6.3 mm² (ostial left anterior descending artery), 7.2 mm² (polygon of confluence [POC]), and 8.2 mm² (proximal left main artery above the POC)

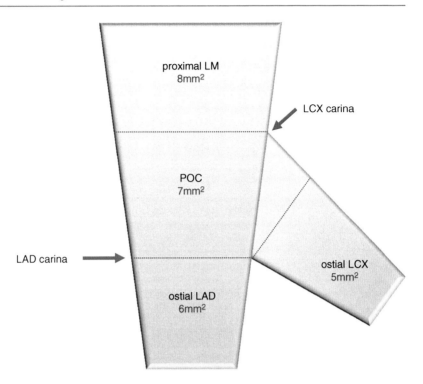

MSA was reported as 8.7 mm² in the MAIN-COMPARE (revascularization for unprotected left main coronary artery stenosis: comparison of percutaneous coronary angioplasty versus surgical revascularization) study [2]. Considering 4 segments of left main bifurcation, the best minimal stent area criteria to predict angiographic restenosis were 5.0 mm² (ostial left circumflex artery), 6.3 mm² (ostial left anterior descending artery), 7.2 mm² (polygon of confluence [POC]), and 8.2 mm² (proximal left main artery above the POC) (Fig. 7.3) [16].

7.4 Evaluation of Stent Expansion (Well Expansion vs. Underexpansion)

In the BMS era, MUSIC study (multicenter ultrasound stenting in coronaries study) defined adequate expansion as >90% of the average reference cross-sectional area (CSA), or >100% of a smaller reference CSA with complete apposition and symmetric expansion [4]. CRUISE (Can Routine Ultrasound Influence Stent Expansion) study showed better stent expansion of IVUS-guided PCI than angiography-guided PCI, especially in terms of target vessel revascularization (TVR), but not in mortality or myocardial infarction [17]. In contrast to the BMS era, early studies of IVUS-guided PCI with DES had no significant benefit in terms of TVR or clinical events. AVIO (Angiography Versus IVUS Optimization) study which defined optimal stent expansion as final minimum stent CSA of at least 70% of the hypothetical CSA of the fully inflated balloon used for post-dilatation did not show any difference in clinical outcome [18]. However, attention should be paid to avoid stent underexpansion. Several evidences indicate that stent underexpansion is one of the major causes of stent failure such as stent restenosis or stent thrombosis (Table 7.1) [14, 19–21]. ADAPT-DES (Assessment of Dual Antiplatelet Therapy With Drug-Eluting Stents) study showed reduction in stent thrombosis, myocardial infarction, and major adverse cardiac events by IVUS-guided optimization of stent expansion and apposition [22]. Representative

Table 7.1 Underexpansion as the predictor of DES thrombosis and restenosis

Study	Stent type	No. of lesion	Minimal stent area
Fujii K, et al. [19]	Sirolimus-eluting stent (Cypher)	15 in ST group vs. 45 in control group	4.3 ± 1.6 mm^2 in ST group vs. 6.2 ± 1.9 mm^2 in control group
Okabe T, et al. [20]	Sirolimus-eluting stent (Cypher), paclitaxel-eluting stent (Taxus)	14 in ST group vs. 30 in control group	4.6 ± 1.1 mm^2 in ST group vs. 5.6 ± 1.7 mm^2 in control group
Liu X, et al. [21]	Sirolimus-eluting stent (Cypher), paclitaxel-eluting stent (Taxus)	20 in ST group vs. 50 in ISR group vs. 50 in control group	3.9 ± 1.0 mm^2 in ST group vs. 5.0 ± 1.7 mm^2 in ISR group vs. 6.0 ± 1.6 mm^2 in control group
Hong MK, et al. [14]	Sirolimus-eluting stent (Cypher)	21 in ISR group vs. 522 in control group	5.1 ± 1.5 mm^2 in ISR group vs. 6.5 ± 1.9 mm^2 in control group

DES drug-eluting stent, *ST* stent thrombosis, *ISR* in-stent restenosis

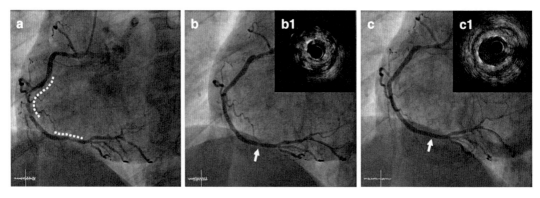

Fig. 7.4 A representative images of stent underexpansion and well expansion. A 53-year-old man was admitted with stable angina. The coronary angiogram (CAG) showed significant stenosis (*dotted line*) on mid and distal right coronary artery (RCA) (**a**). Two drug-eluting stents (Ultimaster 3.0 × 33 mm on mid RCA and Ultimaster 2.75 × 18 mm on distal RCA) were implanted separately and CAG after stent implantation showed stent underexpansion on distal RCA (**b**, *arrow*). Corresponding intravascular ultrasound image showed minimal stent area (MSA) of 2.57 mm^2 (*b1*). After additional dilation with noncompliant balloon, CAG showed well expansion of distal stent (**c**, *arrow*) and MSA was increased as 5.06 mm^2 (*c1*)

IVUS images of underexpansion and well expansion are shown in Fig. 7.4.

7.5 Detection of Stent Edge Dissection

Stent edge dissection is a tear in the plaque parallel to the vessel wall with visualization of blood flow in the false lumen <5 mm to a stent edge. The incidence of edge dissections by IVUS is approximately 10–20% and 40% of the IVUS-identified dissections was not detected by angiography [23–25]. Significant (major) edge dissections, defined by IVUS as lumen area < 4 mm^2 or dissection angle ≥60°, have been associated with early stent thrombosis [26]. However, minor non-flow-limiting dissection at the edge of stent may not be associated with an increased incidence of clinical events although no consensus exists on an optimal strategy. Figure 7.5 is an example of stent edge dissection.

7.6 Detection of Acute Incomplete Stent Apposition

Incomplete stent apposition (ISA), synonymous with stent malapposition, was defined as the absence of contact between at least one strut and

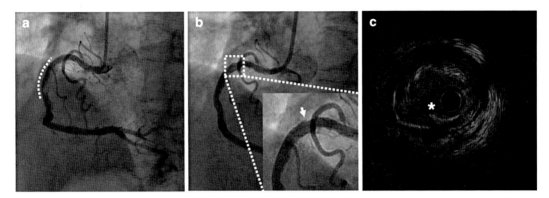

Fig. 7.5 A case of stent edge dissection. A 60-year-old woman with stable angina showed calcified stenotic lesion (*dotted line*) on mid-right coronary artery (**a**). The coronary angiogram after drug-eluting stent implantation showed small dissection on proximal stent edge (**b**, *arrow*). Dissection flap (*asterisk*) was observed by intravascular ultrasound (**c**)

the lumen wall that did not overlap a side branch with evidence of blood speckle behind the strut and can occur acutely after stent implantation (acute ISA) or develop over time (late-acquired ISA). Acute ISA is almost due to suboptimal stent deployment. The frequency of acute ISA has been reported to be nearly 10% and it appears not to be associated with increased cardiac events [27, 28].

7.7 Detection of Tissue Protrusion (Plaque Prolapse and Intra-stent Thrombus)

Tissue protrusion (TP) was defined as a visible tissue extrusion through the stent struts by IVUS (Fig. 7.6) [29, 30]. Although thrombus was characterized by heterogeneous echodensity tissue with a sparkling pattern by IVUS [31], the accurate discrimination of atherosclerotic plaque and thrombus within stent is very difficult because of limited resolution of IVUS. Thus, TP includes plaque and/or thrombus extrusion within stent [32]. The incidence of TP has been reported in various ranges between 20% and 73%, depended on characteristics of enrolled patients (Table 7.2) [29, 30, 32–36]. In fact, TP is likely to develop in patients with acute coronary syndrome, especially ST-segment elevation myocardial infarction owing to a higher chance of thrombus or friable plaque compared to stable patients [32, 35] and receiving longer stent probably caused by unequal distribution of inflation pressure during stent deployment [30, 34]. Other predictors of TP are larger reference lumen area, greater plaque burden, more plaque rupture, attenuated plaque, positive vascular remodeling, and virtual histology thin-cap fibroatheroma by IVUS [30, 32]. The clinical impact of TP remains a controversy. Previous studies suggested that TP after stent implantation may increase the risk of stent thrombosis [26, 37]. Other studies, however, have been failed to show this relationship [29, 32, 38].

Although some investigators demonstrated greater cardiac enzyme elevation after stent implantation in patients with TP, it did not translate into the increased risk of stent thrombosis or periprocedural myocardial infarction [30, 32]. An IVUS substudy from ADAPT-DES reported the 2-year clinical outcomes of TP after stenting. At 2-year clinical follow-up, there was no difference in the rate of major adverse cardiac events between patients with or without TP. Interestingly, patients with TP showed a less frequency of clinically driven target lesion revascularization at 2 years (1.9% vs. 4.0%, $p = 0.008$), probably due to larger minimal stent area at the end of procedure [32]. Taken together, TP may influence the early clinical phase rather than late clinical stage after

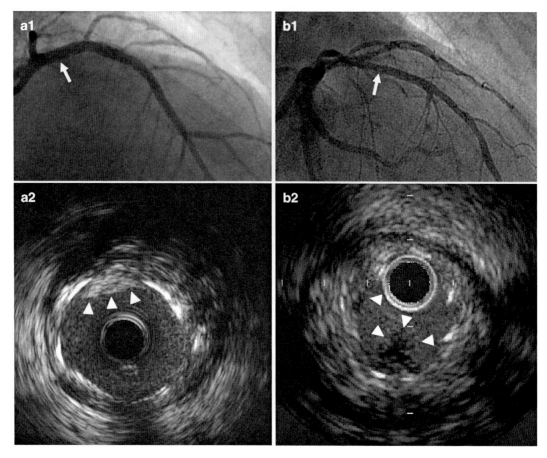

Fig. 7.6 Representative cases of tissue protrusion. A 65-year-old man was admitted with ST-segment elevation myocardial infarction (**a**). The coronary angiogram (CAG) after drug-eluting stent (DES) implantation showed no luminal narrowing within stented segments (*a1, arrow*). Correspondingly, intravascular ultrasound (IVUS) revealed tissue protrusion (plaque and/or thrombus) between stent struts (*a2, arrowheads*). A 55-year-old woman was admitted with ST-segment elevation myocardial infarction (**b**). The CAG after DES implantation showed mild luminal narrowing within stented segments (*b1, arrow*). Correspondingly, IVUS revealed tissue protrusion (most likely thrombus) between stent struts (*b2, arrowheads*)

stent implantation even though its clinical significance is still uncertain.

7.8 Evaluation of Full Lesion Coverage

IVUS can assess plaque amount in atherosclerotic coronary lesion, enabling to determine reference segment during stent implantation. Based on IVUS examination, reference segment is defined as <40% of plaque burden at cross-sectional image adjacent to the lesion [39]. Early IVUS study has demonstrated that angiographically normal looking segments, namely reference vessel segments, have 30–50% of plaque burden at cross-sectional image [40]. Several studies have shown that a reference segment that has >50% of plaque burden at cross-sectional area may increase the risk of target lesion revascularization or restenosis at follow-up after DES implantation (Fig. 7.7) (Table 7.3) [41–43]. Recent study also reported plaque burden with a cutoff value of 54.7% at less than 1 mm from proximal stent edge as a predictor of stent edge restenosis after everolimus-eluting stent implantation [43]. During or after stent deployment, thus, estimation of plaque amount at landing

Table 7.2 Summary of tissue protrusion after stent implantation

Study	Patients/lesions	% of TP	% of ACS (% of STEMI)	Cardiac enzyme elevation	% of peri-procedural MI	% of stent thrombosis	Clinical outcomes (TP vs. non-TP)
Sohn J, et al. [29]	38/40	45%	65.8% (18.4%)	Yes	5.3%	0%	2-year MACE: no difference
Choi SY, et al. [26] (HORIZON-AMI IVUS substudy)	401/401	73.6%	100% (100%)	NA	NA	Early: 3.4%	1-year clinical events: no difference
Hong YJ, et al. [37]	418/418	34%	100% (37.1%)	Yes	NA	Acute: 3.5% Subacute: 4.2%	1-year cardiac death, MI, TVR: no difference
Maehara A, et al. [48]	286/286	27.3%	39.1% (0%)	NA	NA	NA	NA
Qiu F, et al. [32] (ADAPT-DES)	2072/2446	34.3%	58.5% (17.9%)	Yes	1.8%	0.6%	2-year cardiac death, MI, ST: no difference
Shimohama T, et al. [36]	183/199	19.1%	12.7% (NA)	NA	NA	NA	9-month TLR: 3.3%

TP tissue protrusion, *ACS* acute coronary syndrome, *STEMI* ST-segment elevation myocardial infarction, *MACE* major adverse cardiac events, *TVR* target vessel revascularization, *ST* stent thrombosis, *TLR* target lesion revascularization

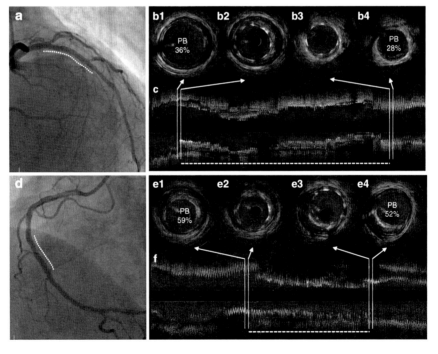

Fig. 7.7 Representative cases of appropriate (**a–c**) and inappropriate (**d–f**) location of drug-eluting stent (DES) based on post-stenting intravascular ultrasound findings. A 49-year-old man with acute myocardial infarction was treated with a second-generation DES 3.0 × 30 mm (*dotted line*) at mid to proximal left anterior descending artery (**a**). There is a well-expanded and apposed struts at the proximal (**b2**) and distal (**b3**) edges of stent. In addition, less than 50% of plaque burden is observed at proximal (**b1**) and distal (**b4**) reference segments, suggesting that the location of deployed stent is appropriate. A 68-year-old man with stable angina was treated with a second-generation DES 3.0 × 16 mm (*dotted line*) at mid right coronary artery (**d**). There is a well-expanded and apposed struts at the proximal (**e2**) and distal (**e3**) edges of stent. However, more than 50% of plaque burden is observed at proximal (**e1**) and distal (**e4**) reference segments, suggesting that the location of deployed stent is inappropriate

Table 7.3 Suggestive IVUS criteria for stent optimization

Completely apposed struts
Apposition of stent struts to the vessel wall, not surrounded by lumen
Well expanded struts
Minimal stent area (MSA) at least
• 5.0–5.5mm^2 (non-LM) & 8.7 mm^2 (LM) for DES
• 6.5–7.5 mm^2 for BMS (not in small vessels)
• >90% of distal reference segment LA or >80% of average reference segment LA
No edge dissection
Post-procedure IVUS for evaluation of edge dissection
Full lesion coverage
Reference site with plaque burden of <50%

IVUS intravascular ultrasound, *LM* left main, *DES* drug-eluting stents, *BMS* bare metal stents, *LA* lumen area

point determined by IVUS can assess future clinical outcomes.

7.9 Evaluation of Plaque Characteristics at Stented or Reference Segments

IVUS can provide qualitative and quantitative change of plaque characteristics at stented segments as well as adjacent reference segments by serial IVUS examination. Analysis of radiofrequency backscatter signals of IVUS allows us to understand whether stent strut is placed underlying necrotic core or not at reference segments due to capability of tissue characterization at adjacent segment to the stent [44]. One investigator reported that a higher frequency of plaque vulnerability behind the stent strut as well as at reference segments in DES-treated lesions compared to BMS by virtual histology IVUS (VH-IVUS) [44]. Another long-term serial VH-IVUS study demonstrated similar change of neointimal tissue characterization beyond 3 years between DES and BMS [45]. On the other hand, a recent study suggested that decrease in plaque located behind the stent area may be associated with neointimal proliferation at follow-up after BMS implantation [46].

7.10 Impact on Final Procedure During Stent Deployment

The most important utility of IVUS after stent implantation is that it can provide information whether additional procedure is needed or not. An IVUS substudy from ADAPT-DES showed that the operator changed the PCI strategy based on IVUS findings in three fourth of 3349 patients including the use of a larger stent or balloon (38%) and a longer stent (22%), higher inflation pressure (23%), additional post-stent dilatation due to underexpansion (13%) or incomplete apposition (7%), and additional stent implantation (8%) [22]. Among them, post-stenting IVUS was performed in 93% of patients (Fig. 7.8). A study by Kim et al. also reported that post-stenting IVUS findings contributed to performance of additional balloon inflation or stent implantation [47].

7.11 Summary

Since stent optimization has been reported to be associated with clinical events, IVUS assessment after stent implantation might be important in a clinical point of view. Although the clinical relevance of stent eccentricity, acute stent malapposition, and tissue protrusion was a matter of debate, numerous studies have shown that smaller MSA, stent underexpansion, and major edge dissection were independent predictors of poor clinical outcomes. Even in the current era of bioresorbable scaffold, improved procedural results under IVUS guidance still contribute to avoidance of early scaffold failure. In conclusion, post-stenting IVUS can offer qualitative as well as quantitative information within and adjacent stented segments that may expand our comprehensive understanding during procedure. Importantly, the major role of IVUS after stent implantation is that IVUS-driven suboptimal procedure results can provide a clue of whether operator should perform additional intervention during stenting procedure for making better acute and long-term clinical outcomes.

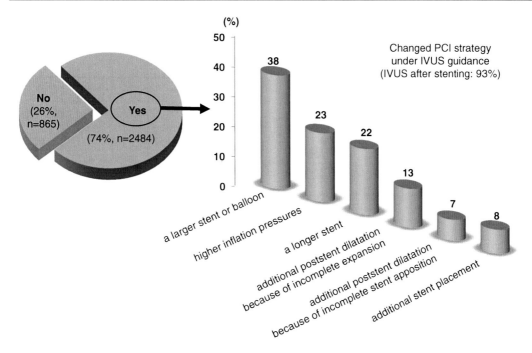

Fig. 7.8 The frequency and detailed information of changed the percutaneous coronary intervention (PCI) strategy after procedural intravascular ultrasound use (data from ADAPT-DES study) [22]. The operator changed the PCI strategy in 74% (2484/3349) of patients to choose (1) a larger stent or balloon (in 38% [943/2484] of cases); (2) higher inflation pressures (in 23% [564/2484] of cases); (3) a longer stent (in 22% [546/2484] of cases); (4) additional post-stent dilatation because of incomplete expansion (in 13% [329/2484]) or incomplete stent apposition (in 7% [166/2484]); and/or (5) additional stent placement (in 8% [197/2484])

References

1. Yoon HJ, Hur SH. Optimization of stent deployment by intravascular ultrasound. Korean J Intern Med. 2012;27(1):30–8.
2. Park SJ, Kim YH, Park DW, Lee SW, Kim WJ, Suh J, et al. Impact of intravascular ultrasound guidance on long-term mortality in stenting for unprotected left main coronary artery stenosis. Circ Cardiovasc Interv. 2009;2(3):167–77.
3. Hong SJ, Kim BK, Shin DH, Nam CM, Kim JS, Ko YG, et al. Effect of intravascular ultrasound-guided vs angiography-guided everolimus-eluting stent implantation: the IVUS-XPL randomized clinical trial. JAMA. 2015;314(20):2155–63.
4. de Jaegere P, Mudra H, Figulla H, Almagor Y, Doucet S, Penn I, et al. Intravascular ultrasound-guided optimized stent deployment. Immediate and 6 months clinical and angiographic results from the multicenter ultrasound stenting in coronaries study (MUSIC study). Eur Heart J. 1998;19(8):1214–23.
5. Brugaletta S, Gomez-Lara J, Diletti R, Farooq V, van Geuns RJ, de Bruyne B, et al. Comparison of in vivo eccentricity and symmetry indices between metallic stents and bioresorbable vascular scaffolds: insights from the ABSORB and SPIRIT trials. Catheter Cardiovasc Interv. 2012;79(2):219–28.
6. Suwannasom P, Sotomi Y, Ishibashi Y, Cavalcante R, Albuquerque FN, Macaya C, et al. The impact of post-procedural asymmetry, expansion, and eccentricity of bioresorbable everolimus-eluting scaffold and metallic everolimus-eluting stent on clinical outcomes in the ABSORB II trial. JACC Cardiovasc Interv. 2016;9(12):1231–42.
7. von Birgelen C, Gil R, Ruygrok P, Prati F, Di Mario C, van der Giessen WJ, et al. Optimized expansion of the Wallstent compared with the Palmaz-Schatz stent: on-line observations with two- and three-dimensional intracoronary ultrasound after angiographic guidance. Am Heart J. 1996;131(6):1067–75.
8. Nakano M, Wagatsuma K, Iga A, Nii H, Amano H, Toda M, et al. Impact of highly asymmetric stent expansion after sirolimus-eluting stent implantation on twelve-month clinical outcomes. J Cardiol. 2007;49(6):313–21.
9. Hwang CW, Wu D, Edelman ER. Physiological transport forces govern drug distribution for stent-based delivery. Circulation. 2001;104(5):600–5.
10. Mintz GS, Weissman NJ. Intravascular ultrasound in the drug-eluting stent era. J Am Coll Cardiol. 2006;48(3):421–9.

11. Morino Y, Honda Y, Okura H, Oshima A, Hayase M, Bonneau HN, et al. An optimal diagnostic threshold for minimal stent area to predict target lesion revascularization following stent implantation in native coronary lesions. Am J Cardiol. 2001;88(3):301–3.
12. Doi H, Maehara A, Mintz GS, Yu A, Wang H, Mandinov L, et al. Impact of post-intervention minimal stent area on 9-month follow-up patency of paclitaxel-eluting stents: an integrated intravascular ultrasound analysis from the TAXUS IV, V, and VI and TAXUS ATLAS workhorse, long lesion, and direct stent trials. JACC Cardiovasc Interv. 2009;2(12):1269–75.
13. Sonoda S, Morino Y, Ako J, Terashima M, Hassan AH, Bonneau HN, et al. Impact of final stent dimensions on long-term results following sirolimus-eluting stent implantation: serial intravascular ultrasound analysis from the sirius trial. J Am Coll Cardiol. 2004;43(11):1959–63.
14. Hong MK, Mintz GS, Lee CW, Park DW, Choi BR, Park KH, et al. Intravascular ultrasound predictors of angiographic restenosis after sirolimus-eluting stent implantation. Eur Heart J. 2006;27(11):1305–10.
15. Song HG, Kang SJ, Ahn JM, Kim WJ, Lee JY, Park DW, et al. Intravascular ultrasound assessment of optimal stent area to prevent in-stent restenosis after zotarolimus-, everolimus-, and sirolimus-eluting stent implantation. Catheter Cardiovasc Interv. 2014;83(6):873–8.
16. Kang SJ, Ahn JM, Song H, Kim WJ, Lee JY, Park DW, et al. Comprehensive intravascular ultrasound assessment of stent area and its impact on restenosis and adverse cardiac events in 403 patients with unprotected left main disease. Circ Cardiovasc Interv. 2011;4(6):562–9.
17. Fitzgerald PJ, Oshima A, Hayase M, Metz JA, Bailey SR, Baim DS, et al. Final results of the can routine ultrasound influence stent expansion (CRUISE) study. Circulation. 2000;102(5):523–30.
18. Chieffo A, Latib A, Caussin C, Presbitero P, Galli S, Menozzi A, et al. A prospective, randomized trial of intravascular-ultrasound guided compared to angiography guided stent implantation in complex coronary lesions: the AVIO trial. Am Heart J. 2013;165(1):65–72.
19. Fujii K, Carlier SG, Mintz GS, Yang YM, Moussa I, Weisz G, et al. Stent underexpansion and residual reference segment stenosis are related to stent thrombosis after sirolimus-eluting stent implantation: an intravascular ultrasound study. J Am Coll Cardiol. 2005;45(7):995–8.
20. Okabe T, Mintz GS, Buch AN, Roy P, Hong YJ, Smith KA, et al. Intravascular ultrasound parameters associated with stent thrombosis after drug-eluting stent deployment. Am J Cardiol. 2007;100(4):615–20.
21. Liu X, Doi H, Maehara A, Mintz GS, Costa Jde R Jr, Sano K, et al. A volumetric intravascular ultrasound comparison of early drug-eluting stent thrombosis versus restenosis. JACC Cardiovasc Interv. 2009;2(5):428–34.
22. Witzenbichler B, Maehara A, Weisz G, Neumann FJ, Rinaldi MJ, Metzger DC, et al. Relationship between intravascular ultrasound guidance and clinical outcomes after drug-eluting stents: the assessment of dual antiplatelet therapy with drug-eluting stents (ADAPT-DES) study. Circulation. 2014;129(4):463–70.
23. Liu X, Tsujita K, Maehara A, Mintz GS, Weisz G, Dangas GD, et al. Intravascular ultrasound assessment of the incidence and predictors of edge dissections after drug-eluting stent implantation. JACC Cardiovasc Interv. 2009;2(10):997–1004.
24. Hong MK, Park SW, Lee NH, Nah DY, Lee CW, Kang DH, et al. Long-term outcomes of minor dissection at the edge of stents detected with intravascular ultrasound. Am J Cardiol. 2000;86(7):791–5. A9
25. Sheris SJ, Canos MR, Weissman NJ. Natural history of intravascular ultrasound-detected edge dissections from coronary stent deployment. Am Heart J. 2000;139(1 Pt 1):59–63.
26. Choi SY, Witzenbichler B, Maehara A, Lansky AJ, Guagliumi G, Brodie B, et al. Intravascular ultrasound findings of early stent thrombosis after primary percutaneous intervention in acute myocardial infarction: a harmonizing outcomes with revascularization and stents in acute myocardial infarction (HORIZONS-AMI) substudy. Circ Cardiovasc Interv. 2011;4(3):239–47.
27. Steinberg DH, Mintz GS, Mandinov L, Yu A, Ellis SG, Grube E, et al. Long-term impact of routinely detected early and late incomplete stent apposition: an integrated intravascular ultrasound analysis of the TAXUS IV, V, and VI and TAXUS ATLAS workhorse, long lesion, and direct stent studies. JACC Cardiovasc Interv. 2010;3(5):486–94.
28. Kimura M, Mintz GS, Carlier S, Takebayashi H, Fujii K, Sano K, et al. Outcome after acute incomplete sirolimus-eluting stent apposition as assessed by serial intravascular ultrasound. Am J Cardiol. 2006;98(4):436–42.
29. Sohn J, Hur SH, Kim IC, Cho YK, Park HS, Yoon HJ, et al. A comparison of tissue prolapse with optical coherence tomography and intravascular ultrasound after drug-eluting stent implantation. Int J Cardiovasc Imaging. 2015;31(1):21–9.
30. Hong YJ, Jeong MH, Ahn Y, Sim DS, Chung JW, Cho JS, et al. Plaque prolapse after stent implantation in patients with acute myocardial infarction: an intravascular ultrasound analysis. JACC Cardiovasc Imaging. 2008;1(4):489–97.
31. Mintz GS, Nissen SE, Anderson WD, Bailey SR, Erbel R, Fitzgerald PJ, et al. American College of Cardiology clinical expert consensus document on standards for acquisition, measurement and reporting of intravascular ultrasound studies (IVUS). A report of the American College of Cardiology Task Force on clinical expert consensus documents. J Am Coll Cardiol. 2001;37(5):1478–92.
32. Qiu F, Mintz GS, Witzenbichler B, Metzger DC, Rinaldi MJ, Duffy PL, et al. Prevalence and clinical

impact of tissue protrusion after stent implantation: an ADAPT-DES intravascular ultrasound substudy. JACC Cardiovasc Interv. 2016;9(14):1499–507.
33. Hong MK, Park SW, Lee CW, Kang DH, Song JK, Kim JJ, et al. Long-term outcomes of minor plaque prolapsed within stents documented with intravascular ultrasound. Catheter Cardiovasc Interv. 2000;51(1):22–6.
34. Kim SW, Mintz GS, Ohlmann P, Hassani SE, Fernandez S, Lu L, et al. Frequency and severity of plaque prolapse within Cypher and Taxus stents as determined by sequential intravascular ultrasound analysis. Am J Cardiol. 2006;98(9):1206–11.
35. Maehara A, Mintz GS, Lansky AJ, Witzenbichler B, Guagliumi G, Brodie B, et al. Volumetric intravascular ultrasound analysis of paclitaxel-eluting and bare metal stents in acute myocardial infarction: the harmonizing outcomes with revascularization and stents in acute myocardial infarction intravascular ultrasound substudy. Circulation. 2009;120(19):1875–82.
36. Shimohama T, Ako J, Yamasaki M, Otake H, Tsujino I, Hasegawa T, et al. SPIRIT III JAPAN versus SPIRIT III USA: a comparative intravascular ultrasound analysis of the everolimus-eluting stent. Am J Cardiol. 2010;106(1):13–7.
37. Hong YJ, Jeong MH, Choi YH, Song JA, Kim DH, Lee KH, et al. Impact of tissue prolapse after stent implantation on short- and long-term clinical outcomes in patients with acute myocardial infarction: an intravascular ultrasound analysis. Int J Cardiol. 2013;166(3):646–51.
38. Jin QH, Chen YD, Jing J, Tian F, Guo J, Liu CF, et al. Incidence, predictors, and clinical impact of tissue prolapse after coronary intervention: an intravascular optical coherence tomography study. Cardiology. 2011;119(4):197–203.
39. Weissman NJ, Palacios IF, Nidorf SM, Dinsmore RE, Weyman AE. Three-dimensional intravascular ultrasound assessment of plaque volume after successful atherectomy. Am Heart J. 1995;130(3 Pt 1):413–9.
40. Mintz GS, Painter JA, Pichard AD, Kent KM, Satler LF, Popma JJ, et al. Atherosclerosis in angiographically "normal" coronary artery reference segments: an intravascular ultrasound study with clinical correlations. J Am Coll Cardiol. 1995;25(7):1479–85.
41. Morino Y, Tamiya S, Masuda N, Kawamura Y, Nagaoka M, Matsukage T, et al. Intravascular ultrasound criteria for determination of optimal longitudinal positioning of sirolimus-eluting stents. Circ J. 2010;74(8):1609–16.
42. Kang SJ, Cho YR, Park GM, Ahn JM, Kim WJ, Lee JY, et al. Intravascular ultrasound predictors for edge restenosis after newer generation drug-eluting stent implantation. Am J Cardiol. 2013;111(10):1408–14.
43. Takahashi M, Miyazaki S, Myojo M, Sawaki D, Iwata H, Kiyosue A, et al. Impact of the distance from the stent edge to the residual plaque on edge restenosis following everolimus-eluting stent implantation. PLoS One. 2015;10(3):e0121079.
44. Kubo T, Maehara A, Mintz GS, Garcia-Garcia HM, Serruys PW, Suzuki T, et al. Analysis of the long-term effects of drug-eluting stents on coronary arterial wall morphology as assessed by virtual histology intravascular ultrasound. Am Heart J. 2010;159(2):271–7.
45. Kitabata H, Loh JP, Pendyala LK, Omar A, Ota H, Minha S, et al. Intra-stent tissue evaluation within bare metal and drug-eluting stents > 3 years since implantation in patients with mild to moderate neointimal proliferation using optical coherence tomography and virtual histology intravascular ultrasound. Cardiovasc Revasc Med. 2014;15(3):149–55.
46. Andreou I, Takahashi S, Tsuda M, Shishido K, Antoniadis AP, Papafaklis MI, et al. Atherosclerotic plaque behind the stent changes after bare-metal and drug-eluting stent implantation in humans: implications for late stent failure? Atherosclerosis. 2016;252:9–14.
47. Kim IC, Yoon HJ, Shin ES, Kim MS, Park J, Cho YK, et al. Usefulness of frequency domain optical coherence tomography compared with intravascular ultrasound as a guidance for percutaneous coronary intervention. J Interv Cardiol. 2016;29(2):216–24.
48. Maehara A, Ben-Yehuda O, Ali Z, Wijns W, Bezerra HG, Shite J, et al. Comparison of stent expansion guided by optical coherence tomography versus intravascular ultrasound: the ILUMIEN II study (observational study of optical coherence tomography [OCT] in patients undergoing fractional flow reserve [FFR] and percutaneous coronary intervention). JACC Cardiovasc Interv. 2015;8(13):1704–14.

Long-Term Complications and Bioresorbable Vascular Scaffolds Evaluation

Kyeong Ho Yun

Stent failure is defined as loss of short-term, long-term safety and efficacy of drug-eluting stent. Intravascular ultrasound can provide important information of stent failure, which mainly consist of stent restenosis and thrombosis. In this chapter, I will describe intravascular ultrasound findings associated with stent failure.

Bioresorbable vascular scaffolds have become an attractive option due to the complete resorption process over a few years. However, current generation devices have been indicated that an optimized implantation strategy is required for prevention of scaffold thrombosis. This chapter will be discussed optimize implantation technique using intravascular ultrasound in brief.

8.1 Long-Term Complications Contributed to Stent Failure

8.1.1 Stent Underexpansion

Minimal stent area after drug-eluting stent implantation is an important predictor for restenosis. Hong et al. reported that final minimum stent area <5.5 mm^2 is an independent predictor of 6-month angiographic restenosis after sirolimus-eluting stent implantation [1]. After that, minimal stent area of each kind of 1st- and 2nd-generation drug-eluting stent for prediction of restenosis was evaluated (Fig. 8.1) [2, 3]. However, sensitivity and specificity of cut-off minimal stent area was low, and cut-off area would not reflect various clinical situations and vessel size, such as long lesion. Recently, The Impact of Intravascular Ultrasound Guidance on Outcomes of Xience Prime Stents in Long Lesions (IVUS-XPL) study, first randomized 2nd-generation intravascular ultrasound study, suggested optimal intravascular criteria for minimal stent area [4]. In this study, the minimal stent area and minimal stent area-to-distal reference segment lumen area ratio that best predicted in patients with adverse events from those without these events were 5.0 mm^2 and 1.0 mm^2, respectively (Fig. 8.2) [5]. Therefore, final minimal stent area should be obtained at least greater than the lumen cross-sectional area at the distal reference segments.

Stent underexpansion is associated with stent thrombosis as well as stent restenosis. Lui et al. reported that underexpansion associated with thrombosis is more severe, diffuse, and proximal in location compared with restenosis [6].

8.1.2 Stent Fracture

Stent fracture is defined as the presence of an angiographically visible interrupted connection

K.H. Yun
Departments of Cardiovascular Medicine, Regional Cardiovascular Center, Wonkwang University Hospital, Iksan, South Korea
e-mail: dryunkh@gmail.com

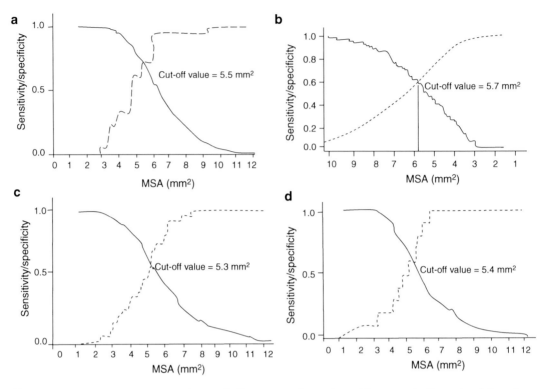

Fig. 8.1 Optimal minimal stent area in the sirolimus-eluting stent (**a**), paclitaxel-eluting stent (**b**), zotarolimus-eluting stent (**c**), and everolimus-eluting stent (**d**) to predict angiographic restenosis (Reproduced from Song et al. 2014, Doi et al. 2009). *MSA* minimal stent area

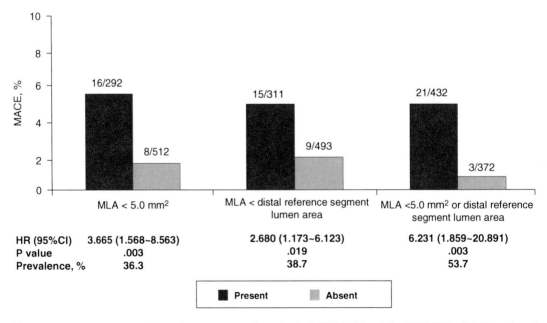

Fig. 8.2 Major adverse cardiovascular events rate for patients with and without minimal lumen area (MLA) < 5.0 mm² or distal reference segment lumen area in the RESET trial and the IVUS-XPL trial (Reproduced from Lee et al. 2016). *MLA* minimal lumen area

of stent struts or fewer visible stent struts at the suspected site than normal looking stented area on intravascular ultrasound. The classification of stent fracture varies from study to study (Table 8.1) [7]. Intravascular ultrasound can identify complete stent fracture (complete strut absent) and partial stent fracture (stent struts absent in ≥1/3 of the vessel wall) (Fig. 8.3) [8]. Most stent fracture occurred in sirolimus-eluting stent, but several cases of stent fracture were also reported in other type of stents. Especially, stent fracture of 2nd generation drug-eluting stent can be associated with longitudinal stent deformation due to their weak compressive force (Fig. 8.4). Stent fracture can be incidental finding in asymptomatic patients, however, it also presents as recurrent angina, myocardial infarction, and even sudden death. The uses of intravascular ultrasound increase the rate of stent fracture detection, and provide associated information regarding neointima formation, vessel remodeling, stent expansion, and aneurysmal formation (Fig. 8.5).

8.1.3 Late Stent Malapposition

Stent malapposition was defined as a separation of at least 1 stent strut not in contact with the intimal surface of the arterial wall that was not overlapping a side branch, was not present immediately after stent implantation, and had evidence of blood speckling behind the strut. Late stent malapposition was defined as stent malapposition developing between 30 days and 1 year, but typically detected on 6-month follow-up intravascular ultrasound [9]. Very late stent malapposition was defined as a late stent malapposition lesion that developed after 1 year (Fig. 8.6). A meta-analysis showed that the risk

Table 8.1 Classification of stent fracture

Type	Description
I	A single-strut fracture
II	2 or more strut fractures without deformation
III	2 or more strut fractures with deformation
IV	Multiple strut fractures with acquired transection but without gap
V	Multiple strut fractures with acquired transection with gap

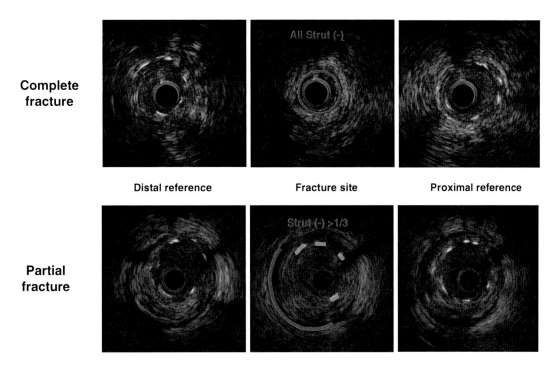

Fig. 8.3 Intravascular ultrasound definition of stent fracture

Fig. 8.4 A case of everolimus-eluting stent fracture. Partial fracture was followed by longitudinal deformation and overlap of fractured edges leading to excessive neointimal hyperplasia

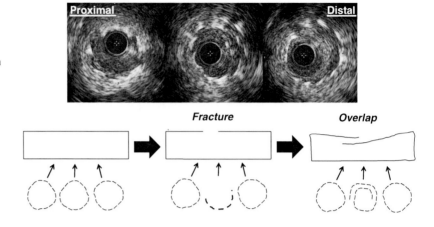

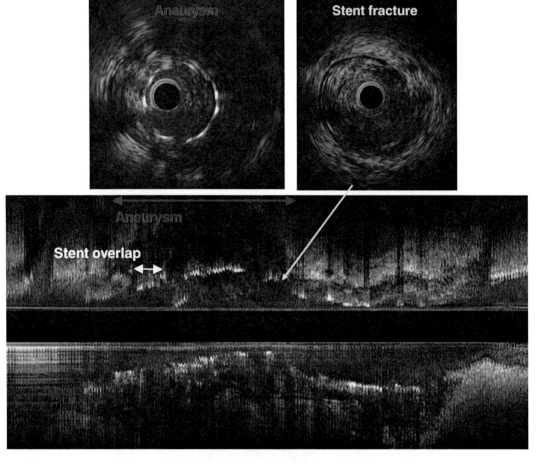

Fig. 8.5 A case of sirolimus-eluting stent fracture associated with coronary aneurysm

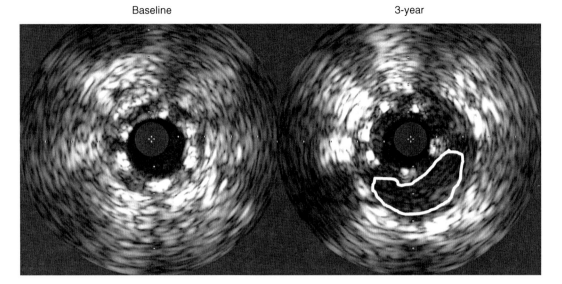

Fig. 8.6 Example of late stent malapposition 3-year after sirolimus-eluting stent implantation

of late stent malapposition was significantly greater after drug-eluting stent than bare-metal stent, whereas other studies showed that primary stenting in acute myocardial infarction was an independent predictor after both drug-eluting stent and bare-metal stent implantation [10, 11]. Late stent malapposition of drug-eluting stent is associated with positive vascular remodeling and vascular remodeling can be progression [9]. Therefore, stent malapposition could continuously progress and new areas of malaposion also could develop in later stage.

The clinical impact of stent malapposition has been a matter of concern and debate. In the harmonizing outcomes with revascularization and stents in acute myocardial infarction (HORIZONS-AMI) trial, stent malapposition was not associated with stent thrombosis (Fig. 8.7) [12]. Hong et al. also reported that late stent malapposition after drug-eluting stent implantation was not a predictor of major adverse cardiac events or stent thrombosis at 3 years after the 6-month intravascular ultrasound [13]. However, Cook et al. reported that late stent malapposition associated with stent thrombosis had higher maximal malapposition area, length, and depth than without stent thrombosis (Fig. 8.8) [14]. The prognostic impact of late stent malapposition on long-term clinical outcomes requires further investigation.

8.1.4 In-Stent Neoatherosclerosis

In-stent neoatherosclerosis has emerged as an important contributing factor to late vascular complications including very late stent thrombosis and late in-stent restenosis. Histologically, neoatherosclerosis is characterized by accumulation of lipid-laden foamy macrophages within the neointima with or without necrotic core formation and/or calcification. The development of neoatherosclerosis may occur in months to years following stent placement. Pathologic and clinical imaging studies have demonstrated that neoatherosclerosis occurs more frequently and at an earlier time point in drug-eluting stent when compared with bare-metal stent [15, 16]. In intravascular ultrasound, in-stent plaque rupture likely accounts for most thrombotic events associated with neoatherosclerosis (Fig. 8.9). However, because intravascular ultrasound has poor resolution (spatial resolution = 150–250 µm),

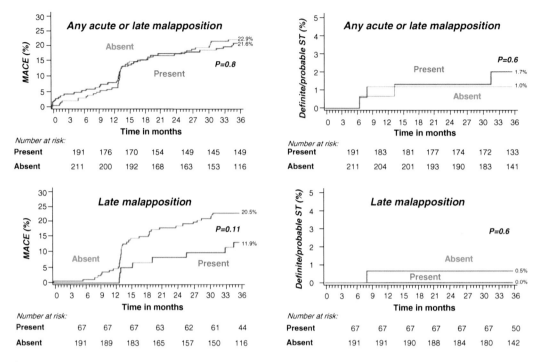

Fig. 8.7 3-year follow-up of HORIZONS-AMI intravascular substudy (Reproduced from Yakushiji T, et al. ACC 2012)

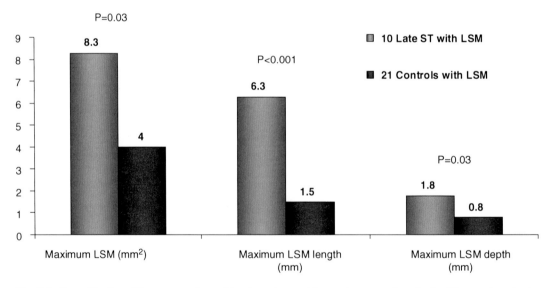

Fig. 8.8 Quantification of late stent malapposition in patients with very late stent thrombosis. *ST* stent thrombosis, *LSM* late stent malapposition. (Modified from Cook S et al. 2007)

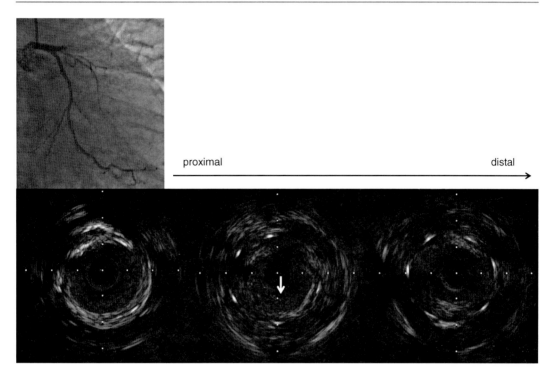

Fig. 8.9 A case of very late stent thrombosis. Intravascular ultrasound after thrombus aspiration showed significant neointimal tissue growth and neointimal flap (*arrow*)

intravascular optical coherence tomography is the best image tool for detection of neoatherosclerosis. The detailed discussion will be described in optical coherence tomography section.

8.2 Intravascular Guided Bioresorbable Vascular Scaffold Implantation

Bioresorbable scaffolds have become an attractive option in the field of percutaneous coronary intervention, due to the advantages associated with the complete resorption process. Recent randomized trials have demonstrated non-inferiority of BRS when compared to contemporary drug-eluting stents [17, 18]. However, recently concerns have been raised regarding a potentially higher incidence of scaffold thrombosis. This may be related to implantation strategy, since consistent optimized implantation strategies were not utilized in most of the prior reports, including relatively low rates of post-dilation and intravascular imaging use [17–20]. Furthermore, a recent report demonstrated that the incidence of scaffold thrombosis could be significantly reduced with an optimal implantation strategy [21].

Several types of bioresorbable scaffold are in development, however, currently the Absorb bioresorbable vascular scaffold (BVS) is only used in Korea. BVS has different implantation strategy compared with metallic stent (Table 8.2).

Table 8.2 Optimal implantation technique for the Absorb bioresorbable vascular scaffold

5P's of optimal implantation technique
1. Prepare the lesion
2. Properly size the vessel
3. Post-dilate with a non-compliant balloon
4. Pay attention to expansion limits
5. Prescribe dual anti-platelet therapy

In BVS implantation, careful device sizing is required to avoid both over- and undersizing. Undersizing can lead to malapposition, which is thought to be a major cause of scaffold thrombosis [22]. Malapposition with undersized BVS may be frequently difficult to correct after deployment due to limited expansion capabilities. On the other hand, an oversized BVS increases the percentage of vessel coverage and strut volume in the vessel lumen more dramatically than current metallic stents, which may increase thrombogenicity and side branch occlusions [23, 24]. Intravascular ultrasound provides accurate size of the vessel/lumen before BVS

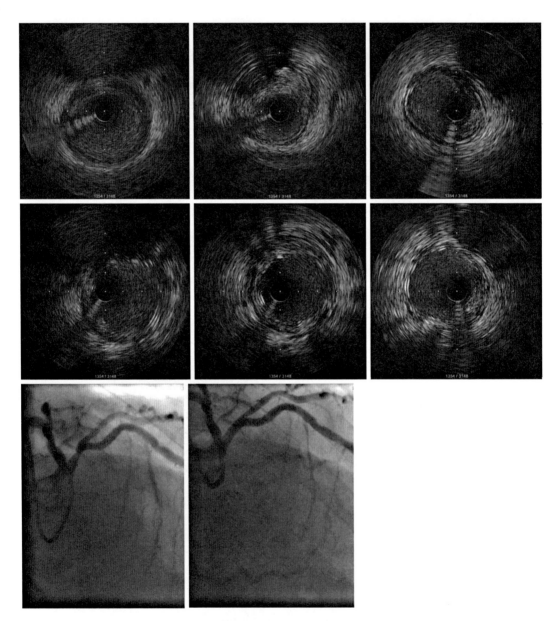

Fig. 8.10 An example of intravascular ultrasound guided Absorb bioresorbable vascular scaffold (BVS) implantation. Reference diameter was 4.0 mm in pre-interventional ultrasound (*upper*). Lesion preparation was performed by 3.75 mm balloon, and then 3.5 × 18mm BVS was implanted at 6 atm. Post-dilation with 3.75 non-compliant balloon was performed at 14 atm. Final IVUS (*lower*) showed good apposition without significant complications

implantation and information regarding suboptimal implantation such as underexpansion, incomplete lesion coverage, and malapposition (Fig. 8.10). Therefore, operators should have low threshold for intravascular imaging use especially during the early experiences for the operator and institution.

The decision on whether to use intravascular ultrasound or optical coherence tomography (OCT) depends on several factors. OCT has higher resolution than ultrasound, while less power of tissue penetration. Furthermore, OCT generally requires an injection of contrast media to obtain the images, which may increase total contrast volume and procedure time. Therefore, intravascular ultrasound is advantageous when confirming vessel diameters especially in large vessels or flow limiting distal vessels, and also can be used widely in any situation and repeatedly even in the presence of chronic kidney disease. Although intravascular ultrasound can also detect the majority of potentially concerning findings after BVS implantation including malapposition, underexpansion, and dissection, OCT can show them more clearly. Moreover, it is frequently difficult for intravascular ultrasound to detect scaffold fracture. Therefore OCT may be advantageous when an operator wants to observe BVS struts with fine detail, although it is still unclear whether it is clinically beneficial to detect findings that intravascular ultrasound cannot [25, 26].

References

1. Hong MK, Mintz GS, Lee CW, Park DW, Choi BR, Park KH, Kim YH, Cheong SS, Song JK, Kim JJ, Park SW, Park SJ. Intravascular ultrasound predictors of angiographic restenosis after sirolimus-eluting stent implantation. Eur Heart J. 2006;27:1305–10.
2. Doi H, Maehara A, Mintz GS, Yu A, Wang H, Mandinov L, Popma JJ, Ellis SG, Grube E, Dawkins KD, Weissman NJ, Turco MA, Ormiston JA, Stone GW. Impact of post-intervention minimal stent area on 9-month follow-up patency of paclitaxel-eluting stents: an integrated intravascular ultrasound analysis from the TAXUS IV, V, and VI and TAXUS ATLAS workhorse, long lesion, and direct stent trials. JACC Cardiovasc Interv. 2009;2:1269–75.
3. Song HG, Kang SJ, Ahn JM, Kim WJ, Lee JY, Park DW, Lee SW, Kim YH, Lee CW, Park SW, Park SJ. Intravascular ultrasound assessment of optimal stent area to prevent in-stent restenosis after zotarolimus-, everolimus-, and sirolimus-eluting stent implantation. Catheter Cardiovasc Interv. 2014;83:873–8.
4. Hong SJ, Kim BK, Shin DH, Nam CM, Kim JS, Ko YG, Choi D, Kang TS, Kang WC, Her AY, Kim YH, Hur SH, Hong BK, Kwon H, Jang Y, Hong MK, Investigators IVUS-XPL. Effect of intravascular ultrasound-guided vs angiography-guided everolimus-eluting stent implantation: the IVUS-XPL randomized clinical trial. JAMA. 2015;314:2155–63.
5. Lee SY, Shin DH, Kim JS, Kim BK, Ko YG, Choi D, Jang Y, Hong MK. Intravascular ultrasound predictors of major adverse cardiovascular events after implantation of everolimus-eluting stents for long coronary lesions. Rev Esp Cardiol (Engl Ed). 2016; doi:10.1016/j.rec.2016.06.019. [Epub ahead of print]
6. Liu X, Doi H, Maehara A, Mintz GS, Costa Jde R Jr, Sano K, Weisz G, Dangas GD, Lansky AJ, Kreps EM, Collins M, Fahy M, Stone GW, Moses JW, Leon MB, Mehran R. A volumetric intravascular ultrasound comparison of early drug-eluting stent thrombosis versus restenosis. JACC Cardiovasc Interv. 2009;2:428–34.
7. Nakazawa G, Finn AV, Vorpahl M, Ladich E, Kutys R, Balazs I, Kolodgie FD, Virmani R. Incidence and predictors of drugeluting stent fracture in human coronary artery a pathologic analysis. J Am Coll Cardiol. 2009;54:1924–31.
8. Lee SH, Park JS, Shin DG, Kim YJ, Hong GR, Kim W, Shim BS. Frequency of stent fracture as a cause of coronary restenosis after sirolimus-eluting stent implantation. Am J Cardiol. 2007;100:627–30.
9. Kang SJ, Mintz GS, Park DW, Lee SW, Kim YH, Lee CW, Han KH, Kim JJ, Park SW, Park SJ. Late and very late drug-eluting stent malapposition: serial 2-year quantitative IVUS analysis. Circ Cardiovasc Interv. 2010;3:335–40.
10. Hassan AK, Bergheanu SC, Stijnen T, van der Hoeven BL, Snoep JD, Plevier JW, Schalij MJ, Jukema JW. Late stent malapposition risk is higher after drug-eluting stent compared with bare-metal stent implantation and associates with late stent thrombosis. Eur Heart J. 2010;31:1172–80.
11. Hong MK, Mintz GS, Lee CW, Park DW, Park KM, Lee BK, Kim YH, Song JM, Han KH, Kang DH, Cheong SS, Song JK, Kim JJ, Park SW, Park SJ. Late stent malapposition after drug-eluting stent implantation: an intravascular ultrasound analysis with long-term follow-up. Circulation. 2006;113:414–9.
12. Guo N, Maehara A, Mintz GS, He Y, Xu K, Wu X, Lansky AJ, Witzenbichler B, Guagliumi G, Brodie B, Kellett MA Jr, Dressler O, Parise H, Mehran R, Stone GW. Incidence, mechanisms, predictors, and clinical impact of acute and late stent malapposition after primary intervention in patients with acute myocardial infarction: an intravascular ultrasound substudy of the harmonizing outcomes with revascularization and

stents in acute myocardial infarction (HORIZONS-AMI) trial. Circulation. 2010;122:1077–84.
13. Hong MK, Mintz GS, Lee CW, Park DW, Lee SW, Kim YH, Kang DH, Cheong SS, Song JK, Kim JJ, Park SW, Park SJ. Impact of late drug-eluting stent malapposition on 3-year clinical events. J Am Coll Cardiol. 2007;50:1515–6.
14. Cook S, Wenaweser P, Togni M, Billinger M, Morger C, Seiler C, Vogel R, Hess O, Meier B, Windecker S. Incomplete stent apposition and very late stent thrombosis after drug-eluting stent implantation. Circulation. 2007;115:2426–34.
15. Kang SJ, Mintz GS, ParkDW LSW, Kim YH, Lee CW, Han KH, Kim JJ, Park SW, Park SJ. Tissue characterization of in-stent neointimal using intravascular ultrasound radiofrequency data analysis. Am J Cardiol. 2010;106:1561–5.
16. Otsuka F, Vorpahl M, Nakano M, Foerst J, Newell JB, Sakakura K, Kutys R, Ladich E, Finn AV, Kolodgie FD, Virmani R. Pathology of second-generation everolimus-eluting stents versus first-generation sirolimus- and paclitaxel-eluting stents in humans. Circulation. 2014;129:211–23.
17. Ellis SG, Kereiakes DJ, Metzger DC, Caputo RP, Rizik DG, Teirstein PS, Litt MR, Kini A, Kabour A, Marx SO, et al. Everolimus-eluting bioresorbable scaffolds for coronary artery disease. N Engl J Med. 2015;373:1905–15.
18. Serruys PW, Chevalier B, Dudek D, Cequier A, Carrie D, Iniguez A, Dominici M, van der Schaaf RJ, Haude M, Wasungu L, et al. A bioresorbable everolimus-eluting scaffold versus a metallic everolimus-eluting stent for ischaemic heart disease caused by de-novo native coronary artery lesions (ABSORB II): an interim 1-year analysis of clinical and procedural secondary outcomes from a randomised controlled trial. Lancet. 2015;385:43–54.
19. Capodanno D, Gori T, Nef H, Latib A, Mehilli J, Lesiak M, Caramanno G, Naber C, Di Mario C, Colombo A, et al. Percutaneous coronary intervention with everolimus-eluting bioresorbable vascular scaffolds in routine clinical practice: early and midterm outcomes from the European multicentre GHOST-EU registry. EuroIntervention. 2014;10:1144–53.
20. Gori T, Schulz E, Hink U, Kress M, Weiers N, Weissner M, Jabs A, Wenzel P, Capodanno D, Munzel T. Clinical, angiographic, functional, and imaging outcomes 12 months after implantation of drug-eluting bioresorbable vascular scaffolds in acute coronary syndromes. JACC Cardiovasc Interv. 2015;8:770–7.
21. Puricel S, Cuculi F, Weissner M, Schmermund A, Jamshidi P, Nyffenegger T, Binder H, Eggebrecht H, Munzel T, Cook S, et al. Bioresorbable coronary scaffold thrombosis: multicenter comprehensive analysis of clinical presentation, mechanisms, and predictors. J Am Coll Cardiol. 2016;67:921–31.
22. Raber L, Brugaletta S, Yamaji K, O'Sullivan CJ, Otsuki S, Koppara T, Taniwaki M, Onuma Y, Freixa X, Eberli FR, et al. Very late scaffold thrombosis: intracoronary imaging and histopathological and spectroscopic findings. J Am Coll Cardiol. 2015;66:1901–14.
23. Kawamoto H, Jabbour RJ, Tanaka A, Latib A, Colombo A. The bioresorbable scaffold: will oversizing affect outcomes? JACC Cardiovasc Interv. 2016;9:299–300.
24. Kolandaivelu K, Swaminathan R, Gibson WJ, Kolachalama VB, Nguyen-Ehrenreich KL, Giddings VL, Coleman L, Wong GK, Edelman ER. Stent thrombogenicity early in high-risk interventional settings is driven by stent design and deployment and protected by polymer-drug coatings. Circulation. 2011;123:1400–9.
25. Lotfi A, Jeremias A, Fearon WF, Feldman MD, Mehran R, Messenger JC, Grines CL, Dean LS, Kern MJ, Klein LW, et al. Expert consensus statement on the use of fractional flow reserve, intravascular ultrasound, and optical coherence tomography: a consensus statement of the Society of Cardiovascular Angiography and Interventions. Catheter Cardiovasc Interv. 2014;83:509–18.
26. Waksman R, Kitabata H, Prati F, Albertucci M, Mintz GS. Intravascular ultrasound versus optical coherence tomography guidance. J Am Coll Cardiol. 2013;62:S32–40.

Near-Infrared Spectroscopy

Byoung-joo Choi

Nowadays, various newly developed intracoronary imaging techniques have provided unique information on the coronary plaque and are widely used either for clinical decision-making or for research purposes (Table 9.1). However, there is still unmet need for the characterization of atheromatous plaque, especially for in vivo measurement of lipid burden within coronary artery wall. Near-infrared spectroscopy (NIRS) uses properties of the light reflection and absorption in each specific chemical component and provides us information on the presence of lipid core plaque in the coronary artery wall. This chapter will review the basic mechanism, validation, and techniques of NIRS followed by the results of early clinical studies.

Table 9.1 Comparison of different intravascular imaging modalities

Imaging modality	Resolution	Cap thickness	Lipid core	Calcium	Thrombus	Macrophage	Neovascularization
IVUS	100 μm	+	+	++	+	−	−
OCT	10 μm	+++	++	++	++	+	++
VH	100 μm	+	+	++	+	−	−
NIRS	−	+	+++	−	−	−	−
Angioscopy	−	+	+	−	+++	−	−

IVUS intravascular ultrasound, *OCT* optical coherence tomography, *VH* virtual histology, *NIRS* near-infrared spectroscopy

B.-j. Choi
Department of Cardiology, Ajou University School of Medicine, Suwon, South Korea
e-mail: bjchoi@ajou.ac.kr

9.1 Basic Mechanism

Spectroscopy is well established and widely accepted method to identify unknown chemicals in a variety of industries and scientific studies. Basically, spectroscopy employs the mechanism that light reflection (scattering) and absorption vary at different wavelengths according to each chemical component or substance [1, 2]. Organic component in the atheromatous plaque (collagen, cholesterol, etc.), when near-infrared (wavelength 780–2500 nm) light is shed on them, provides unique spectral signature (there are particular and specific peaks and trough patterns according to each chemical substances) that can be used as "chemical thumbprint" [3]. All these information are integrated with grayscale intravascular ultrasound (IVUS) images and displayed into a single picture (Fig. 9.1).

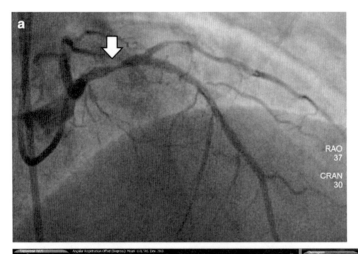

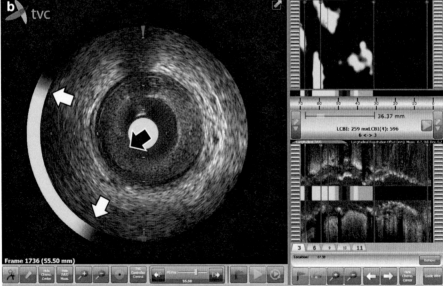

Fig. 9.1 Representative case of near-infrared spectroscopy (NIRS) in patient with acute coronary syndrome. The coronary angiogram shows significant stenosis at the proximal segment of the left anterior descending artery (*white arrow*) (**a**). NIRS shows large lipid burden within coronary artery wall (**b**). The cross-sectional image of NIRS clearly reveals lipid accumulation is present from 7 o'clock to 10 o'clock (*white arrow*), while concomitant intravascular ultrasound (IVUS) image demonstrates the presence of plaque rupture (*black arrow*) at the same location. In this case the identification of lipid by IVUS image is not feasible

9.2 Validation

NIRS system was rigorously validated with 84 human autopsy hearts in a prospective and double-blind manner to assess the accuracy in detecting the lipid core plaque (LCP) [4]. In order to develop quantitative index for the validation, an LCP of interest was defined as a lipid core >60° in circumferential extent, >200 μm thickness, and with a mean fibrous cap thickness <450 μm. The algorithm of NIRS system prospectively identified LCP with a receiver-operator characteristic area of 0.80 (95% confidence interval [CI]: 0.76–0.85). The lipid core burden index detected the presence or absence of any fibroatheroma with an area under the curve of 0.86 (95% CI: 0.81–0.91). This study successfully demonstrated good agreement between NIRS system and histopathology in coronary autopsy specimens. Clinical verification of NIRS system was performed by SPECTACL (Spectroscopic Assessment of Coronary Lipid) study. This study showed that spectral data obtained from patients by NIRS system were similar with those from autopsy specimens [5]. Furthermore, high reproducibility of NIRS system for the detection of LCP was demonstrated by Garcia et al. [6].

9.3 NIRS System and Measurement

NIRS system (TVC®, InfraReDx, Burlington, MA, USA) consists of 3.2F catheter, which uses 0.014-in. coronary guidewire system and pullback devices (Fig. 9.2). Mechanical pullback and rotation are performed at a speed of 0.5 cm/s and 240 rotation/m. The NIRS system acquires approximately 1000 NIRS measurement/12.5 cm of artery scanned and determines the presence of

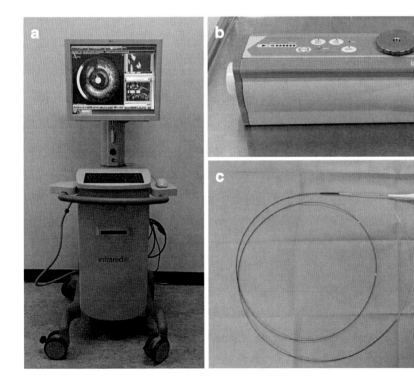

Fig. 9.2 Near-infrared (NIR) spectroscopy system (TVC®, InfraReDx, Burlington, MA, USA). The system consists of a console (**a**), a mechanical rotation pullback device (**b**), and a 3.2F imaging catheter (**c**). The disposable imaging catheter uses traditional 0.014-in. monorail system and contains an optical fiber to deliver NIR light from a console as well as intravascular ultrasound (IVUS) imaging system. The console integrates NIR information with IVUS image using predictive algorithm

lipid core plaque (LCP) at each interrogated location in the artery using a predictive algorithm. The calculated data are displayed in a two-dimensional map of the vessel ("chemogram") (Fig. 9.3a). The x-axis of the chemogram represents pullback position in millimeter scale, and the y-axis represents circumferential position in degrees (0–360°); a color scale from red to yellow indicates increasing probability that a LCP is present.

The block chemogram is a summary measurement of the probability that a LCP of 2-mm pullback interval is analyzed and displayed in a color map (Fig. 9.3b). The block chemogram uses the same color scale as the chemogram, but the display is summed up to four discrete colors to facilitate visual interpretation (red, $p < 0.57$; orange, $0.57 \leq p \leq 0.84$; tan, $0.84 \leq p \leq 0.98$; yellow, $p > 0.98$, algorithm probability that a LCP is present in that 2-mm block). Lipid core burden index (LCBI) is defined as the fraction of valid pixel in the chemogram that exceed a LCP probability of 0.6, multiplied by 1000 (Fig. 9.4). LCBI provides a summary measurement of the LCP presence in the entire scanned segment. The maxLCBI$_{4mm}$ is defined as the maximum value of LCBI for any of the 4-mm segment in the interrogated region and used as the index representing the size of the LCP (Fig. 9.5).

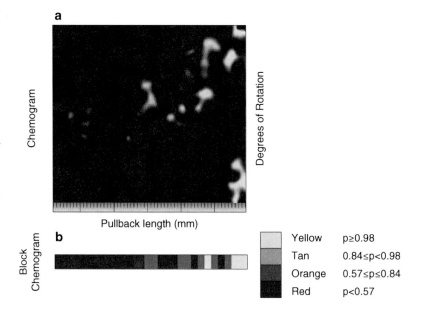

Fig. 9.3 An example of chemogram and block chemogram. (**a**) The color of chemogram from red to yellow indicates the increasing probability that a lipid core plaque (LCP) is present at this location. (**b**) Each color of the block chemogram is determined by 90th percentile value of the chemogram within a 2-mm segment. Four colors of the block chemogram represent chance of a LCP at this location (*red, $p < 0.57$; orange, $0.57 \leq p \leq 0.84$; tan, $0.84 \leq p < 0.98$; yellow, $p \geq 0.98$*)

Fig. 9.4 Lipid core burden index (LCBI). LCBI is defined as cholesterol-positive signals which exceed an LCP probability of 0.6 within the region of interest divided by total valid signals multiplied by 1000 (‰)

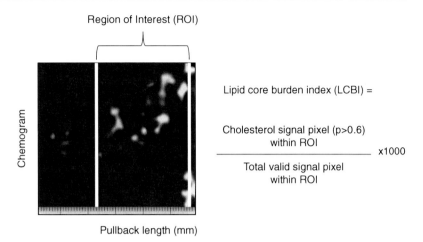

9.4 Clinical Studies

9.4.1 Prediction of Periprocedural MI

NIRS is able to identify high risk of periprocedural myocardial infarction (MI). Goldstein JA et al. observed 62 patients with stable cardiac biomarker who underwent coronary stenting [7]. Periprocedural MI was observed in 50% of patients with a maxLCBI4mm ≥ 500. On the other hand, periprocedural MI occurred only in 4.2% of patients with maxLCBI4mm < 500 ($p = 0.0002$). Quantification of LCP measured as maxLCBI4mm ≥ 500 was associated with increased risk of periprocedural MI, which is completely in accordance with traditional studies with IVUS or virtual histology (Fig. 9.6). The CANARY (Coronary Assessment by NIR of Atherosclerotic Rupture-Prone Yellow) study [8] enrolled 85 stable angina patients in a prospective and multicenter manner. NIRS performed prior to PCI

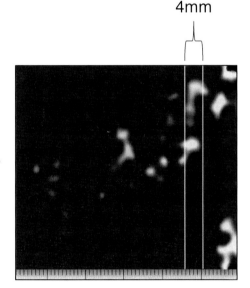

Fig. 9.5 maxLCBI4mm. maxLCBI4mm is defined as the maximum value of lipid core burden index for any of the 4-mm segment. It represents the angular size of the LCP

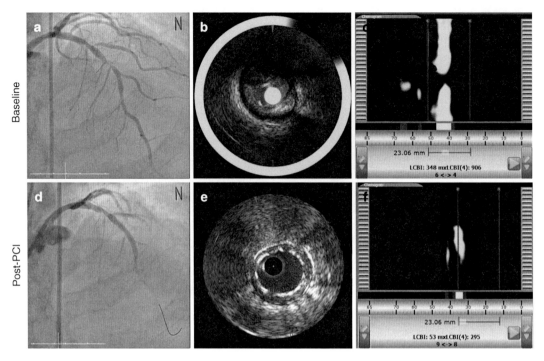

Fig. 9.6 Representative case of periprocedural myocardial infarction (MI) predicted by near-infrared spectroscopy. (**a**) Baseline angiogram shows discrete tight stenosis at the middle segment of the *left* anterior descending artery. (**b**) Baseline intravascular ultrasound (IVUS) shows significantly narrowed lumen with 1.75 mm^2 of minimal lumen area due to a large eccentric echo-attenuated plaque. The plaque burden was 86.9%. (**c**) The baseline chemogram displays "yellow" lipid-rich plaque extending almost 330° of the vessel circumference with maxLCBI4mm 906 (between *blue* lines), which highly suggests the development of periprocedural MI or no-reflow phenomenon. (**d**) Post-PCI angiogram shows no-reflow phenomenon. Cardiac biomarkers taken after the procedure were significantly elevated. (**e**) Post-PCI IVUS shows multiple stent struts well expanded and opposed to the vessel wall. Final minimal stent area is 5.8 mm^2. (**f**) The post-PCI chemogram displays significantly reduced and partly relocated lipid core area (*yellow*) after stenting. The maxLCBI4mm is 295 (between *blue* lines)

showed maxLCBI4mm was significantly higher (481.5 vs. 371.5, $p = 0.05$). However, among the randomized lesions with maxLCBI4mm ≥600, there was no difference of periprocedural MI with vs. without the use of distal protection filter (35.7% vs. 23.5%, respectively; relative risk, 1.52; 95% confidence interval: 0.50–4.60, $p = 0.69$). It is unclear whether this result is due to the limitation of NIRS predicting periprocedural MI or that of distal protection device preventing periprocedural MI. Further investigations will be needed to clarify this issue.

9.4.2 PCI Guidance

Visual assessment of coronary angiogram is commonly used to determine stent length. However, in terms of full lesion coverage, it is frequently inaccurate. IVUS can provide us more precise information than angiogram on lesion length by showing intravascular plaque morphology. Further, NIRS system substantiates another potential that it can give us additional information by showing the extent of lipid within coronary artery wall. Dixon et al. [9] observed that LCP extended beyond the angiographic margin of the lesion in 16% of PCI lesions. Whether LCP extending beyond the stent edges produces adverse outcome is unclear and requires further investigation. However, it is not difficult to expect that incomplete lesion coverage may increase the risk of stent edge problems such as restenosis requiring additional PCI or myocardial infarction. Strategy of PCI optimization with NIRS currently may be implicative.

9.4.3 Prediction of Outcome

Prospective identification of both vulnerable plaque and patient has been an important issue. However, only a small number of prospective outcome studies (Table 9.2), which assessed non-culprit lesions with intravascular imaging modalities, have been available. Most of them used IVUS or virtual histology (VH-IVUS) and have been describing several well-established features of vulnerable plaque (Table 9.2). Now accumulating data suggest that NIRS can identify vulnerable or rupture-prone plaque and predict outcome of the patients. Madder et al. reported maxLCBI4mm was 5.8-fold higher in STEMI culprit segments than in non-culprit segments of the STEMI culprit vessel (median [interquartile range (IQR)]: 523 [445–821] vs. 90 [6–265]; $p < 0.001$) [15]. A threshold of maxLCBI4mm ≥400 distinguished STEMI culprit (sensitivity, 85%; specificity, 98%). Oemrawsingh RM et al. observed non-culprit coronary arteries in 203 patients who were referred for coronary angiography [14]. About half (46%) of the patients had acute coronary syndrome. A fourfold increase in major adverse cardiac and cerebrovascular events during 1-year follow-up was observed in patients with LCBI above the median (16.7% vs. 4.0% event rate [adjusted hazard ratio, 4.04; 95% confidence interval, 1.33–12.29; $p = 0.01$]). Furthermore, the majority of event in this study was unplanned revascularization, which suggest NIRS is able to identify "active phase" or "rapid growing" plaque as well as rupture-prone plaque. Similarly, Madder et al. reported that in their 121 registry patients analysis maxLCBI4mm ≥400 in a non-stented segment at baseline is significantly associated with

Table 9.2 Imaging predictors in non-culprit lesion for clinical outcomes

Study	Patients	Method	Outcome	Results
Ohtani et al. [10]	552 pts	Angioscopy	7.1% ACS events at 57.3 ± 22.1-month FU	Number of yellow plaques (adjusted HR1.23[1.03–1.45], $p = 0.02$)
Prospect Stone et al. [11]	697 ACS pts	3-vessel VH-IVUS	11.6% MACE (cardiac death, cardiac arrest, MI, or rehospitalization) at 3.4-year FU	PB ≥ 70% (HR 5.03[2.51–10.11], $p < 0.001$), MLA ≤ 4.0 mm^2 (HR3.21[1.61–6.42], $p = 0.001$), VH-TCFA (HR3.35[1.77–6.36], $p < 0.001$)
Calvert et al. [12]	931 non-culprit lesions in 170 pts (41% ACS)	3-vessel VH-IVUS	1.4% MACE (death, MI, or unplanned revascularization) at 625-day FU	VH-TCFA (HR7.53, $p = 0.038$) and PB > 70% (HR 8.13, $p = 0.011$) remodeling index (HR2686 [1.94–3.72×10], $p = 0.032$)
Atheroremo-IVUS Cheng et al. [13]	581 pts (54% ACS)	VH-IVUS	7.8% MACE (mortality, ACS, or unplanned revascularization) at 1-year FU	VH-TCFA (adjusted HR1.98[1.09–3.60], $p = 0.026$) PB ≥ 70% (adjusted HR2.90[1.15–5.49], $p = 0.021$)
Atheroremo-NIRS Oemrawsingh et al. [14]	203 pts (47% ACS)	1-vessel NIRS	13.7% MACE (all-cause mortality, nonfatal ACS, stroke, and unplanned revascularization) at 1-year FU	LCBI ≥ 43.0 (median) (adjusted HR4.04[1.33–12.29], $p = 0.01$)

ACS acute coronary syndrome, *FU* follow-up, *HR* hazard ratio, *pts* patients, *VH-IVUS* virtual histology intravascular ultrasound, *MACE* major adverse cardiac event, *MI* myocardial infarction, *PB* plaque burden, *MLA* minimal luminal area, *VH-TCFA* virtual histology thin-capped fibroatheroma, *NIRS* near-infrared spectroscopy, *LCBI* lipid core burden index

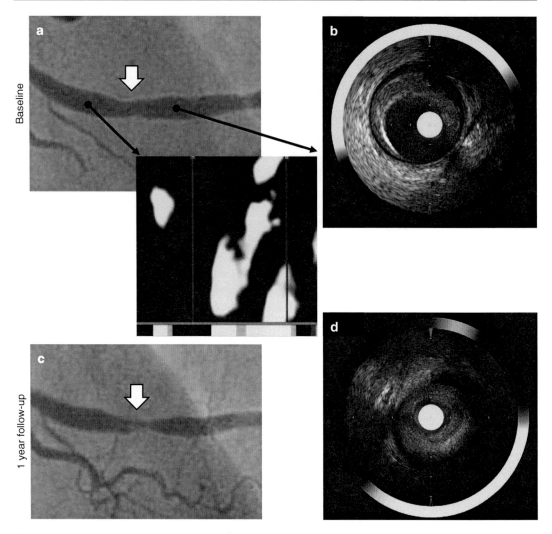

Fig. 9.7 Representative case of plaque progression predicted by near-infrared spectroscopy (NIRS). (**a**) Baseline angiogram shows insignificant stenosis (*white arrow*) at the middle segment of the *right* coronary artery. Concomitant NIRS scan displays the presence of large lipid core in the coronary artery wall (maxLCBI4mm is 483), which highly suggest the future cardiac event. (**b**) Baseline intravascular ultrasound (IVUS) shows an eccentric plaque with 8.2 mm^2 of minimal lumen area (MLA). The plaque burden is 58%. (**c**) The 1-year follow-up coronary angiogram shows definite "progression of plaque" with significant luminal narrowing (*white arrow*). (**d**) Follow-up IVUS shows narrowing of MLA (2.1 mm) and increased plaque burden (88%) compared with baseline images

adverse cardiac events during follow-up (HR 10.2, 95%CI 3.4–30.6, $P < 0.001$) [16]. NIRS is able to predict outcome in patients with coronary artery disease (Fig. 9.7).

9.4.4 Endothelial Dysfunction

Although the mechanism of exacerbating atherosclerosis by endothelial dysfunction has been extensively investigated in vitro and animal studies, in vivo demonstration using intravascular imaging technique such as IVUS has failed to substantiate this association. Choi B et al. reported that there was a significant correlation between LCBI ($r = -0.460$, $p = 0.008$), LCBI divided by lesion length ($r = -0.453$, $p = 0.009$), and maxLCBI4mm ($r = -0.431$, $p = 0.014$) and the degree of epicardial endothelial function [17]. NIRS system was sensitive enough to detect the early changes of atherosclerosis according to the degree of endothelial dysfunction, which suggest it may serve as an important tool for assessing atherosclerosis and pathogenic mechanism of it.

9.5 Limitation

The NIRS system only provides two-dimensional information of cholesterol accumulation and does not provide information on the depth of the cholesterol within the coronary artery wall. IVUS may therefore be used for additional evaluation of plaque structure. False-positive reading of NIRS could be caused by fibroatheromas too small or with caps too thick to meet criteria for the LCP of interest or by lesions containing significant lipid but not having necrotic core (intimal xanthoma and pathologic intimal thickening).

Conclusion

The new lipid-identification methodology with NIR spectroscopy seems to be of value to research as well as clinical decision-making. Initial studies successfully demonstrated its ability and potentials. Several ongoing clinical trials may confirm its clinical usefulness and future applications.

References

1. Moreno PR, Muller JE. Identification of high-risk atherosclerotic plaques: a survey of spectroscopic methods. Curr Opin Cardiol. 2002;17:638–47.
2. Hall JW, Pollard A. Near-infrared spectrophotometry: a new dimension in clinical chemistry. Clin Chem. 1992;38:1623–31.
3. Caplan JD, Waxman S, Nesto RW, Muller JE. Near-infrared spectroscopy for the detection of vulnerable coronary artery plaques. J Am Coll Cardiol. 2006;47:C92–6.
4. Gardner CM, Tan H, Hull EL, Lisauskas JB, Sum ST, Meese TM, Jiang C, Madden SP, Caplan JD, Burke AP, Virmani R, Goldstein J, Muller JE. Detection of lipid core coronary plaques in autopsy specimens with a novel catheter-based near-infrared spectroscopy system. JACC Cardiovasc Imaging. 2008;1:638–48.
5. Waxman S, Dixon SR, L'Allier P, Moses JW, Petersen JL, Cutlip D, Tardif JC, Nesto RW, Muller JE, Hendricks MJ, Sum ST, Gardner CM, Goldstein JA, Stone GW, Krucoff MW. In vivo validation of a catheter-based near-infrared spectroscopy system for detection of lipid core coronary plaques: initial results of the SPECTACL study. JACC Cardiovasc Imaging. 2009;2:858–68.
6. Garcia BA, Wood F, Cipher D, Banerjee S, Brilakis ES. Reproducibility of near-infrared spectroscopy for the detection of lipid core coronary plaques and observed changes after coronary stent implantation. Catheter Cardiovasc Interv. 2010;76:359–65.
7. Goldstein JA, Maini B, Dixon SR, Brilakis ES, Grines CL, Rizik DG, Powers ER, Steinberg DH, Shunk KA, Weisz G, Moreno PR, Kini A, Sharma SK, Hendricks MJ, Sum ST, Madden SP, Muller JE, Stone GW, Kern MJ. Detection of lipid-core plaques by intracoronary near-infrared spectroscopy identifies high risk of periprocedural myocardial infarction. Circ Cardiovasc Interv. 2011;4:429–37.
8. Stone GW, Maehara A, Muller JE, Rizik DG, Shunk KA, Ben-Yehuda O, Genereux P, Dressler O, Parvataneni R, Madden S, Shah P, Brilakis ES, Kini AS, Investigators C. Plaque characterization to inform the prediction and prevention of periprocedural myocardial infarction during percutaneous coronary intervention: the CANARY trial (coronary assessment by near-infrared of atherosclerotic rupture-prone yellow). JACC Cardiovasc Interv. 2015;8:927–36.
9. Dixon SR, Grines CL, Munir A, Madder RD, Safian RD, Hanzel GS, Pica MC, Goldstein JA. Analysis of target lesion length before coronary artery stenting using angiography and near-infrared spectroscopy versus angiography alone. Am J Cardiol. 2012;109:60–6.
10. Ohtani T, Ueda Y, Mizote I, Oyabu J, Okada K, Hirayama A, Kodama K. Number of yellow plaques detected in a coronary artery is associated with future risk of acute coronary syndrome: detection of vulnerable patients by angioscopy. J Am Coll Cardiol. 2006;47:2194–200.
11. Stone GW, Maehara A, Lansky AJ, de Bruyne B, Cristea E, Mintz GS, Mehran R, McPherson J, Farhat N, Marso SP, Parise H, Templin B, White R, Zhang Z, Serruys PW, Investigators P. A prospective natural-history study of coronary atherosclerosis. N Engl J Med. 2011;364:226–35.
12. Calvert PA, Obaid DR, O'Sullivan M, Shapiro LM, McNab D, Densem CG, Schofield PM, Braganza D, Clarke SC, Ray KK, West NE, Bennett MR. Association between IVUS findings and adverse outcomes in patients with coronary artery disease: the VIVA (VH-IVUS in vulnerable atherosclerosis) study. JACC Cardiovasc Imaging. 2011;4:894–901.
13. Cheng JM, Garcia-Garcia HM, de Boer SP, Kardys I, Heo JH, Akkerhuis KM, Oemrawsingh RM, van Domburg RT, Ligthart J, Witberg KT, Regar E, Serruys PW, van Geuns RJ, Boersma E. In vivo detection of high-risk coronary plaques by radiofrequency intravascular ultrasound and cardiovascular outcome: results of the ATHEROREMO-IVUS study. Eur Heart J. 2014;35:639–47.

14. Oemrawsingh RM, Cheng JM, Garcia-Garcia HM, van Geuns RJ, de Boer SP, Simsek C, Kardys I, Lenzen MJ, van Domburg RT, Regar E, Serruys PW, Akkerhuis KM, Boersma E, Investigators A-N. Near-infrared spectroscopy predicts cardiovascular outcome in patients with coronary artery disease. J Am Coll Cardiol. 2014;64:2510–8.
15. Madder RD, Goldstein JA, Madden SP, Puri R, Wolski K, Hendricks M, Sum ST, Kini A, Sharma S, Rizik D, Brilakis ES, Shunk KA, Petersen J, Weisz G, Virmani R, Nicholls SJ, Maehara A, Mintz GS, Stone GW, Muller JE. Detection by near-infrared spectroscopy of large lipid core plaques at culprit sites in patients with acute ST-segment elevation myocardial infarction. JACC Cardiovasc Interv. 2013;6:838–46.
16. Madder RD, Husaini M, Davis AT, VanOosterhout S, Khan M, Wohns D, McNamara RF, Wolschleger K, Gribar J, Collins JS, Jacoby M, Decker JM, Hendricks M, Sum ST, Madden S, Ware JH, Muller JE. Large lipid-rich coronary plaques detected by near-infrared spectroscopy at non-stented sites in the target artery identify patients likely to experience future major adverse cardiovascular events. Eur Heart J Cardiovasc Imaging. 2016;17:393–9.
17. Choi BJ, Prasad A, Gulati R, Best PJ, Lennon RJ, Barsness GW, Lerman LO, Lerman A. Coronary endothelial dysfunction in patients with early coronary artery disease is associated with the increase in intravascular lipid core plaque. Eur Heart J. 2013;34:2047–54.

Part II

OCT

10

Physical Principles and Equipment of Intravascular Optical Coherence Tomography

Jinyong Ha

Optical coherence tomography (OCT) is an emerging imaging modality analogous to intravascular ultrasound imaging but uses light instead of sound. The integration of a fiber-optic probe with frequency domain OCT enables video images that display the location and changes of coronary plaques and stent apposition in live patients. This chapter details the basic principles of intravascular optical coherence tomography (IV-OCT) in clinical practice. The system architecture and catheter structure consisting of an optical probe and a protective sheath are discussed in detail. Also, recent technology advances in IV-OCT are briefly introduced.

10.1 Introduction to OCT

Optical coherence tomography (OCT) is a high-resolution imaging modality that provides real-time cross-sectional images of tissue microstructures using the near-infrared light [1]. As OCT has common features with ultrasound imaging and microscopy in medical applications, it has been clinically adopted in ophthalmology, dermatology, and cardiology [2–4]. In most tissues, OCT imaging plays an important role in filling a gap between microscopy and ultrasound in comparison with resolution and imaging depth as shown in Fig. 10.1 [5]. Microscopy performs very high-resolution (~1 μm) imaging of en face tissue plane, but imaging depth in biological tissues is limited up to only a few hundred micrometers due to the signal attenuation from large optical scattering. The resolution of medical ultrasound imaging varies 0.1–1 mm depending on the sound wave frequency. It is, however, possible to see internal organs even if the imaging depth is limited to only millimeter ranges at high frequencies of ultrasound waves [6].

Compared with ultrasound imaging, OCT has the same operation principle, echo signal detection, but utilizes infrared light instead of ultrasound. In general, imaging is performed by measuring the magnitude and time delay of backscattered or backreflected signal from internal biological tissues. As a sound wave travels at 340 m/s in air, the echo signal can be measured with a time resolution of ~100 ns, which is within the limits of the electronic detection process. However, it is impossible to electrically measure echoes of backscattered light due to the light speed of 3×10^8 m/s in air, and optical interferometric techniques were then proposed [7–9]. Optical interferometers are widely used in science and engineering to measure small displacements and spatial irregularities by measuring interference patterns. To achieve microscale resolutions of the optical sectioning ability, low-coherence interferometry, using the short

J. Ha
Department of Optical Engineering, Sejong University, Seoul, South Korea
e-mail: jinyong.ha@gmail.com

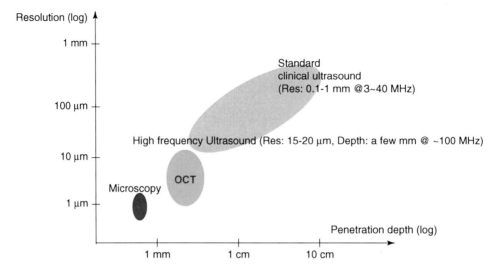

Fig. 10.1 Image resolution and penetration depth for OCT, microscopy, and ultrasound. OCT plays a role in filling a gap between microscopy and ultrasound in medical imaging. The image resolution in OCT is about 1–15 μm and the imaging depth is limited to 2–3 mm. Microscopy performs very high-resolution (~1 μm) imaging of en face tissue plane, but the imaging depth in biological tissues is limited up to only a few hundred micrometers due to the signal attenuation from large optical scattering. The resolution of medical ultrasound imaging varies 0.1–1 mm depending on the sound wave frequency, but it is possible to see internal organs even if imaging depth is limited to only millimeter ranges at high frequencies of ultrasound waves

coherence length of a broadband light source, is required. Low-coherence interferometry used in OCT is a unique solution to measure the echo signal of backscattered light with a very high signal-to-noise ratio, which is termed system sensitivity. The incident beam of light source is divided into a reference beam reflected from a mirror and a sample beam illuminated on biological tissues, and the sum of two beams is directed to a photodetector that measures the intensity of the combined signals [10].

10.1.1 Time Domain OCT and Frequency Domain OCT

OCT systems mainly consist of a light source, a coupler- or a circulator-based interferometer, and a photodetector. OCT is categorized into time domain OCT (TD-OCT) and frequency domain OCT (FD-OCT) as illustrated in Fig. 10.2 [11, 12]. TD-OCT employs a broadband light source such as a superluminescent diode and a scanning reference arm. The path length difference between the sample and reference arms of the interferometer is modulated by scanning a reference path length. The envelope of interference fringes is then extracted as a function of time, which means that image data is finally generated in a time domain. A single travel of the reference mirror creates a depth profile or an A-scan. Development of the high-speed scanning delay line of the reference arm allows OCT imaging speeds of several thousand axial profiles per second and video frame rate [13, 14]. FD-OCT additionally requires a spectrometer as a photodetector or wavelength-swept laser as light source without scanning a reference length. The former case is referred to as spectral domain OCT (SD-OCT), whereas the latter is termed swept-source OCT (SS-OCT) or alternatively optical frequency domain imaging (OFDI) [15–19]. SS-OCT has the advantages of easy implementation of polarization diverse detection as well as large depth range over SD-OCT. In the case of SS-OCT, a narrow instantaneous linewidth over a broad spectral range is tuned in wavelength as a function of time, and all echo signals from different depths are measured simultaneously. Thus, system sensitivity and imaging speeds can be dramatically improved [11, 19].

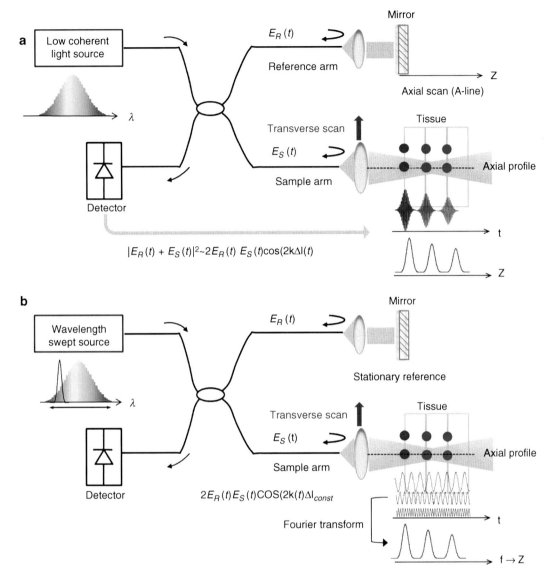

Fig. 10.2 Comparison between TD-OCT and FD-OCT. (**a**) Time domain optical coherence tomography (TD-OCT) system configuration consisting of a low-coherent light source, an interferometer, and a photodetector. To generate axial profiles, the reference arm is scanned as a function of time. (**b**) Frequency domain OCT (FD-OCT) system configuration. A wavelength-swept source is utilized and the reference arm is stationary. Interferometric patterns are measured as a function of wavelength and time, rather than as a function of time alone. The delay of echo signals from different positions in tissue results in different frequency modulations which are measured by Fourier transform

10.2 Intravascular OCT System

Intravascular OCT (IV-OCT) is a catheter-based imaging modality using a fiber-optic probe (Fig. 10.3). In general, in vivo intracoronary imaging is challenging because a suitable OCT catheter and contrast agent flushing protocols to remove blood need to be developed. IV-OCT additionally requires a catheter system that is composed of a rotary junction and a catheter as shown in Fig. 10.4 [20, 21]. A rotary junction plays the important role of pulling back and

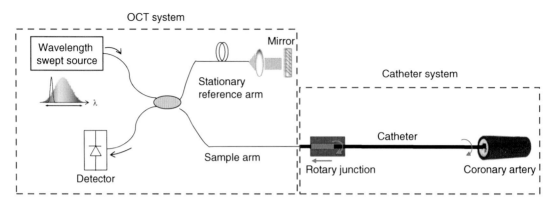

Fig. 10.3 Configuration of intravascular OCT. Intravascular OCT (IV-OCT) additionally requires a catheter system, which includes a rotary junction and a catheter. The rotary junction connects an IV-OCT platform to a catheter

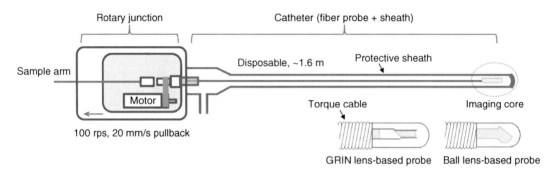

Fig. 10.4 Rotary junction and catheter structure. To create cross-sectional OCT images, a catheter needs to be rotated and pulled back by a rotary junction. The rotary junction also couples OCT light from the IV-OCT system to the fiber-optic probe in the protective sheath. To focus light beam, a GRIN lens or a polished ball lens is utilized. To provide the stable torque transmission from the proximal end to the distal tip of a fiber probe, the fiber probe is inserted into a hollow torque cable and fixed

rotating a catheter as well as optically connecting an OCT system console with a catheter. A catheter consists of an optical probe and a protective sheath. An optical probe is composed of a single mode fiber, a small lens, and a hollow torque cable. The torque cable has multiple threads and layers to accurately transmit the proximal end rotation to the distal tip of a fiber probe in curved environments. As a rotating optical fiber is prone to fragility, it is inserted into a torque cable and fixed. To focus light on the vessel, an angle-polished ball lens or a gradient index (GRIN) lens is utilized. An inserted catheter is rotated to create two-dimensional cross-sectional images of coronary artery while it is pulled back to generate multiple frames (Fig. 10.5). IV-OCT creates original rectangular OCT images that are converted from polar to Cartesian coordinates for display as shown in Fig. 10.6. To examine the coronary artery by IV-OCT, blood in the artery must be removed to avoid massive optical scattering and attenuation by red blood cells. The first catheter-based imaging of a human artery ex vivo was conducted by Tearney et al. [22, 23]. This study reported that OCT images were capable of differentiating the intima, media, and adventitia of the artery. The first in vivo IV-OCT imaging in human patients was conducted by Jang et al. who demonstrated a comparison of OCT with IVUS images of tissue prolapse in a stent [24].

The first commercial IV-OCT product was the M2 OCT system (LightLab Imaging, Inc., Westford, MA, USA, now part of St. Jude Medical, Inc.) with regulatory approval in Europe and Japan in 2004, and the M3 system was launched 3 years later in Japan. Since both M2

10 Physical Principles and Equipment of Intravascular Optical Coherence Tomography

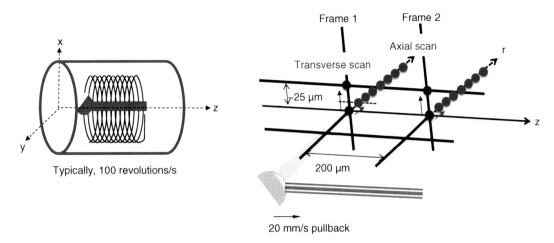

Fig. 10.5 Image data acquisition by helical scanning. Image data in IV-OCT are acquired by helical scanning of a catheter. This helical scanning consists of transverse scanning and pullback motion

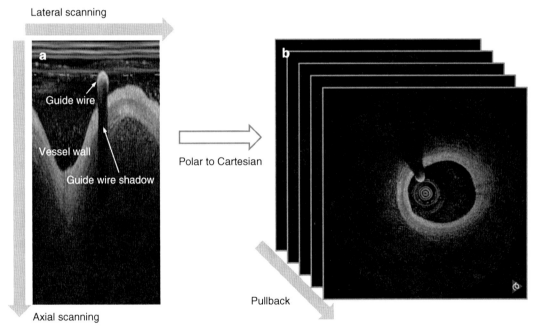

Fig. 10.6 Converting image in polar to Cartesian coordinate. IV-OCT images are generated by axial, lateral scannings and pullback motion. The original polar OCT images are transformed into Cartesian coordinates. (**a**) Rectangular OCT image in the polar domain naturally generated by IV-OCT. (**b**) Cross-sectional OCT image converted to Cartesian domain for visualization of coronary artery

and M3 systems were based on TD-OCT technology, the imaging speeds were limited (frame rate, 15.6/s for the M2 vs. 20/s for the M3, and pullback speed, 3.0 mm/s vs. 2.0 mm/s) [25, 26]. The balloon occlusion with lactated Ringer's solution or normal flushing was performed from proximal location to the lesion during the OCT imaging. Dramatic advancements in the wavelength-swept source in FD-OCT enabled much faster frame rates (100 frames/s) and pullback speeds (5–20 mm/s), resulting in non-occlusive OCT imaging with flushing of viscous

Table 10.1 Comparison of TD-OCT and FD-OCT

Imaging method	TD-OCT	FD-OCT
Axial resolution	15 μm	10–15 μm
Lateral resolution	90 μm	20–40 μm
Catheter profile	Maximum outer diameter of 0.019″	2.4-F to 3.2-F
Frame rate	16–20 frames/s (typ. 15.4 frames/s)	100–160 frames/s
Pullback speed	0.5–2 mm/s for occlusion 2–4 mm/s for non-occlusion	10–40 mm/s (typ. 20 mm/s)
Balloon occlusion	Highly recommended	Not required

contrast [4, 27]. The commercial version of FD-OCT, a Dragonfly imaging catheter and C7-XR OCT system (St. Jude Medical/LightLab Imaging, Westford, MA, USA), was launched with the world's first regulatory approval in 2010 [26]. The feasibility and safety of the Terumo OFDI system allowing 160 frames/s were evaluated in human patients in 2011 [28], and the first product of a LUNAWAVE OFDI system (Terumo Corporation, Tokyo, Japan) was then launched in more than 30 countries in Europe in 2013. The main specifications of TD-OCT and FD-OCT systems are summarized in Table 10.1 [4, 26, 28].

10.3 Image Quality of OCT

Since image quality is determined by imaging modality resolution, developing a high-resolution optical imaging system has been one of main research topics, involving high-speed and penetration depth imaging modalities. The image resolution of OCT is divided into axial resolution and lateral or transverse resolution. The axial resolution is the smallest distance between two objects that can be resolved along the axis of the incident beam. It is independent of the lens design and proportionally dependent on the center wavelength of light source and inversely proportional to the source bandwidth [29]. The lateral resolution in OCT imaging is the minimum resolvable distance between two objects which lie perpendicular to the OCT beam at the same depth position. It is mainly dependent on the focusing lens in the imaging core [30]. Another important parameter to determine the image quality is the depth of focus (DOF). DOF is twice the Rayleigh range, defined as the axial distance from the position of the minimum spot size (d) to the position of $\sqrt{2}d$. A trade-off exists between lateral resolution and DOF. Thus, increasing the lateral resolution to acquire better image quality results in decreasing the DOF. Since the relative position of a catheter from the vessel wall significantly varies during IV-OCT imaging, great variations of the lateral resolution induce distortion of images. The axial resolution for IV-OCT is typically ~10 μm at a center wavelength (~1300 nm) in the light source, and the lateral resolution is between 20 and 40 μm and the DOF is ~1.3 mm [31, 32].

10.3.1 Image Distortion in IV-OCT

Image artifacts in IV-OCT are mainly caused by catheter motion and cardiac dynamics. To create OCT cross-sectional images of vessels, a catheter is inserted into the coronary artery and rotated with an automatic pullback. Here, the torque applied at the proximal end of the catheter is not evenly transmitted to the distal imaging core, and thus the catheter is nonuniformly rotated since a catheter is placed through tortuous vessels or a crimped imaging sheath or a tight hemostatic value. This distortion is referred to as nonuniform rotational distortion, which also occurs in IVUS imaging [32]. Cardiac dynamics causes image distortion. During the cardiac cycle, heart motion directly affects the catheter motion in both the radial and longitudinal direction as shown in Fig. 10.7 [33, 34].

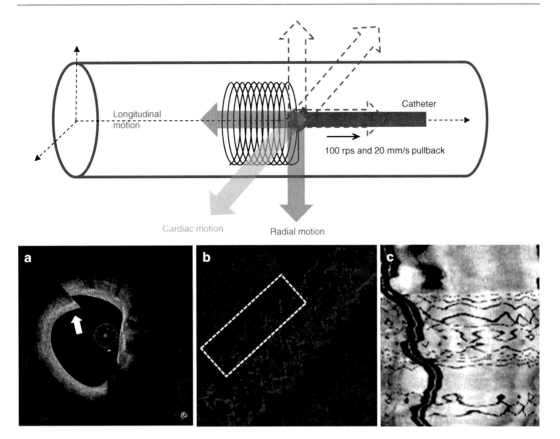

Fig. 10.7 Representative motion artifacts caused by cardiac motion. A catheter is significantly fluctuated in the coronary artery due to cardiac dynamics. Cardiac motion mainly generates the probe motion in the radial and longitudinal directions. During the cardiac cycle, the probe oscillates in both directions. (**a**) Cross-sectional view of coronary artery, (**b**) longitudinal cutaway view of the stented vessel caused by the radial motion artifact and (**c**) unfolded longitudinal view of the stented vessel due to the longitudinal motion artifact

As a result, an axial discontinuity in the cross-sectional image of coronary artery may appear at the transition region between the first and the last A-scan, and the repeated images of cross-sectional coronary artery can be generated by the longitudinal motion of a catheter that is pulled back and forth. In addition, there is also image distortion by saturation due to high optical backscattering, shadowing effect caused by blood inside of a catheter or stent struts, and suboptimal flushing [32].

10.4 Advances in IV-OCT Technology

Imaging speed, resolution, and penetration depth are most importantly considered when the performance of imaging modalities is evaluated. High-speed imaging may provide motion-artifact-free images and reduce the contrast flush volume. Recently, an ECG-triggered high-speed OCT system demonstrated a cardiac motion-free imaging at a rate of 500 frames/s and a pullback speed of 100 mm/s, and the in vivo imaging experiment was conducted in a beating swine heart [35]. It was also reported that micro-OCT (μOCT) with enhanced lateral resolution of 1 μm proved an ability to observe cells, extracellular components, and endothelial coverage in situ although a μOCT catheter needs to be developed [36]. However, it may be difficult to improve penetration depth in tissue without increasing the center wavelength of light source, which decreases the axial resolution.

There has been a great deal of interest in identifying high-risk plaques by characterizing tissue components. Polarization-sensitive OCT (PS-OCT) is a functional extension and provides the property of tissue birefringence which may be associated with collagen and smooth muscle cell content [37, 38]. Recently, an in vivo human pilot study demonstrated the feasibility and robustness of intravascular PS-OCT by achieving improved tissue characterization such as plaque rupture [39, 40]. For other light-based approaches to detecting high-risk plaques, multimodality OCT combined with near-infrared fluorescence (NIRF) or near-infrared autofluorescence (NIRAF) has been demonstrated. Multimodality IV-OCT and NIRF imaging system accurately identified lipid-rich inflamed plaques using a FDA-approved indocyanine green (ICG) in rabbit models [41]. A first-in-human IV-OCT and NIRAF study was conducted with a 2.6-F coronary catheter. This study showed that an elevated NIRAF signal was focally associated with a high-risk morphological phenotype as determined by IV-OCT [42]. It was recently noted that the most advanced IVUS-OCT system at a rate of 72 frames/s was successfully demonstrated in a rabbit artery in vivo. The accurate registration between IVUS and OCT data sets showed great potential to accelerate the clinical adoption for accurate identification of vulnerable plaques in humans [43].

10.5 Summary

IV-OCT has a light-based imaging modality using an ultrathin catheter. Development of the high-speed imaging technology based on FD-OCT has enabled real-time non-occlusive coronary artery imaging. As IV-OCT has a great potential for further understanding and treatment for atherosclerotic coronary artery disease, new technologies will be constantly developed and revolutionized by taking multidisciplinary approaches.

References

1. Huang D, Swanson EA, Lin CP, Schuman JS, Stinson WG, Chang W, et al. Optical coherence tomography. Science. 1991;254(5035):1178–81.
2. Fercher AF, Hitzenberger CK, Drexler W, Kamp G, Sattmann H. In-vivo optical coherence tomography. Am J Ophthalmol. 1993;116(1):113–5.
3. Gladkova ND, Petrova GA, Nikulin NK, Radenska-Lopovok SG, Snopova LB, Chumakov YP, et al. In vivo optical coherence tomography imaging of human skin: norm and pathology. Skin Res Technol. 2000;6(1):6–16.
4. Tearney GJ, Waxman S, Shishkov M, Vakoc BJ, Suter MJ, Freilich MI, et al. Three-dimensional coronary artery microscopy by intracoronary optical frequency domain imaging. JACC Cardiovasc Imaging. 2008;1(6):752–61.
5. Drexler W, Fujimoto JG. Optical coherence tomography : technology and applications. Berlin: Springer; 2008. xxix, 1346 p.
6. Szabo TL. Diagnostic ultrasound imaging: inside out. Burlington: Elsevier; 2004.
7. Beaud P, Schutz J, Hodel W, Weber HP, Gilgen HH, Salathe RP. Optical reflectometry with micrometer resolution for the investigation of integrated optical-devices. IEEE J Quantum Electron. 1989;25(4):755–9.
8. Takada K, Yokohama I, Chida K, Noda J. New measurement system for fault location in optical waveguide devices based on an interferometric-technique. Appl Opt. 1987;26(9):1603–6.
9. Youngquist RC, Carr S, Davies DEN. Optical coherence-domain Reflectometry—a new optical evaluation technique. Opt Lett. 1987;12(3):158–60.
10. Huang D, Wang J, Lin CP, Puliafito CA, Fujimoto JG. Micron-resolution ranging of cornea anterior chamber by optical reflectometry. Lasers Surg Med. 1991;11(5):419–25.
11. Choma MA, Sarunic MV, Yang CH, Izatt JA. Sensitivity advantage of swept source and Fourier domain optical coherence tomography. Opt Express. 2003;11(18):2183–9.
12. Leitgeb R, Hitzenberger C, Fercher A. Performance of fourier domain vs. time domain optical coherence tomography. Opt Express. 2003;11(8):889–94.
13. Rollins AM, Kulkarni MD, Yazdanfar S, Ungarunyawee R, Izatt JA. In vivo video rate optical coherence tomography. Opt Express. 1998;3(6):219–29.

14. Tearney GJ, Bouma BE, Fujimoto JG. High-speed phase- and group-delay scanning with a grating-based phase control delay line. Opt Lett. 1997;22(23):1811–3.
15. Chinn SR, Swanson EA, Fujimoto JG. Optical coherence tomography using a frequency-tunable optical source. Opt Lett. 1997;22(5):340–2.
16. Fercher AF, Hitzenberger CK, Kamp G, Elzaiat SY. Measurement of intraocular distances by backscattering spectral interferometry. Opt Commun. 1995;117(1–2):43–8.
17. Golubovic B, Bouma BE, Tearney GJ, Fujimoto JG. Optical frequency-domain reflectometry using rapid wavelength tuning of a Cr4+:forsterite laser. Opt Lett. 1997;22(22):1704–6.
18. Yun SH, Tearney GJ, Bouma BE, Park BH, de Boer JF. High-speed spectral-domain optical coherence tomography at 1.3 mu m wavelength. Opt Express. 2003;11(26):3598–604.
19. Yun SH, Tearney GJ, de Boer JF, Iftimia N, Bouma BE. High-speed optical frequency-domain imaging. Opt Express. 2003;11(22):2953–63.
20. Tearney GJ, Boppart SA, Bouma BE, Brezinski ME, Weissman NJ, Southern JF, et al. Scanning single-mode fiber optic catheter-endoscope for optical coherence tomography. Opt Lett. 1996;21(7):543–5.
21. Yaqoob Z, Wu JG, McDowell EJ, Heng X, Yang CH. Methods and application areas of endoscopic optical coherence tomography. J Biomed Opt. 2006;11(6):063001.
22. Tearney GJ, Brezinski ME, Boppart SA, Bouma BE, Weissman N, Southern JF, et al. Catheter-based optical imaging of a human coronary artery. Circulation. 1996;94(11):3013.
23. Tearney GJ, Jang IK, Kang DH, Aretz HT, Houser SL, Brady TJ, et al. Optical coherence tomography of human coronary arteries: a new imaging modality to visualize different components of plaques. J Am Coll Cardiol. 2000;35(2):52a–3a.
24. Jang IK, Tearney G, Bouma B. Visualization of tissue prolapse between coronary stent struts by optical coherence tomography—comparison with intravascular ultrasound. Circulation. 2001;104(22):2754.
25. Inami S, Wang Z, Ming-Juan Z, Takano M, Mizuno K. Current status of optical coherence tomography. Cardiovasc Interv Ther. 2011;26(3):177–85.
26. Terashima M, Kaneda H, Suzuki T. The role of optical coherence tomography in coronary intervention. Korean J Intern Med. 2012;27(1):1–12.
27. Yun SH, Tearney GJ, Vakoc BJ, Shishkov M, Oh WY, Desjardins AE, et al. Comprehensive volumetric optical microscopy in vivo. Nat Med. 2006;12(12):1429–33.
28. Okamura T, Onuma Y, Garcia-Garcia HM, van Geuns RJ, Wykrzykowska JJ, Schultz C, et al. First-in-man evaluation of intravascular optical frequency domain imaging (OFDI) of Terumo: a comparison with intravascular ultrasound and quantitative coronary angiography. EuroIntervention. 2011;6(9):1037–45.
29. Swanson EA, Huang D, Hee MR, Fujimoto JG, Lin CP, Puliafito CA. High-speed optical coherence domain reflectometry. Opt Lett. 1992;17(2):151–3.
30. Saleh BEA, Teich MC. Fundamentals of photonics. 2nd ed. Hoboken: Wiley; 2007. xix, 1175 p.
31. Lowe HC, Narula J, Fujimoto JG, Jang IK. Intracoronary optical diagnostics current status, limitations, and potential. JACC Cardiovasc Interv. 2011;4(12):1257–70.
32. Tearney GJ, Regar E, Akasaka T, Adriaenssens T, Barlis P, Bezerra HG, et al. Consensus standards for acquisition, measurement, and reporting of intravascular optical coherence tomography studies: a report from the international working group for intravascular optical coherence tomography standardization and validation. J Am Coll Cardiol. 2012;59(12):1058–72.
33. Ha JY, Shishkov M, Colice M, Oh WY, Yoo H, Liu L, et al. Compensation of motion artifacts in catheter-based optical frequency domain imaging. Opt Express. 2010;18(11):11418–27.
34. Ha J, Yoo H, Tearney GJ, Bouma BE. Compensation of motion artifacts in intracoronary optical frequency domain imaging and optical coherence tomography. Int J Cardiovasc Imaging. 2012;28(6):1299–304.
35. Jang SJ, Park HS, Song JW, Kim TS, Cho HS, Kim S, et al. ECG-triggered, single cardiac cycle, high-speed, 3D, intracoronary OCT. JACC Cardiovasc Imaging. 2016;9(5):623–5.
36. Liu LB, Gardecki JA, Nadkarni SK, Toussaint JD, Yagi Y, Bouma BE, et al. Imaging the subcellular structure of human coronary atherosclerosis using micro-optical coherence tomography. Nat Med. 2011;17(8):1010–U132.
37. Kuo WC, Chou NK, Chou C, Lai CM, Huang HJ, Wang SS, et al. Polarization-sensitive optical coherence tomography for imaging human atherosclerosis. Appl Opt. 2007;46(13):2520–7.
38. Nadkarni SK, Pierce MC, Park BH, de Boer JF, Whittaker P, Bouma BE, et al. Measurement of collagen and smooth muscle cell content in atherosclerotic plaques using polarization-sensitive optical coherence tomography. J Am Coll Cardiol. 2007;49(13):1474–81.
39. van der Sijde JN, Karanasos A, Villiger M, Bouma BE, Regar E. First-in-man assessment of plaque rupture by polarization-sensitive optical frequency domain imaging in vivo. Eur Heart J. 2016;37(24):1932.
40. Villiger M, Karanasos A, Ren J, Lippok N, Shishkov M, van Soest G, et al., editors. Intravascular polarization sensitive optical coherence tomography in human patients. Conference on Lasers and Electro-Optics. San Jose: Optical Society of America; 2016.
41. Lee S, Lee MW, Cho HS, Song JW, Nam HS, Oh DJ, et al. Fully integrated high-speed intravascular optical coherence tomography/near-infrared fluorescence structural/molecular imaging in vivo using a clinically available near-infrared fluorescence-emitting indocyanine green to detect inflamed lipid-rich atheromata in coronary-sized vessels. Circ Cardiovasc Interv. 2014;7(4):560–9.

42. Ughi GJ, Wang H, Gerbaud E, Gardecki JA, Fard AM, Hamidi E, et al. Clinical characterization of coronary atherosclerosis with dual-modality OCT and near-infrared autofluorescence imaging. JACC Cardiovasc Imaging. 2016;9(11):1304–14.

43. Li J, Ma T, Mohar D, Steward E, Yu M, Piao Z, et al. Ultrafast optical-ultrasonic system and miniaturized catheter for imaging and characterizing atherosclerotic plaques in vivo. Sci Rep. 2015;5:18406.

Image Acquisition Techniques

Ki-Seok Kim

In recent times, clinical usefulness of optical coherence tomography (OCT) is showing greater potential in the intracoronary imaging field. In this chapter, we will discuss about basic characteristics and imaging acquisition technique during Frequency-domain OCT (FD-OCT) examination.

11.1 Introduction

Optical coherence tomography (OCT) is a catheter-based invasive coronary imaging system. Using light source instead of ultrasound, OCT provided high-resolution coronary plaque image and state of deployed stent. Naohiro Tanno and James G. Fujimoto developed OCT in 1991 [1], and they first performed OCT on the human retina. Intravascular OCT was performed in 2002 (Fig. 11.1) [2, 3]. Intravascular OCT requires a single fiber-optic wire that both emits light and records the reflection while being simultaneously rotated and pulled back along the coronary artery [4]. As compared to intravascular ultrasound (IVUS), OCT provided ten times higher resolution (10–15 versus 100 μm). OCT cannot be able to make an image through the blood and more shorter penetration into the tissue (2 versus 1 cm) compared to IVUS [2, 5]. However, OCT provided high-resolution image of coronary plaque and detailed information of coronary atherosclerosis (Table 11.1), which may aid in future diagnosis and treatment of coronary artery disease.

K.-S. Kim
Division of Cardiology, Department of Internal Medicine, Jeju National University College of Medicine, Jeju, South Korea
e-mail: kiseok@jejunu.ac.kr

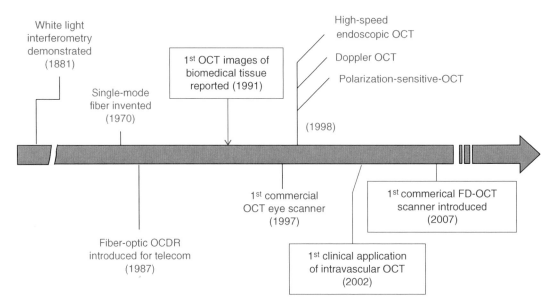

Fig. 11.1 History of the OCT development. Naohiro Tanno and James G. Fujimoto developed OCT in 1991, and they first performed OCT on the human retina. Intravascular OCT was performed in 2002. The first commercial second-generation FD-OCT introduced in 2007, which overcomes the limitation of TD-OCT system

Table 11.1 Performance comparison between intravascular ultrasound (IVUS) and frequency-domain optical coherence tomography (FD-OCT)

	IVUS	FD-OCT
Axial resolution (μm)	100–200	12–15
Beam width	200–300	20–40
Frame rate (frames/s)	30	100
Pullback speed (mm/s)	0.5–1	20
Scan diameter (mm)	15	10
Tissue penetration (mm)	10	1.0–2.0
Line per frame	256	500
Lateral sampling (μm)	225	19
Frame rate (frames/s)	Not required	Required

11.2 History of OCT Development

Intracoronary OCT catheter is connected to a rotary junction, which uses a motor to rotate the optical fiber in the catheter and couples light from this rotating fiber to light from the reference arm [6]. The rotary junction mounted to an automated pullback device (Fig. 11.2). There are two types of OCT system: time domain and frequency domain. The first-generation OCT is time domain (TD-OCT) which requires balloon occlusion in the proximal

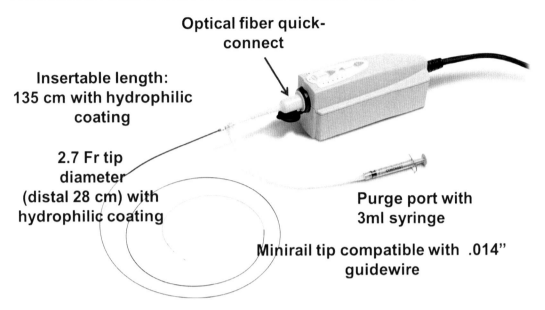

Fig. 11.2 Dragonfly OCT catheter and DOC system. Intracoronary OCT catheter connected to a rotary junction, which uses a motor to rotate the optical fiber in the vessel, which provides blood clearing during image generation. The problem of TD-OCT was prolonged examination time, shorter lengths of imaging segment, and intermediate imaging quality [7]. The first commercial second-generation FD-OCT was introduced in 2007, which overcomes the limitation of the TD-OCT system (Tables 11.2 and 11.3) [8]. catheter and couples light from this rotating fiber to light from reference arm. The rotary junction mounts to an automated pullback device

Table 11.2 Difference of time-domain versus frequency-domain optical coherence tomography (OCT)

	TD-OCT	FD-OCT
Scan method	Mechanically scans a reference mirror	Electronically scans the laser wavelength
Imaging speed	Slow	Fast
Image quality	Moderate	Exceptional

TD-OCT Time-domain OCT, *FD-OCT* frequency-domain OCT

Table 11.3 Performance comparison between TD-OCT and FD-OCT

	TD-OCT	FD-OCT
Axial resolution (µm)	12–15	15–20
Frame rate (frames/s)	100	15–20
Pullback speed (mm/s)	20	2–3
Scan diameter (mm)	10	6.8
Tissue penetration (mm)	1.0–2.0	1.0–2.0
Line per frame	500	200
Lateral sampling (µm)	19	39

TD-OCT Time-domain OCT, *FD-OCT* frequency-domain OCT, *s* second

11.3 Principle of FD-OCT Image Acquisition

Using the FD-OCT system, the OCT probe is first positioned over a regular guidewire, distal to the region of interest. Identification of the pullback starting point is a simple task as a dedicated marker identifies the exact position of the OCT beam, located at 20 mm proximal to the marker itself. When the OCT catheter is positioned and blood clearance is visually obtained distally through the contrast injection, the acquisition of a rapid OCT image sequence with fast pullback can be automatically commenced by injecting a bolus of solution through the guiding catheter, with the pullback speed of 20 mm/s (Fig. 11.3). The infusion rate of contrast is usually set to 3–4 ml/s for the left coronary artery and 2–3 ml/s for the right coronary, but can be modified based on the vessel runoff and size. This contrast agent is recommended for low arrhythmogenic potential and high viscosity, which help to prolong imaging time [9]. Most expert users advocate the use of automated contrast injection to optimize image quality. The pullback can start automatically when blood clearance is distally recognized or can be manually activated. An acquisition speed of 20 mm/s enables the acquisition of 200 cross-sectional image frames over a 5 cm length of artery in 2.5 s with a total infused volume of 14 ml of contrast [4]. This may represent a concrete advantage of FD-OCT for use in percutaneous coronary interventions (PCI), allowing quick evaluation of the stent and of the landing zones and avoiding geographical miss. The FD-OCT pullback speed is too fast to interpret the run during the acquisition, but the recorded images are stored digitally and can be reviewed in a slow playback loop [10].

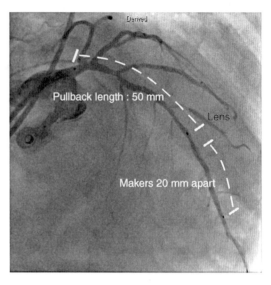

Fig. 11.3 OCT imaging catheter insertion and positioning. Using the FD-OCT system, the OCT probe is first positioned over a regular guidewire, distal to the region of interest. Identification of the pullback starting point is a simple task as a dedicated marker identifies the exact position of the OCT beam, located at 20 mm proximal to the marker itself

11.4 FD-OCT Image Acquisition Protocol

The St. Jude Medical OCT system and the Dragonfly intravascular imaging catheter are used to perform OCT intravascular imaging after intracoronary injection of 200 µg of nitroglycerin through conventional 6 Fr guiding catheters without side hole. A 0.014 in guidewire is positioned distal to the region of interest. The Dragonfly catheter is wiped proximal to the shaft to activate hydrophilic coating and gently purge the catheter with 100% contrast until three drops exit the

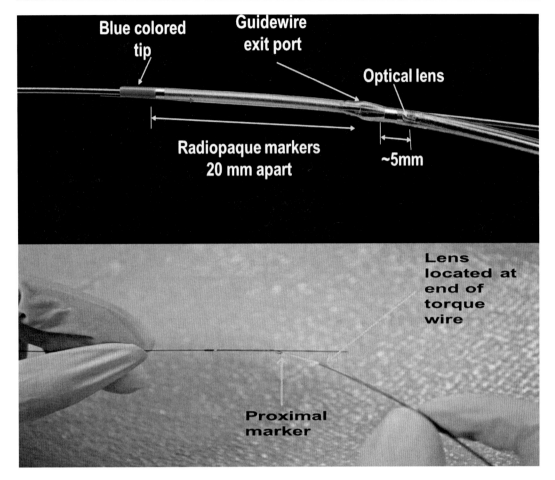

Fig. 11.4 OCT imaging catheter preparation. The Dragonfly catheter is wiped proximal to the shaft to activate hydrophilic coating and gently purge catheter with 100% contrast until three drops exit the catheter tip. The guidewire is then back-loaded through the blue tip and out of the exit port on the Dragonfly catheter. A slight bend is recommended to help ease the guidewire out of the exit portal

catheter tip. The guidewire is then back-loaded through the blue tip and out of the exit port on the Dragonfly catheter. A slight bend is recommended to help ease the guidewire out of the exit portal (Fig. 11.4). The Dragonfly catheter is advanced until the proximal radiopaque marker is distal to the target lesion. A test injection of 1–2 cc of 100% contrast is used to ensure guide catheter positioning. Before pullback procedure, purging is necessary to remove residual blood in the catheter lumen (Fig. 11.5a, b). During live scan, use a puff of contrast to evaluate clarity (Fig. 11.5c, d). Once the pullback is enabled on the system, the coronary blood flow is replaced by continuous flushing of 100% contrast media using a power injector or manual injection. The system labeling suggests power injector settings of 14 ml of total volume at 4 ml/s rate at 300 psi and 0 rise. We recommend these settings for the left anterior descending (LAD) and left circumflex (LCX) arteries and 12 ml of total volume at 4 ml/s rate at 300 psi and 0 rise for the right coronary artery (RCA). We find these settings to provide consistent, high-quality images. Measurements are performed using the system after proper calibration settings of the Z-offset [11].

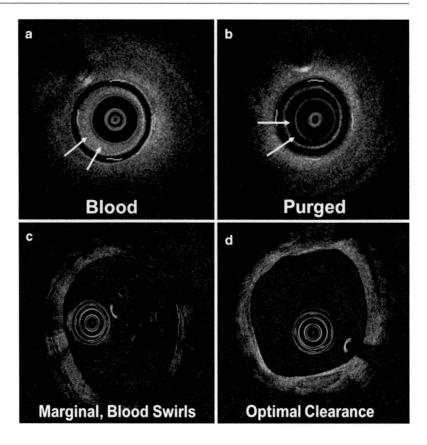

Fig. 11.5 OCT imaging catheter preparation before pullback procedure. Before pullback procedure, purging is necessary to remove residual blood in the catheter lumen (**a**, **b**). During live scan, use a puff of contrast to evaluate clarity (**c**, **d**)

11.5 FD-OCT Imaging Acquisition: Tips and Trick

In our experience, the fiber-optic OCT catheter is softer and less amenable to pulling than the IVUS catheter, and even the diameter (2.7 Fr) is less than IVUS. Before the operator advances the fiber-optic catheter, diffuse, long, relatively calcified, or bending lesions should be well prepared to avoid breaking the fiber-optic catheter. Moreover, OCT should be used to carefully coaxially guide the catheter position and measure firm catheter engagement in the coronary ostium to prevent residual blood attenuation. Vessel sizes range from 2.0 to 3.75 mm in diameter, which is ideal for OCT imaging. Thus, operators should be aware of "out-of-screen" loss of image, which is a result of the vessel size being larger than the scan diameter (field of view) of OCT, and foldover artifacts. So far, ostial lesions of the main trunk are still a limitation of OCT due to poor blood washing and catheter engagement (Table 11.4).

Table 11.4 Summary of optical coherence tomography mage acquisitioni

Do not use less than 6 Fr guiding catheter and side-hole guiding catheter
Use soft standard 0.014″ coronary guidewire
Diffuse, long, calcified, or bending lesion should be carefully prepared
Ideal vessel size is 2.0–3.75 mm in diameter
To ensure firm catheter engagement with good coaxial alignment can avoid blood attenuation
Inject non-diluted iodine contrast at rates of 3–5 ml/s in 4–5 s

11.6 Artifact of OCT

Residual blood attenuates the OCT light beam and may defocus the beam if red cell density is high. This will reduce brightness of the vessel wall, especially at large radial distances from the image wire. Blood swirls are caused by turbulent flow between flushing contrast fluid and blood. Flush fluid dose not filling the vessel lumen or end of bolus flush injection (Fig. 11.6a). Blood Speckling occurred by red blood cell (RBC) mixed into flush fluid or diluted with saline make less viscosity, which does not remove all RBS during image formation (Fig. 11.6b).

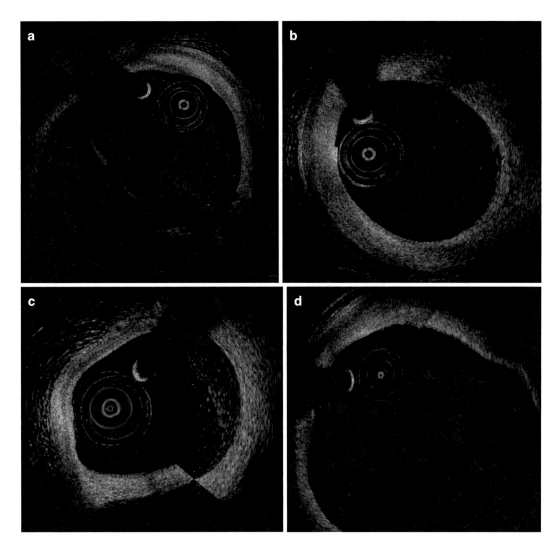

Fig. 11.6 Imaging artifact during pullback procedure. Blood swirls is caused by turbulent flow between flushing contrast fluid and blood. Flush fluid is not filled in the vessel lumen or the bolus injection of contrast is ended (**a**). Blood speckling occurs by red blood cell (RBC) mixed into flush fluid or diluted with saline make less viscosity, which does not remove all RBS during image formation (**b**). Sew-up artifact is the result of rapid artery or imaging wire movement in single-frame imaging formation, leading to single-point misalignment of the lumen border (**c**). Fold-over artifact is more specific to the new generation of FD-OCT. Typical examples are side branch and large vessels (**d**)

Sew-up artifact is the result of rapid artery or imaging wire movement in single-frame imaging formation, leading to single-point misalignment of the lumen border (Fig. 11.6c).

Fold-over artifact is more specific to the new generation of FD-OCT. It is the consequence of the "phase wrapping" or "alias" along the Fourier transformation when structure signals are reflected from outside the system's field of view. Typical examples are side branch and large vessels (Fig. 11.6d).

11.7 Summary

The advanced FD-OCT system provides more detailed coronary plaque information to plan an appropriate PCI procedure. A recent clinical trial (ILUMIEN III) has shown that OCT-guided PCI is not inferior to IVUS-guided PCI. Precise FD-OCT catheter manipulation and imaging acquisition technique will provide coronary vessel information and improve PCI outcomes.

References

1. Huang D, Swanson EA, Lin CP, Schuman JS, Stinson WG, Chang W, et al. Optical coherence tomography. Science. 1991;25:1178–81.
2. Jang IK, Bouma BE, Kang DH, Park SJ, Park SW, Seung KB, et al. Comparison with intravascular ultrasound. J Am Coll Cardiol. 2002;39:604–9.
3. Kawase Y, Hoshino K, Yoneyama R, McGregor J, Hajjar RJ, Jang IK, et al. In vivo volumetric analysis of coronary stent using optical coherence tomography with a novel balloon occlusion-flushing catheter: a comparison with intravascular ultrasound. Ultrasound Med Biol. 2005;31:1343–9.
4. Prati F, Jenkins MW, Di Giorgio A, Rollins AM. Intracoronary optical coherence tomography, basic theory and image acquisition techniques. Int J Cardiovasc Imaging. 2011;27:251–8.
5. Bezerra HG, Costa MA, Guagliumi G, Rollins AM, Simon DI. Intracoronary optical coherence tomography: a comprehensive review clinical and research applications. JACC Cardiovasc Interv. 2009;2:1035–46.
6. Brezinski ME, Tearney GJ, Bouma BE, Izatt JA, Hee MR, Swanson EA, et al. Optical coherence tomography for optical biopsy. Properties and demonstration of vascular pathology. Circulation. 1996;93:1206–13.
7. Prati F, Cera M, Ramazzotti V, Imola F, Giudice R, Albertucci M. Safety and feasibility of a new non-occlusive technique for facilitated intracoronary optical coherence tomography (OCT) acquisition in various clinical and anatomical scenarios. EuroIntervention. 2007;3:365–70.
8. Choma M, Sarunic M, Yang C, Izatt J. Sensitivity advantage of swept source and Fourier domain optical coherence tomography. Opt Express. 2003;11:2183–9.
9. Prati F, Cera M, Ramazzotti V, Imola F, Giudice R, Giudice M, et al. From bench to bedside: a novel technique of acquiring OCT images. Circ J. 2008;72:839–43.
10. Barlis P, Gonzalo N, Di Mario C, Prati F, Buellesfeld L, Rieber J, et al. A multicentre evaluation of the safety of intracoronary optical coherence tomography. EuroIntervention. 2009;5:90–5.
11. Prati F, Regar E, Mintz GS, Arbustini E, Di Mario C, Jang IK, et al. Expert review document on methodology, terminology, and clinical applications of optical coherence tomography: physical principles, methodology of image acquisition, and clinical application for assessment of coronary arteries and atherosclerosis. Eur Heart J. 2010;31:401–15.

12

Interpretation of Optical Coherence Tomography: Quantitative Measurement

So-Yeon Choi

Obtaining of good-quality image is essential to make accurate measurements. The image should be correctly calibrated for z-offset, the zero-point setting of the system before measurements. The definition of lesion, reference, and stented segment from the Journal of American College of Cardiology intravascular ultrasound (IVUS) consensus document has been adopted for optical coherence tomography (OCT) [1]. For standardization of OCT measurement, expert review documents and consensus standards have been published previously [2–4]. Studies regarding the accuracy and the reproducibility of qualitative and quantitative OCT measurements have been published previously [5–7].

12.1 Border Identification

The borders of the lumen, external elastic membrane (EEM), internal elastic membrane (IEM), plaque, and stent could be demarcated in OCT cross-sectional images similar to IVUS. In normal vessel without any plaque, OCT may discriminate IEM which is defined as the border between the intima and media and EEM which is defined as the border between the media and the adventitia. Measurements that EEM uses are likely closer to those of IVUS, whereas IEM measurements that use the IEM more closely approximate the pathologic definition of atherosclerosis as a disease of the intima. However, because of low penetration depth and rapid attenuation of its signal, OCT could not visualize IEM or EEM border in most diseased segments. The border measurements should not be made in cross-sectional images that contain artifacts that obscure a significant portion (>90°) of the image or over regions that contain side branches. The differences between OCT and IVUS measurements were demonstrated in Table 12.1 and Fig. 12.1.

12.2 Lesion Assessment

12.2.1 Reference Segment

Reference Assessment *Proximal or distal reference* is defined as the sites with the largest lumen proximal or distal to a stenosis within the same

S.-Y. Choi
Department of Cardiology, Ajou University Medical Center, Ajou University College of Medicine, Suwon, South Korea
e-mail: sychoimd@outlook.com

Table 12.1 Comparison of major quantitative measurements between optical coherence tomography and intravascular ultrasound

	OCT	IVUS
Lesion		
Lumen area	+	+
Vessel area	−/+	+
Plaque burden	−/+	+
Area stenosis	+	+
Stent		
Stent area	+	+
Vessel remodeling	−/+	+

IVUS intravascular ultrasound; *OCT* optical coherence tomography

segment with no major intervening branches (usually within 10 mm of the stenosis).

Reference Lumen and EEM Assessment *Proximal or distal mean reference lumen diameter* is the mean value of the shortest and the longest lumen diameter through the center of mass of the lumen at proximal or distal reference site. *Proximal or distal mean reference EEM diameter* is the mean value of the shortest and the longest EEM diameter through the center of mass of the lumen at proximal or distal reference site.

Average reference lumen diameter is the average value of mean lumen diameter at the proximal and distal reference sites. *Average reference EEM diameter* is the average value of mean EEM diameter at the proximal and distal reference sites. Both average reference lumen diameter and average reference EEM diameter are useful parameters for stent sizing during PCI.

Average reference EEM CSA, which is a useful parameter for evaluation of lesion severity in terms of stenosis, is the average value of EEM CSA at the proximal and distal reference sites.

Recent in the OPINION study, which had a randomized controlled design to compare the benefit of OCT guidance with IVUS guidance during percutaneous coronary intervention (PCI), OCT reference site was defined as the most normal-looking site with free of lipidic plaque (defined as signal-poor region with diffuse border) at a cross-section adjacent to the target lesion [8]. In other randomized controlled study, the ILUMIEN III: OPTIMIZE PCI study comparing OCT guidance, IVUS guidance, or angiography-guided stent implantation, proximal and distal reference mean EEM diameters and the smaller of these diameters to determine stent diameter or the proximal and distal lumen diameters were used if the EEM could not be visualized [9].

12.2.2 Lesion Segment

Lumen Measurements *Lumen CSA* is the area bounded by the luminal border. *Minimum lumen diameter* is the shortest diameter through the center of mass of the lumen. *Maximum lumen diameter* is the longest diameter through the center of mass of the lumen. *Lumen eccentricity* is calculated as (maximum lumen diameter minus minimum lumen diameter) divided by maximum lumen diameter.

OCT-measured lumen CSA is well correlated with IVUS-measured lumen CSA. In both phantom models and in vivo study comparing quantitative coronary analysis (QCA) for angiography vs IVUS vs OCT measurements, OCT was most precise to the real value, and IVUS measurement was 8% larger than OCT measurement [6]. The mean minimum lumen diameter (MLD) measured by QCA was 5% smaller than that measured by OCT, and the minimum lumen diameter measured by IVUS was 9% greater than that measured by FD-OCT [6].

Previously several studies regarding IVUS criteria for defining the functional significance evaluated with fractional flow reserve (FFR) demonstrated that MLD had a good correlation with the FFR values, but the utility of IVUS MLA as an alternative to FFR to guide intervention in intermediate lesions may be limited in accuracy and vessel dependent [10–13]. Anatomical measurements of coronary stenosis obtained by OCT show significant correlation with FFR. OCT-derived parameters were smaller than those reported in previous IVUS studies (Table 12.2) [14, 15]. Recent study assessing computational fractional flow reserve from OCT in patient with intermediate stenosis showed promising approach of it in assessment not only of anatomic information but also of the functional significance of intermediate stenosis [16].

12 Interpretation of Optical Coherence Tomography: Quantitative Measurement

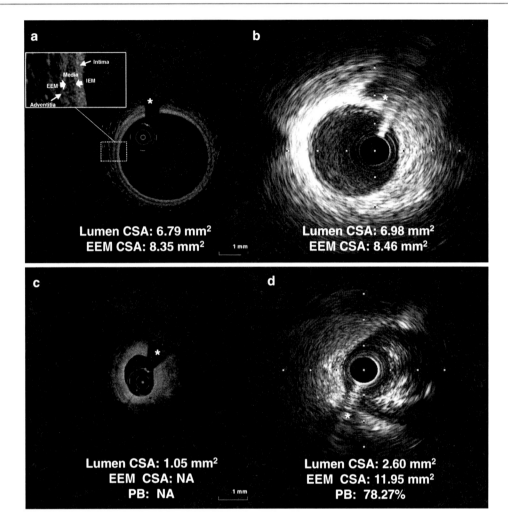

Fig. 12.1 Comparision of border detection between optical coherence tomography (OCT) and intravascular ultrasound (IVUS). Normal artery wall shows a 3-layered architecture, comprising a high backscattering, thin intima, a low backscattering media, a heterogeneous and/or high backscattering adventitia in both OCT (**a**) and IVUS (**b**). OCT could visualize internal elastic membrane (IEM) and external elastic membrane (EEM) (bold arrow heads) (inset, x3). The OCT-derived EEM or IEM measurement could not be made in cross-sectional image that contains diseased vessel (**c**) whereas IVUS demonstrate EEM border well (**d**). * represents wire artifact. *CSA* cross sectional area; *PB* plaque burden

Table 12.2 OCT-derived minimal lumen area predicting for physiologic significance assessed by fractional flow reserve

Study	Patients	FFR value	OCT	IVUS
Gonzalo et al. [14]	61 intermediate lesions in 56 patients	FFR < 0.8	1.95 mm² (AUC, 0.74; 95% CI, 0.61–0.84; sensitivity, 82%; specificity, 63%)	2.36 mm² (AUC, 0.63; 95% CI, 0.47–0.77, sensitivity, 67%; specificity 65%)
Shiono et al. [15]	62 intermediate lesions in 59 patients	FFR < 0.75	1.91 mm² (sensitivity, 94%; specificity, 77%)	NA

AUC area under curve; *CI* confidence interval; *FFR* fractional flow reserve; *IVUS* intravascular ultrasound; *OCT* optical coherence tomography

EEM Measurements *EEM CSA* is the area bounded by EEM border as a surrogated parameter for vessel area. A discrete interface at the border between the media and the adventitia is almost invariably present within OCT images and corresponds closely to the location of the EEM. Because of low penetration depth of OCT signal and rapid OCT signal attenuation within plaque, EEM circumference and area mostly cannot be measured reliably especially in lesion segment. If low signal involves a relatively small arc (<90°), planimetry of the circumference can be performed by extrapolation from the closest identifiable EEM borders, although measurement accuracy and reproducibility will be reduced.

Plaque (or Atheroma) Measurement *Plaque (or atheroma) CSA* is the EEM CSA minus the lumen CSA. *Maximum plaque (or atheroma) thickness* is the largest distance from the intimal leading edge to the EEM along any line passing through the center of mass of the lumen. *Minimum plaque (or atheroma) thickness* is the shortest distance from the intimal leading edge to the EEM along any line passing through the center of mass of the lumen. *Plaque (or atheroma) eccentricity* is calculated as (maximum plaque thickness minus minimum plaque thickness) divided by maximum plaque thickness. If EEM area could not be obtained, plaque measurement is not available.

Plaque Burden *Plaque (or atheroma) burden* is assessed as plaque CSA divided by the EEM CSA. This parameter can only be defined when the EEM can be demonstrated. The plaque burden is distinct from the luminal area stenosis. The former represents the area within the EEM occupied by atheroma regardless of lumen compromise. The latter is a measure of luminal compromise relative to a reference lumen analogous to the angiographic diameter stenosis. If EEM area cannot be obtained, plaque burden cannot be assessed.

Lumen Area Stenosis *Lumen area stenosis* is assessed as reference lumen CSA minus minimum lumen CSA divided by reference lumen CSA.

Plaque Component and Other Measurements The presence of specific component within the plaque or over the plaque, such as calcium, lipid, or thrombus, could be assessed as quantitative measurements like angle, depth, thickness, or area. *Angle or arc* could be measured using the center of mass of the lumen as the angle point. *Depth* is the distance between the lumen and the leading edge of the plaque feature. *Thickness* is usually assessed as the thickest distance between the inner and outer surfaces of the plaque component (valid only if the deep boundary can be identified). *Area* of some component could be described as the CSA of the plaque component (valid only if the deep boundary can be identified).

Fibrous cap thickness can be measured by the thickness of a cap present over OCT-delineated lipid or necrotic core either at the single cross-section where the fibrous cap thickness is considered minimal or from multiple samples (three or more). Although studies have been performed to compare the OCT measurement of fibrous cap thickness with histologic measurements of cap thickness, it was generally considered that this area needs further validation, as the boundary between the cap and the necrotic core is not always straightforward to precisely determine.

Remodeling An index of remodeling can be assessed as lesion EEM CSA/reference EEM CSA, if the EEM CSA is identified in OCT image.

Because of its limited tissue penetration, OCT does not appear to be suited to study vessel remodeling.

12.3 Stent Measurements

OCT has been considered as an useful intracoronary imaging modality for the lesion assessment, stent sizing, and stent optimization during PCI (Figs. 12.2 and 12.3). The Clinical usefulness of OCT-guided PCI will be discussed in next chapter (Chap. 13).

OCT is capable of visualizing the vascular response between stent strut and vessel wall, and

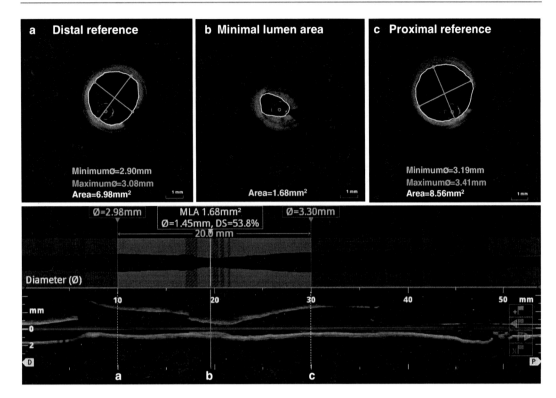

Fig. 12.2 Pre-percutaneous coronary intervention optical coherence tomography measurements. Both proximal and distal reference was obtained at normal looking segments from longitudinal and cross-sectional images. Mean reference lumen diameter (ø) was assessed from the each cross-sectional images (**a** and **c**) and the lesion length was the distance between proximal and distal reference segments. Minimum lumen area (*MLA*) was measured by detection of lumen border at the most narrowest site (**b**)

identifying tissues surrounding stent struts. Most metallic stent struts have strong reflection to optic signal creating a bright hyperintense signal at the surface of strut (blooming appearance) with a shadow that obscures deeper structure within the vessel. The polymeric struts of bioabsorbable vascular scaffolds are transparent to the optic signal, allowing visualization of the vascular wall structure behind the struts without shadowing (Fig. 12.4).

Strut assessment is limited by the axial resolution of the OCT system, and OCT could not allow the visualization of a single layer of endothelial cells. Furthermore the biological and clinical significance of some OCT-derived stent measurements within stent segment has not been fully understood. Recently a retrospective data evaluating OCT measurements to predict very late stent thrombosis demonstrated that malapposition, neoatherosclerosis, uncovered struts, and stent underexpansion, without differences between patients treated with early- and new-generation drug-eluting stents, were leading OCT findings associated with very late stent thrombosis in descending order [17].

12.3.1 Stented Segment

Stent Area Measurements *Stent CSA* is the area bounded by the stent border. *Minimum stent diameters* are the shortest diameter through the center point of the stent. *Maximum stent diameters* are the shortest and the longest diameter through the center point of the stent. *Stent eccentricity (symmetry)* is calculated as (maximum stent diameter minus minimum stent diameter) divided by maximum stent diameter.

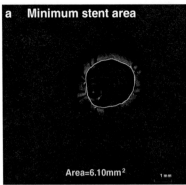
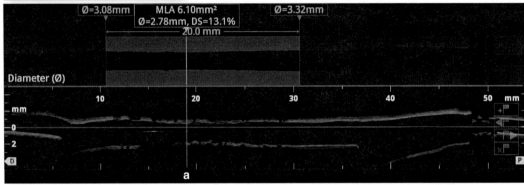

Fig. 12.3 Post-percutaneous coronary intervention optical coherence tomography measurements. The minimum stent area was obtained by detection of stent border at the most narrowest cross-sectional area (a). *MLA* minimum lumen area

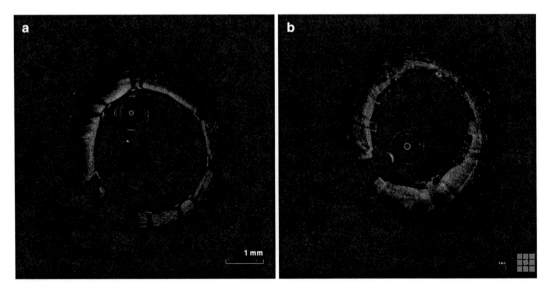

Fig. 12.4 Comparison of stent struts detection between metal stent and bioabsorbable vascular scaffolds (BVS) assessed by optical coherence tomography. Metallic drug eluting stent struts have strong reflection to optic signal creating a bright hyper-intense signal at the surface of strut (blooming appearance) with a shadow that obscures deeper structure within the vessel (**a**). BVS are transparent to the optic signal, allowing visualization of the vascular wall structure behind the struts without shadowing (**b**)

Stent expansion is calculated as the minimum stent CSA compared with the predefined reference area which can be the average reference lumen area or EEL area if possible. An underexpanded stent has an in-stent minimal lumen area less than 90% of the average reference lumen area. In the CLI-THRO study, which compared OCT parameters between in patient with subacute stent thrombosis and in those without, stent thrombus group had smaller OCT stent CSA (5.6 ± 2.6 vs 6.8 ± 1.7 mm^2, $p = 0.03$) and higher incidence of stent underexpansion (42.8% vs 16.7%, $p = 0.05$) when compared with control group [18].

12.3.2 Stent Strut Measurements

OCT has been considered as the most useful intracoronary imaging modality to assess immediate- and long- term vascular response after stent implantation. Stent strut measurements can be obtained at a cross-section level or can be evaluated at the strut level analysis. (Fig. 12.5). The assessment of stent struts requires strict interval ranging from every 0.5 mm to 1 mm to obtain high rate of reproducibility. Stent strut maps can be computed with the *x* axis representing the length of the stent (millimeters) and the *y* axis representing the circumference (0–360°).

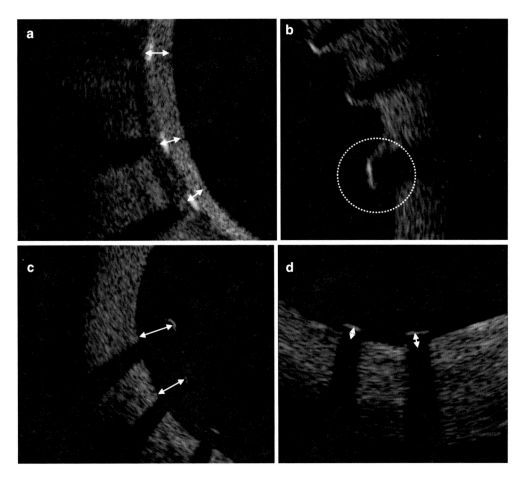

Fig. 12.5 Example of stent strut coverage and apposition assessment. (**a**) shows three covered struts and (**b**) represents an uncovered strut. In (**c**) and (**d**) examples of malapposed and apposed struts are, respectively, presented

A contour plot optical coherence tomography analysis for evaluating stent strut may provide more useful information to understanding the serial changes in strut coverage [19].

Stent Apposition Incomplete stent apposition or malapposition occurs if there is a separation of a stent strut from the vessel wall. Malapposition is defined as a measured distance greater than the strut thickness for stent materials (metal or metal plus polymer). *Malapposition distance* could be measured as the distance between the luminal surface of the covering tissue and the luminal surface of the strut. The area between the endoluminal midpoint of the struts and the vessel wall was measured as *malapposition area*.

Acute, late-persistent, and late-acquired stent malapposition assessed by OCT has relatively high incidences, but their clinical importance and the mechanism have been shown different. The clinical outcome of acute malapposition is favorable, but late malapposition has been considered as a predictor of stent thrombosis [20].

Strut Coverage and Neointima Measurements
Strut coverage thickness is the distance between the luminal surface of the covering tissue and the luminal surface of the strut. *Percentage of uncovered stent struts* is calculated as the number of struts without distinct overlying tissue, in which the luminal reflection of the strut surface is directly interfacing with the lumen, divided by total number of analyzable struts.

Variable thickness of stent struts which consisted of metal and polymer should be considered to determine whether struts are "covered" or "uncovered." OCT cannot visualize a single layer of endothelium over the strut, or it does not demonstrate accurate nature of tissue. In a case-controlled study, the presence of uncovered stent struts assessed by OCT was associated with late stent thrombosis after DES implantation [21]. Won et al. showed that the best cutoff value of percentage of uncovered struts for predicting major safety events (a composite occurrence of cardiovascular death, myocardial infarction, and stent thrombosis) was 5.9% using the maximal χ [2] method (area under the receiver-operating characteristic curve, 0.779; 95% confidence interval, 0.648–0.910; $p = 0.019$, a sensitivity of 83.3% and a specificity of 70.3%) [22].

Neointima area is defined as stent CSA minus lumen CSA. *Percent neointima area* is defined as (neointima area divided by stent CSA) X 100. The qualitative assessment of the neointima pattern is assessed at the site of the largest cross-sectional area of neointima within stent. OCT has been considered as the best tool to evaluate tissue characterization of neointima, and it also could discriminate neoatherosclerosis from intimal hyperplasia by OCT qualitative measurements.

The rates of stent strut coverage or the characteristics of neointima assessed by OCT surveillance differed according to stent type. The clinical implications of these differences require further study but may imply on the differences in rates of stent thrombosis observed in clinical trials with different stent types [23–26].

12.4 Length and Volume Measurements

OCT image acquisition is performed using motorized transducer pullback commonly at 100 frames/s with an automatic pullback speed of 20 mm/s. Longitudinal view is obtained automatically, and length measurements can be assessed from longitudinal view or calculated as the number of seconds by the pullback speed. This approach can be used to determine the length of a lesion, stenosis, stent, or any other longitudinal features (calcium, lipid, thrombus, etc.). OCT offered more accuracy than IVUS in longitudinal geometric measurement of coronary artery [27].

Lesion length is determined as the distance from distal to proximal reference site using the OCT automated lumen detection feature. Stent length is determined as the distance from distal to proximal edge of stent using the OCT automated lumen detection feature. Length measurements of any longitudinal features can be performed using motorized transducer pullback (number of seconds x pullback speed).

Volume measurements are calculated by Simpson's rule and area measurements from every single frame usually at 0.5–1 mm.

References

1. Mintz GS, Nissen SE, Anderson WD, Rosenfield K, Bailey SR, Siegel RJ, et al. American College of Cardiology clinical expert consensus document on standards for acquisition, measurement and reporting of intravascular ultrasound studies (IVUS): a report of the American College of Cardiology task force on clinical expert consensus documents developed in collaboration with the European Society of Cardiology endorsed by the Society of Cardiac Angiography and Interventions. J Am Coll Cardiol. 2001;37:1478–92.
2. Prati F, Regar E, Mintz GS, Arbustini E, Di Mario C, Jang IK, et al. Expert review document on methodology, terminology, and clinical applications of optical coherence tomography: physical principles, methodology of image acquisition, and clinical application for assessment of coronary arteries and atherosclerosis. Eur Heart J. 2010;31:401–15.
3. Prati F, Guagliumi G, Mintz GS, Costa M, Regar E, Akasaka T, et al. Expert review document part 2: methodology, terminology and clinical applications of optical coherence tomography for the assessment of interventional procedures. Eur Heart J. 2012;33:2513–20.
4. Tearney GJ, Regar E, Akasaka T, Adroaemssems T, Barlis P, Bezerra HG, et al. Consensus standards for acquisition, measurement, and reporting of intravascular optical coherence tomography studies: a report from the international working group for intravascular optical coherence tomography standardization and validation. J Am Coll Cardiol. 2012;59:1058–72.
5. Gerbaud E, Weisz G, Tanaka A, Kashiwagi M, Shimizu T, Wang L, et al. Multi-laboratory inter-institute reproducibility study of IVOCT and IVUS assessments using published consensus document definitions. Eur Heart J Cardiovasc Imaging. 2016;17:756–64.
6. Kubo T, Akasaka T, Shite J, Suzuki T, Uemura S, Yu B, et al. OCT compared with IVUS in a coronary lesion assessment: the OPUS-CLASS study. JACC Cardiovasc Imaging. 2013;6:1095–104.
7. Tanimoto S, Rodriguez-Granillo G, Barlis P, de Winter S, Bruining N, Hamers R, et al. A novel approach for quantitative analysis of intracoronary optical coherence tomography: high inter-observer agreement with computer-assisted contour detection. Catheter Cardiovasc Interv. 2008;72:228–35.
8. Kubo T, Shinke T, Okamura T, Hibi K, Nakazawa G, Morino Y, et al. Optical frequency domain imaging vs. intravascular ultrasound in percutaneous coronary intervention (OPINION trial): study protocol for a randomized controlled trial. J Cardiol. 2016;68:455–60.
9. Ali ZA, Maehara A, Généreux P, Shlofmitz RA, Fabbiocchi F, Nazif TM, et al. Optical coherence tomography compared with intravascular ultrasound and with angiography to guide coronary stent implantation (ILUMIEN III: OPTIMIZE PCI): a randomised controlled trial. Lancet. 2016;388:2618–28.
10. Ben-Dor I, Torguson R, Deksissa T, Bui AB, Xue Z, Satler LF, et al. Intravascular ultrasound lumen area parameters for assessment of physiological ischemia by fractional flow reserve in intermediate coronary artery stenosis. Cardiovasc Revasc Med. 2012;13:177–82.
11. Kang SJ, Lee JY, Ahn JM, Mintz GS, Kim WJ, Park DW, et al. Validation of intravascular ultrasound-derived parameters with fractional flow reserve for assessment of coronary stenosis severity. Circ Cardiovasc Interv. 2011;4:65–71.
12. Koo BK, Yang HM, Doh JH, Choe J, Lee SY, Yoon CH, et al. Optimal intravascular ultrasound criteria and their accuracy for defining the functional significance of intermediate coronary stenoses of different locations. JACC Cardiovasc Interv. 2011;4:803–11.
13. Waksman R, Legutko J, Singh J, Orlando Q, Marso S, Schloss T, et al. FIRST: fractional flow reserve and intravascular ultrasound relationship study. J Am Coll Cardiol. 2013;61:917–23.
14. Gonzalo N, Escaned J, Alfonso F, Nolte C, Rodriques V, Jimenez-Quevedo P, et al. Morphometric assessment of coronary stenosis relevance with optical coherence tomography: a comparison with fractional flow reserve and intravascular ultrasound. J Am Coll Cardiol. 2012;59:1080–9.
15. Shiono Y, Kitabata H, Kubo T, Masuno T, Ohta S, Ozaki Y, et al. Optical coherence tomography-derived anatomical criteria for functionally significant coronary stenosis assessed by fractional flow reserve. Circ J. 2012;76:2218–25.
16. Ha J, Kim JS, Lim J, Kim G, Lee S, Lee JS, et al. Assessing computational fractional flow reserve from optical coherence tomography in patients with intermediate coronary stenosis in the left anterior descending artery. Cir Cardiovasc Interv. 2016;9(8):e003613. doi:10.1161/CIRCINTERVENTIONS.116.003613.
17. Taniwaki M, Radu MD, Zaugg S, Amabile N, Carcia-Carcia HM, Yamaji K, et al. Mechanisms of very late drug-eluting stent thrombosis assessed by optical coherence tomography. Circulation. 2016;133:650–60.
18. Prati F, Kodama T, Romagnoli E, Gatto L, Di Vito L, Ramazzotti V, et al. Suboptimal stent deployment is associated with subacute stent thrombosis: optical coherence tomography insights from a multicenter matched study. From the CLI foundation investigators: the CLI-THRO study. Am Heart J. 2015;169:249–56.

19. Kim JS, Ha J, Kim BK, Shin DH, Ko YG, Choi D, et al. The relationship between post-stent strut apposition and follow-up strut coverage assessed by a contour plot optical coherence tomography analysis. JACC Cardiovasc Interv. 2014;7:641–51.
20. Im E, Kim BK, Ko YG, Shin DH, Kim JS, Choi D, et al. Incidences, predictors, and clinical outcomes of acute and late stent malapposition detected by optical coherence tomography after drug-eluting stent implantation. Circ Cardiovasc Interv. 2014;7:88–96.
21. Guagliumi G, Sirbu V, Musumeci G, Gerber R, Biondi-Zoccai G, Ikejima H, et al. Examination of the in vivo mechanisms of late drug-eluting stent thrombosis: findings from optical coherence tomography and intravascular ultrasound imaging. JACC Cardiovasc Interv. 2012;5:12–20.
22. Won H, Shin DH, Kim BK, Mintz GS, Kim JS, Ko YG, et al. Optical coherence tomography derived cut-off value of uncovered stent struts to predict adverse clinical outcomes after drug-eluting stent implantation. Int J Cardiovasc Imaging. 2013;29:1255–63.
23. Kim JS, Jang IK, Fan C, Kim TH, Kim JS, Park SM, et al. Evaluation in 3 months duration of neointimal coverage after zotarolimus-eluting stent implantation by optical coherence tomography: the ENDEAVOR OCT trial. JACC Cardiovasc Interv. 2009;2:1240–7.
24. Lee KS, Lee JZ, Hsu CH, Husnain M, Riaz H, Riaz IB, et al. Temporal trends in strut-level optical coherence tomography evaluation of coronary stent coverage: a systematic review and meta-analysis. Catheter Cardiovasc Interv. 2016;88:1083–93.
25. Matsumoto D, Shite J, Shinke T, Otake H, Tanino Y, Ogasawara D, et al. Neointimal coverage of sirolimus-eluting stents at 6-month follow-up: evaluated by optical coherence tomography. Eur Heart J. 2007;28:961–7.
26. Toledano Delgado FJ, Alvarez-Ossorio MP, de Lezo Cruz-Conde JS, Bellido FM, Romero Moreno MA, Femandez-Aceytuno AM, et al. Optical coherence tomography evaluation of late strut coverage patterns between first-generation drug-eluting stents and everolimus-eluting stent. Catheter Cardiovasc Interv. 2014;84:720–6.
27. Liu Y, Shimamura K, Kubo T, Tanaka A, Kitabata H, Ino Y, et al. Comparison of longitudinal geometric measurement in human coronary arteries between frequency-domain optical coherence tomography and intravascular ultrasound. Int J Cardiovasc Imaging. 2014;30:271–7.

13

Qualitative Assessments of Optical Coherence Tomography

Ae-Young Her and Yong Hoon Kim

Optical coherence tomography (OCT) is a recently developed high-resolution imaging technique that has a homoaxial resolution of 10 μm and a lateral resolution of 20 μm, which is about 10 times higher than that of any clinically available diagnostic imaging modality [1]. Therefore, OCT is capable of resolving microstructural features of atherosclerotic plaques such as thin fibrous cap, lipid core, and intracoronary thrombus, which are thought to be responsible for plaque vulnerability [2, 3]. In this chapter, the qualitative assessments of characterizing atherosclerotic plaques and plaque rupture triggering coronary thrombosis in native lesion will be introduced.

13.1 Qualitative Assessment of Atherosclerosis

Table 13.1 summarizes the appearance of atherosclerotic components by OCT and intravascular ultrasound (IVUS). The OCT interpretation of the images was based on the results of in vitro studies [4–7].

All plaques identified by OCT are characterized by the loss of the layered structure observed in normal vessels or vessels with intimal hyperplasia. As the various components of atherosclerotic plaques have different optical properties, OCT makes it possible to differentiate them to a great extent. Identification of plaque components by OCT depends on the penetration depth of the incident light beam into the vessel wall. The depth of penetration is greatest for fibrous tissue and least for thrombi with calcium and lipid tissue having intermediate values [3, 8, 9].

Calcifications within plaques are identified by the presence of well-delineated, low-backscattering heterogeneous regions (Figs. 13.1 and 13.2) [3, 6–9]. Superficial microcalcifications, considered to be a distinctive feature of plaque vulnerability, are revealed as small superficial calcific deposit. The contrast between calcifications and the surrounding vessel wall is often well-defined in IVUS images. However, the bright IVUS signal from calcifications can cause difficulty in accurate assessment of neighboring plaque composition due to saturation artifact. In contrast, OCT images allow improved evaluation of the extent of calcifications within plaques and visualization of plaque microstructure adjacent to calcifications. *Fibrous plaques* are typically rich in collagen or muscle cells and consist of homogeneous high-backscattering area (Figs. 13.2 and 13.3) [3, 6–9]. *Necrotic lipid pools* are less well-delineated than calcifications and exhibit decreased signal density and more heterogeneous backscattering than fibrous plaques (Fig. 13.2) [3, 6–9]. The strong contrast

A.-Y. Her • Y.H. Kim (✉)
Division of Cardiology, Department of internal Medicine, Kangwon National University School of Medicine, Chuncheon, South Korea
e-mail: yhkim02@kangwon.ac.kr

Table 13.1 Image features of optical coherence tomography vs. intravascular ultrasound by histopathologic findings [2, 4–7]

Histopathologic findings	Image features	
	OCT	IVUS
Calcification	Heterogeneous	Very high reflectivity
	Sharply well-delineated	Shadowing
	Low reflectivity	
	Low attenuation	
Fibrous plaque	Homogeneous	Homogeneous
	High reflectivity	High reflectivity
	Low attenuation	
Lipid pool	Homogeneous	Low backscatter
	Less well-delineated	
	High reflectivity	
	High attenuation	
White thrombus	Medium reflectivity	
	Low attenuation	
Red thrombus	Medium reflectivity	Medium-high reflectivity
	High attenuation	

OCT optical coherence tomography, *IVUS* intravascular ultrasound

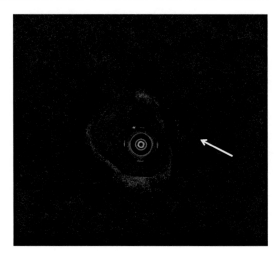

Fig. 13.1 Example of calcifications within plaque. It is identified by well-delineated, low-backscattering heterogeneous regions (*arrow*)

OCT images, correct image interpretation depends on the experience of the observer and on the penetration depth into the tissue [3, 11]. Alternatively, a variety of approaches for quantitative tissue analysis are under development to further improve diagnostic accuracy in an observer-independent way [12].

13.2 Plaque Rupture and Intracoronary Thrombosis

In the presence of thrombosis, three distinct OCT morphologies can be detected: (1) either massive thrombosis or any amount of red thrombus that does not permit assessment of vessel and plaque morphology; (2) thrombosis with signs of ulceration underneath, or (3) thrombosis with apparently normal endothelial lining underneath that may be indicative of erosion [7]. However, a firm diagnosis of erosion cannot be made without knowledge of the morphological or functional alterations of a single endothelial layer that cannot be assessed directly with OCT.

Plaque dissections are common findings associated with ruptured plaques visualized by OCT. They are identified as rims of tissue protruding into the lumen [13] (Fig. 13.6). *Plaque ulceration or rupture* can be detected by OCT as a ruptured fibrous

between lipid-rich cores and fibrous regions in OCT images allows fibrous caps to be easily identified.

Intracoronary thrombi might take a critical role in the pathogenesis and the clinical manifestations of acute myocardial infarction (AMI). But coronary angiography and IVUS cannot reliably identify thrombus, and OCT is able to visualize the intracoronary thrombus clearly [8]. Thrombi are identified by the masses protruding into the vessel lumen discontinuous from the surface of the vessel wall. *White thrombi* consist mainly of platelets and white blood cells and are characterized by a signal-rich, low-backscattering billowing projections protruding into the lumen (Figs. 13.4 and 13.5). *Red thrombi* consist mainly of red blood cells, and relevant OCT images are characterized as high-backscattering protrusions with signal-free shadowing (Figs. 13.2 and 13.5) [10].

Although previous results have reported good accuracy as well as inter- and intra-observer agreement for visual plaque characterization by

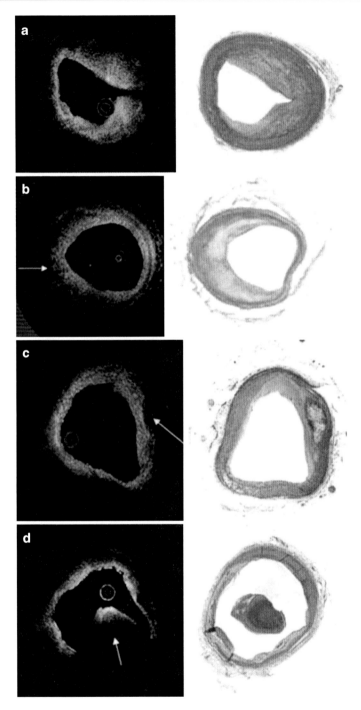

Fig. 13.2 Optical coherence tomography examples of plaque composition (*left panels*) and corresponding histology (*right panels*). (**a**) Optical coherence tomography image of plaque consists of predominantly fibrotic plaque documented by histology. (**b**) Optical coherence tomography image of a plaque with a lipid pool (*arrow*) documented by histology. (**c, d**) Optical coherence tomography image of a calcific component (*arrow* in **c**) and thrombus (*arrow* in **d**)

cap that connects the lumen with the lipid pool. These ulcerated or ruptured plaques may occur with or without a superimposed thrombus. The use of thrombolysis, IIb–IIIa glycoprotein inhibitors, or other antithrombotic drugs facilitates clot degradation and in some circumstances may lead to complete disappearance (Fig. 13.7).

Identification of *erosion* as a mechanism of plaque instability is a challenge even for a technique with a resolution below 20 μm. Validation studies combining OCT with techniques providing a functional assessment of the endothelium may be able to give us more information on vessel thrombosis induced by erosion. OCT is able to evaluate the plaque erosion clearly, and the prevalence of plaque erosion was 23% in patients with AMI in the previous study [8]. Plaque erosion is characterized by loss of the endothelial lining with lacerations of the superficial intimal layers and without "trans-cap" ruptures (Fig. 13.8).

13.3 Controversial Points in Optical Coherence Tomography of Atherosclerosis

Neovascularization or angioneogenesis within coronary atheromas has been known to accelerate coronary atherosclerosis via various mechanisms, including transportation of nourishment to the intima, stimulation of vascular inflammation, and microvascular hemorrhage or leakage. OCT imaging can provide cross-sectional in vivo images of neovascular microchannel formation at micrometer resolution. Despite the lack of specific validation studies, there is a general consensus that microvessels in plaque appear as thin black holes with a diameter of 50–100 μm that are present for at least 3–4 consecutive frames in pullback images [14] (Fig. 13.9).

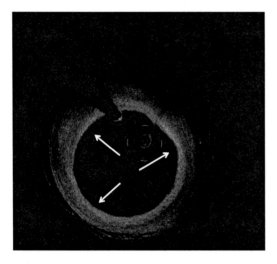

Fig. 13.3 Example of fibrous plaque. Optical coherence tomography has the potential to identify dense fibrotic tissue (*arrows*)

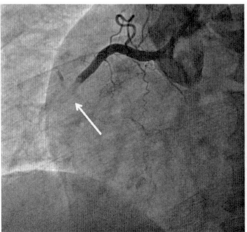

Fig. 13.4 Example of white thrombus. Culprit lesion in the right coronary artery (*arrow* in the *left panel*). White thrombus is platelet rich and exhibits a low signal attenuation (*arrowhead* in the *right panel*)

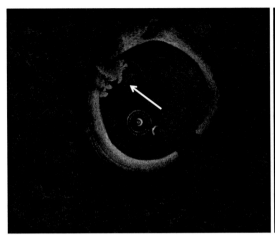 

Fig. 13.5 Example of white and red thrombus. White thrombus is characterized by a signal-rich, low-backscattering protrusions (*arrow* in the *left panel*), while red thrombus, due to presence of red blood cell components, causes a marked signal attenuation with high-backscattering protrusions (*arrow* in the *right panel*)

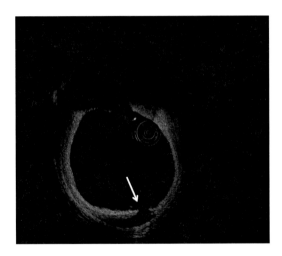

Fig. 13.6 Example of plaque dissection. Optical coherence tomography image shows the plaque dissection with a clear rim of tissue protrusions (*arrow*)

The identification of *plaque hemorrhage* and the link between plaque hemorrhage and plaque vulnerability are major issues that will require additional validation studies. Further studies will be required to assess the ability of OCT to separate recent from old hemorrhagic areas and to clarify if hemosiderin can be distinguished from calcific components.

OCT has the potential to identify *inflammatory cells* such as clusters of lymphocytes. Previous studies showed that application of OCT algorithm can identify inflammatory cells with high specificity and sensitivity [12, 15]. Although this dedicated algorithm may be instrumental to identify and possible quantify plaque inflammation, it is felt that the applicability of these results to real-time imaging is questionable because the detection and quantification algorithms developed in these studies depend on accurate selection of the region of interest.

13.4 Summary

The high resolution of OCT when compared with IVUS enables the identification of the main plaque components that include lipid pools, calcium, fibrotic tissue, and thrombus. However, identification of individual plaque components by OCT requires experience; in other words, a careful analysis of optical properties of the plaque components must be done. The application of new software-based algorithm or other optical tissue properties should improve the characterization of atherosclerotic coronary plaques and provide a more objective assessment.

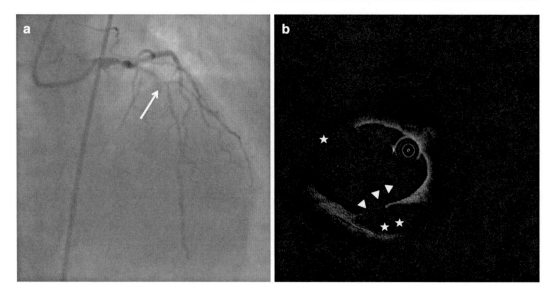

Fig. 13.7 Clinical example of plaque rupture. (**a**) Coronary angiography shows culprit lesion (*arrow*) in the left anterior descending artery in a patient with non-ST elevation myocardial infarction. (**b**) Optical coherence tomography image reveals the ruptured thin cap fibroatheroma (*arrowheads*) in the shoulder region of the plaque. Remnants of the necrotic core are in direct contact with the blood stream (*double stars*). Star indicates guidewire artifacts

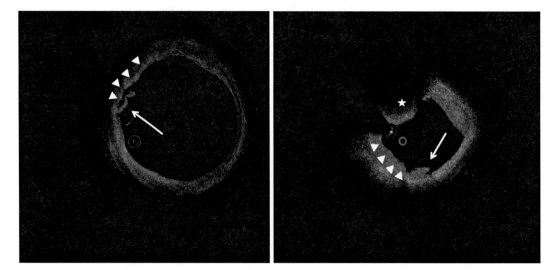

Fig. 13.8 Examples of typical plaque erosion. Plaque erosion in images of optical coherence tomography. Erosion located on the surface of a plaque (*arrowheads*) with intraluminal white thrombi (*arrows*) and red thrombus (*star*)

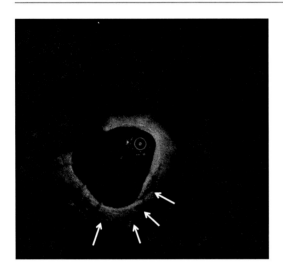

Fig. 13.9 Example of neovascularization. Presence of thin black holes (*arrows*) at optical coherence tomography image. These holes have a diameter of 50–100 μm and are likely due to plaque angioneogenesis

References

1. Huang D, Swanson EA, Lin CP, Schuman JS, Stinson WG, Chang W, et al. Optical coherence tomography. Science. 1991;254:1178–81.
2. Jang IK, Tearney GJ, MacNeill B, Takano M, Moselewski F, Iftima N, et al. In vivo characterization of coronary atherosclerotic plaque by use of optical coherence tomography. Circulation. 2005;111:1551–5.
3. Yabushita H, Bouma BE, Houser SL, Aretz HT, Jang IK, Schlendorf KH, et al. Characterization of human atherosclerosis by optical coherence tomography. Circulation. 2002;106:1640–5.
4. Brezinski ME, Tearney GJ, Bouma BE, Izatt JA, Hee MR, Swanson EA, et al. Optical coherence tomography for optical biopsy. Properties and demonstration of vascular pathology. Circulation. 1996;93:1206–13.
5. Fujimoto JG, Boppart SA, Tearney GJ, Bouma BE, Pitris C, Brezinski ME. High resolution in vivo intra-arterial imaging with optical coherence tomography. Heart. 1999;82:128–33.
6. Jang IK, Bouma BE, Kang DH, Park SJ, Park SW, Seung KB, et al. Visualization of coronary atherosclerotic plaques in patients using optical coherence tomography: comparison with intravascular ultrasound. J Am Coll Cardiol. 2002;39:604–9.
7. Prati F, Regar E, Mintz GS, Arbustini E, Di Mario C, Jang IK, et al. Expert review document on methodology, terminology, and clinical applications of optical coherence tomography: physical principles, methodology of image acquisition, and clinical application for assessment of coronary arteries and atherosclerosis. Eur Heart J. 2010;31:401–15.
8. Kubo T, Imanishi T, Takarada S, Kuroi A, Ueno S, Yamano T, et al. Assessment of culprit lesion morphology in acute myocardial infarction: ability of optical coherence tomography compared with intravascular ultrasound and coronary angioscopy. J Am Coll Cardiol. 2007;50:933–9.
9. Kume T, Akasaka T, Kawamoto T, Watanabe N, Toyota E, Neishi Y, et al. Assessment of coronary arterial plaque by optical coherence tomography. Am J Cardiol. 2006a;97:1172–5.
10. Kume T, Akasaka T, Kawamoto T, Ogasawara Y, Watanabe N, Toyota E, et al. Assessment of coronary arterial thrombus by optical coherence tomography. Am J Cardiol. 2006b;97:1713–7.
11. Manfrini O, Mont E, Leone O, Arbustini E, Eusebi V, Virmani R, et al. Sources of error and interpretation of plaque morphology by optical coherence tomography. Am J Cardiol. 2006;98:156–9.
12. Tearney GJ, Yabushita H, Houser SL, Aretz HT, Jang IK, Schlendorf KH, et al. Quantification of macrophage content in atherosclerotic plaques by optical coherence tomography. Circulation. 2003;107:113–9.
13. Prati F, Cera M, Ramazzotti V, Imola F, Giudice R, Albertucci M. Safety and feasibility of a new non-occlusive technique for facilitated intracoronary optical coherence tomography (OCT) acquisition in various clinical and anatomical scenarios. EuroIntervention. 2007;3:365–70.
14. Taruya A, Tanaka A, Nishiguchi T, Matsuo Y, Ozaki Y, Kashiwagi M, et al. Vasa vasorum restructuring in human atherosclerotic plaque vulnerability: a clinical optical coherence tomography study. J Am Coll Cardiol. 2015;65:2469–77.
15. Raffel OC, Tearney GJ, Gauthier DD, Halpern EF, Bouma BE, Jang IK. Relationship between a systemic inflammatory marker, plaque inflammation, and plaque characteristics determined by intravascular optical coherence tomography. Arterioscler Thromb Vasc Biol. 2007;27:1820–7.

Clinical Evidence of Optical Coherence Tomography-Guided Percutaneous Coronary Intervention

14

Seung-Yul Lee, Yangsoo Jang, and Myeong-Ki Hong

Although the guideline of the European Society of Cardiology has recommended that optical coherence tomography (OCT) may be considered in selected patients during percutaneous coronary intervention (PCI) [1], data regarding OCT guidance are limited (Table 14.1). In this chapter, clinical evidences and benefits of OCT-guided PCI will be discussed.

14.1 Lesional Assessment

The accurate measurement of lumen dimensions is important for assessing the severity of coronary stenoses. In the study comparing the lumen measurement obtained ex vivo in human coronary arteries using intravascular ultrasound, OCT and histomorphometry, and in vivo in patients using intravascular ultrasound and OCT with and without balloon occlusion, both intravascular ultrasound and OCT overestimated the lumen area compared with histomorphometry

S.-Y. Lee
Department of internal Medicine, Sanbon Hospital, Wonkwang University College of Medicine, Gunpo, South Korea

Y. Jang • M.-K. Hong (✉)
Division of Cardiology, Severance Cardiovascular Hospital, Yonsei University College of Medicine, Seoul, South Korea
e-mail: mkhong61@yuhs.ac

Table 14.1 Recent recommendations regarding the usage of optical coherence tomography (OCT) during percutaneous coronary intervention

Recommendations	Class of recommendation	Level of evidence
European Society of Cardiology [1]		
OCT to assess mechanisms of stent failure	IIa	C
OCT in selected patients to optimize stent implantation	IIb	C
ACCF/AHA/SCAI [2]		
Not documented		

ACCF American College of Cardiology Foundation; *AHA* American Heart Association; *SCAI* Society for Cardiovascular Angiography and Interventions

(mean difference 0.8 mm^2 for OCT and 1.3 mm^2 for intravascular ultrasound) [3]. The lumen dimensions in vivo obtained using intravascular ultrasound were larger than those obtained using OCT (mean difference 1.67 mm^2 for intravascular ultrasound relative to OCT with balloon occlusion and 1.11 mm^2 relative to OCT without balloon occlusion) [3]. OCT has a moderate diagnostic efficiency in identifying hemodynamically severe coronary stenoses with fractional flow reserve (FFR) ≤ 0.80 measured with pressure wire (sensitivity = 82%, specificity = 63%) [4]. The optimal cutoff value associated with FFR ≤ 0.80 was 1.95 mm^2 of minimal lumen area [4]. Like intravascular ultrasound, the

Table 14.2 Optical coherence tomographic criteria for defining severe coronary stenosis evaluated by fractional flow reserve (FFR)

Authors	No. of lesions	FFR	Minimal lumen area	Sensitivity	Specificity
Shiono et al. [5]	62	0.75	1.91 mm^2	94%	77%
Gonzalo et al. [4]	61	0.80	1.95 mm^2	82%	63%
Pawlowski et al. [6]	71	0.80	2.05 mm^2	75%	90%
Reith et al. [7]	62	0.80	1.59 mm^2	76%	79%

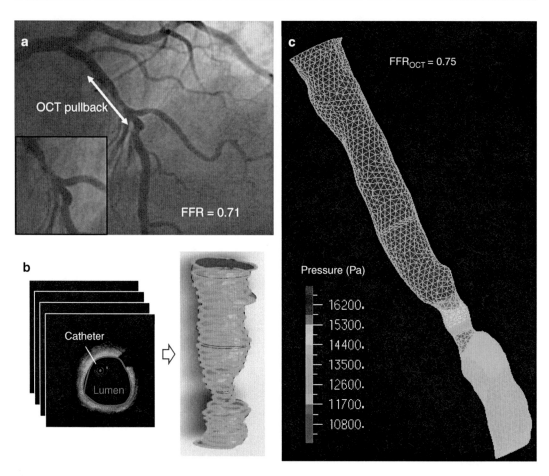

Fig. 14.1 Representative images of computational flow dynamics model and fractional flow reserve (FFR) simulation. Coronary angiography (**a**) showed a moderate stenosis (*arrows*) of the proximal segment of left anterior descending artery. The measured FFR of the lesion was 0.71, indicating functionally significant stenosis. After three-dimensional reconstruction (**b**) was performed using optical coherence tomography (OCT), computational flow dynamics model was applied to the acquired geometry (**c**). The calculated FFR of the lesion was 0.75

assessment of OCT is not specific for identifying severe stenoses, thus limiting the positive predictive value [4]. Table 14.2 summarizes the cutoff values of OCT-derived minimal lumen area that correspond to functionally significant stenosis. Recently, an OCT study reported that there was a moderate correlation between OCT-derived FFR measurements using computational fluid dynamics algorithm and direct FFR measurements using pressure wire ($r = 0.72$, $p < 0.001$) in patients with intermediate coronary stenosis in the left anterior descending coronary

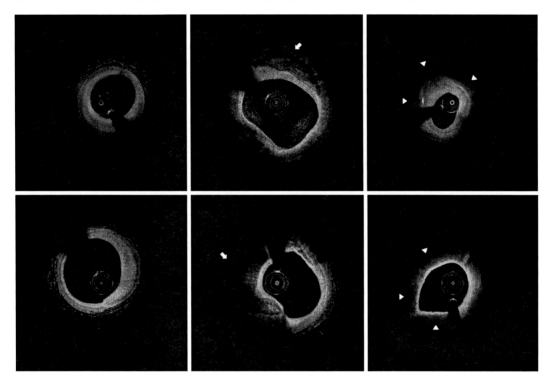

Fig. 14.2 Morphological features of fibrotic (*left panel*), fibrocalcific (*middle panel*), and lipid-rich (*right panel*) plaques at the segment with minimal lumen area. Fibrotic plaques had high backscattering and relatively homogeneous optical signal. Fibrocalcific plaques showed signal-poor heterogeneous region with well-delineated borders, being consistent with calcium (*arrows*). Lipid-rich plaques demonstrated signal-poor regions with poorly delineated borders, indicating lipid (*arrowheads*)

artery, as represented in Fig. 14.1 [8]. This OCT approach without use of pressure wire may be useful for evaluating the simultaneous functional and anatomic severity of coronary stenosis [8]. However, further studies are required to establish its feasibility and effectiveness. In contrast to determination of functionally significant severity of coronary artery, the OCT examination is reliably sensitive and specific for characterizing different types of atherosclerotic plaques: fibrous, fibrocalcific, and lipid-rich plaques (Fig. 14.2) [9, 10]. Morphological features detected by OCT were associated with the occurrence of post-interventional complications. The presence of thin-cap fibroatheroma identified by OCT was a predictor of post-PCI myocardial infarction [11]. Figure 14.3 represents a typical case that showed post-PCI myocardial infarction in patient treated with elective stent implantation.

14.2 Stent Optimization

The optimal OCT criteria for stent deployment have not been established yet. In the CLI-OPCI (Centro per la Lotta contro l'Infarto-Optimisation of Percutaneous Coronary Intervention) study, the reference lumen narrowing had to be greater than 4 mm^2, and the stent-lumen distance, namely, malapposed distance, had to be less than 200 μm for optimal stenting [12]. However, this study was retrospective, and the decision as to whether to perform further actions if the OCT criteria were not satisfied was left at the operator's discretion [12]. In the multicenter, random-

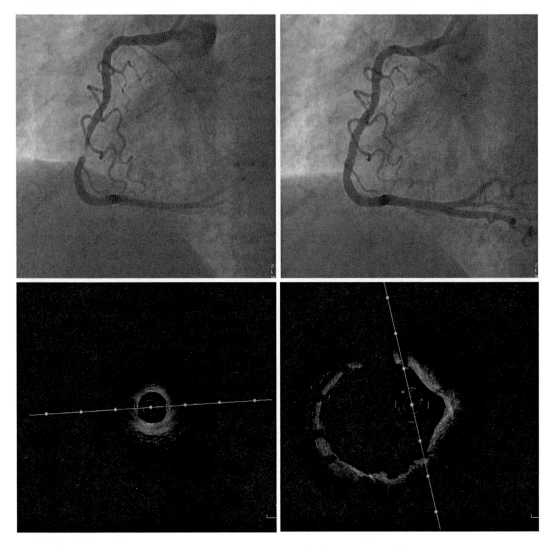

Fig. 14.3 A case showing post-interventional myocardial infarction after successful stent implantation. There was a tight narrowing at midportion of the right coronary artery (*left upper panel*). Pre-intervention optical coherence tomography (OCT) examination showed smallest lumen area with large amounts of lipid pool (*left lower panel*). Stent implantation was successfully performed without residual stenosis on angiogram (*right upper panel*) and with larger stent lumen area on post-intervention OCT examination (*right lower panel*). The level of CK-MB was elevated from 2.1 ng/mL pre-intervention to 22.7 ng/mL post-intervention

ized DOCTORS (Does Optical Coherence Tomography Optimize Results of Stenting) study [13], the guidelines for the procedural strategy incorporating OCT information were as follows: (1) additional balloon overdilations were to be performed in case of stent underexpansion (the ratio of in-stent minimal lumen area to reference lumen area was ≤ 80%), (2) management of malapposition or edge dissection was at the operator's discretion, and (3) additional stent implantations were to be performed to rectify incomplete lesion coverage. These methods of stent optimization led to a larger minimum lumen area compared with immediate post-stenting and subsequently improved the functional outcome assessed by FFR after PCI [13]. Table 14.3 summarizes the considerations for stent optimization using OCT.

Table 14.3 Criteria for optimal stent implantation

Comments
Achievement of adequate stent expansion (minimum lumen area or minimum stent area > 4–5 mm^2 or 80% of reference lumen area)
Avoidance of large stent malapposition (> 200 μm)
Complete lesion coverage with minimal residual plaque burden
No procedure-related complications (edge dissection, thrombosis, and others)

14.3 Clinical Benefits

The CLI-OPCI study firstly evaluated 1-year clinical outcomes in matched patients between angiographic guidance alone and angiographic plus OCT guidance. The use of OCT was associated with a lower risk of cardiac death or myocardial infarction (odds ratio = 0.49, $p = 0.037$) [12]. This observational study suggested the potential usefulness of OCT-guided PCI compared to conventional therapy. The ILUMIEN I (Observational Study of OCT in Patients Undergoing FFR and PCI) was a prospective, nonrandomized, observational study of PCI procedural practice in a total of 418 patients (with 467 stenoses) undergoing intra-procedural pre- and post-PCI FFR and OCT [14]. Based on pre-PCI OCT findings, the procedure was altered in 57% of all stenoses by selecting different stent lengths, and further stent optimization based on post-PCI OCT findings was done in 27% of all stenoses using additional post-dilation or implantation of new stents [14]. With the decreases of stent malapposition, underexpansion, and edge dissection, the change in treatment strategy appeared to be associated with reduced rates of periprocedural myocardial infarction [14]. Although intriguing, these results need confirmation in randomized controlled trial to firmly establish the clinical benefit of OCT-guided PCI.

Several benefits including adequate stent expansion, improved strut coverage, or FFR after PCI were also noted in patients receiving OCT-guided PCI. According to the OCT substudy of the thrombectomy versus PCI alone (TOTAL) trial, OCT-guided primary PCI for ST segment elevation myocardial infarction was associated with a larger final stent minimum lumen diameter compared to angiographic guidance (2.99 ± 0.48 mm versus 2.79 ± 0.47 mm, $p < 0.0001$) [15]. Although this study was statistically underpowered to detect a difference in clinical outcomes in OCT-guided patients, these findings suggested that OCT had the potential to improve clinical outcomes in patients undergoing PCI [15]. The ILUMIEN II study retrospectively compared OCT guidance with intravascular ultrasound guidance in propensity scores matched population and demonstrated that stent expansion was comparable between OCT- and intravascular ultrasound-guided patients [16]. Recently, the ILUMIEN III randomized trial tested whether or not OCT-based stent sizing strategy would result in a minimum stent area similar to or better than that achieved with intravascular ultrasound guidance and better than that achieved with angiography guidance alone [17]. In this trial, stent diameter was determined according to measurements of the external elastic lamina in the proximal and distal reference segments, and stent length was determined as the distance from distal to proximal reference site using the OCT automated lumen detection feature [17]. After stent implantation, high-pressure or larger noncompliant balloon inflation was performed to achieve a minimum stent area of at least 90% in both the proximal and distal halves of the stent relative to the closest reference segment [17]. Regarding minimum stent area, OCT guidance was non-inferior to intravascular ultrasound guidance, but not superior. OCT guidance was also not superior to angiography guidance [17]. Accordingly, these data warrant a large-scale randomized trial to establish whether or not OCT guidance results in superior clinical outcomes to angiography guidance [17].

Another study investigated the impact of OCT guidance on follow-up stent strut coverage after drug-eluting stent implantation. In this randomized trial, OCT-guided PCI significantly reduced the incidence of uncovered stent struts at 6 months compared to angiography-guided PCI (the percentage of uncovered struts, 1.6% versus 4.5%, $p = 0.0004$) [18]. This finding was accompanied with lower percentage of malapposed struts at 6 months in patients undergoing OCT guidance (0.19% versus 0.98%, $p = 0.027$) [18].

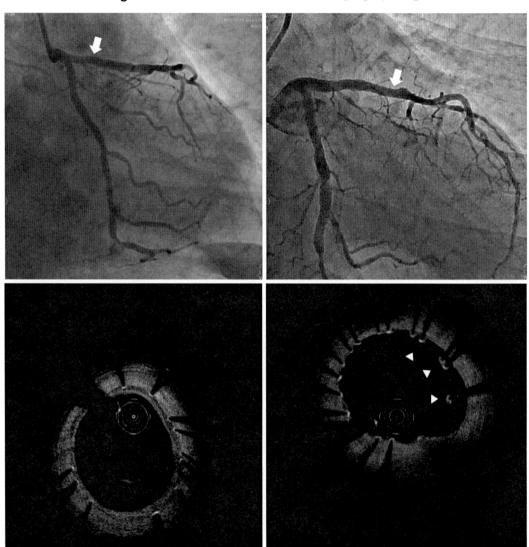

Fig. 14.4 Six-month follow-up after drug-eluting stent implantation. The shown optical coherence tomography (OCT) images were taken at the narrowest site on coronary angiography (*arrows*). Stent struts were almost covered in patients undergoing OCT guidance (*left panel*), while uncovered struts (*arrowheads*) were frequently noted in patients receiving angiographic guidance at stent implantation (*right panel*)

Given that delayed healing of implanted stents has been associated with stent thrombosis [19], this study demonstrates the possible benefit of OCT-guided PCI. Figure 14.4 shows representative cases of OCT versus angiographic guidance.

The use of OCT was related to the improved result of functional status after PCI. A randomized DOCTORS study involving 240 patients with non-ST segment elevation acute coronary syndromes compared OCT-guided PCI with angiography-guided PCI [13]. The OCT-guided group showed a higher value of post-procedural FFR compared with angiography-guided group (0.94 ± 0.04 versus 0.92 ± 0.05, $p = 0.005$) [13]. The OCT evaluation after stent implantation led to the more frequent use of post-stent overdilation

Table 14.4 Clinical benefits of optical coherence tomography (OCT) guidance during percutaneous coronary intervention

Authors	Main findings
Prati et al. [12]	Lower risk of cardiac death or myocardial infarction at 1 year, compared with angiographic guidance
Wijns et al. [14]	OCT-derived changes in treatment strategy were associated with the decrease of periprocedural myocardial infarction
Sheth et al. [15]	Larger minimum lumen diameter at post-intervention, compared with angiographic guidance
Maehara et al. [16]	Stent expansion was comparable between OCT and intravascular ultrasound guidance
Ali et al. [17]	OCT guidance using a specific reference segment external elastic lamina-based stent optimization strategy resulted in similar minimum stent area to that of intravascular ultrasound guidance
Kim et al. [18]	Lower rates of malapposed or uncovered struts at 6 months, compared with angiographic guidance
Meneveau et al. [13]	Higher fractional flow reserve at post-intervention, compared with angiographic guidance

in the OCT-guided group versus the angiography-guided group (43% versus 12.5%, $p < 0.0001$) with lower residual stenosis (7.0 ± 4.3% versus 8.7 ± 6.3%, $p = 0.01$) [13]. However, this functional benefit will translate into the clinical benefit remains to be determined [13]. Nevertheless, it has been previously shown that patients with a post-stent FFR of ≥ 0.90 had event rates of 4.9–6.2% at 6 months, compared with 20.3% in patients with post-stent FFR < 0.90 [20].

Based on the present studies, Table 14.4 summarizes the clinical benefits of OCT-guided PCI.

14.4 Specific Considerations

Left Main Diseases The use of OCT in left main diseases is challenging due to vessel size and anatomical access. OCT cannot adequately evaluate the large-sized vessel and aorto-ostial involvement because it needs to engage the guide catheter for removing the blood by contrast flushing. Although an observational study showed that frequency-domain OCT assessment of non-ostial left main diseases was feasible and provided high-quality imaging [21], data regarding OCT measurements of stenotic severity of left main diseases are currently limited. However, the OCT evaluation may be useful to optimize or guide PCI. Stent underexpansion or malapposition can be corrected to minimize restenosis or to facilitate the strut coverage of implanted stent. In a study comparing frequency-domain OCT with intravascular ultrasound, the OCT achieved imaging completeness less often, whereas it was more sensitive in detecting malapposition and edge dissections [22]. If kissing balloon angioplasty or two-stent techniques are necessary, the position of recrossing guidewire or the degree of stent distortion can be assessed by OCT, possibly improving the outcomes of stent therapy.

Bifurcated Lesions Based on high resolution, OCT provides additional information for treating bifurcated lesions. Three-dimensional OCT evaluation showed morphologic characteristics in jailed side-branch ostium. In lesions treated with single stent, the shape of the side-branch ostium changed from circular to elliptical after stent implantation [23]. The elliptical change of the side-branch ostium led to a larger minimal lumen area measured by OCT compared to minimal lumen area calculated by quantitative coronary angiography [23]. Given that three-dimensional OCT analysis may predict FFR more accurately rather than quantitative coronary angiography [24], the three-dimensional reconstruction using the most recent OCT system (ILUMIEN OPTIS OCT, St. Jude Medical) may be helpful to assess the stenotic severity of the side-branch ostium. OCT often detects vessel injuries or stent complications in bifurcated interventions, and it triggers additional procedures [25, 26]. In bifurcated lesions treated by provisional stenting, stent malapposition was more common at the proximal segment of main vessel and tissue prolapse at the distal segment of main vessel [25]. Stent malapposition was also associated with the location of wire recrossing at the side-branch ostium [26]. The OCT evaluation revealed that the wire passage via distal cell of the side-branch ostium reduced the rate of strut malapposition [26]. Thus, patients who were treated using OCT-guided recrossing had a lower number of malapposed stent struts compared to those treated with angiography guidance alone [26]. Figure 14.5 shows the case of ostial lesion of proximal segment of left anterior descending artery which was treated with single-stent implantation crossover the ostium of left circumflex artery and three-dimensional OCT reconstruction.

Safety and Feasibility Compared to previous OCT models, frequency-domain OCT system becomes more practical and less procedurally demanding [27, 28]. In a single-center registry, frequency-domain OCT was safe and feasible for PCI guidance [28]. The mean time of frequency-domain OCT pullback (from the setup to the completion of the pullback) was 2.1 min [28]. The procedure was almost successful, and major complications in terms of death, myocardial infarction, emergency revascularization, embolization, life-threatening arrhythmia, coronary dissection, prolonged and severe vessel spasm, and contrast-induced nephropathy were not noted [28]. In the randomized DOCTORS trial, there was no significant difference in the rate of procedural complications including periprocedural myocardial infarction and acute kidney injury between OCT and angiographic guidance [13]. However, the duration of OCT-guided procedures was longer than in those guided by angiography

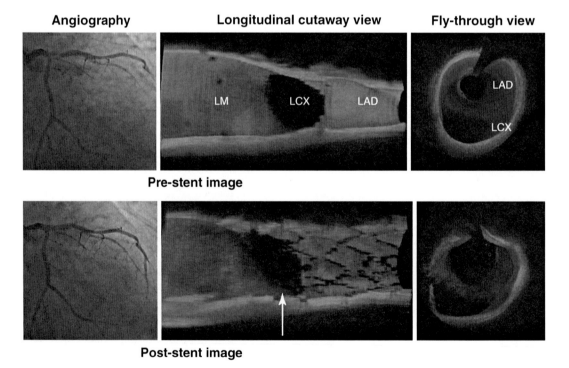

Fig. 14.5 Pre- and post-intervention angiogram and three-dimensional optical coherence tomographic (OCT) reconstructions. After stent implantation for ostial lesion of left anterior descending (LAD) artery, three-dimensional OCT reconstruction image clearly shows that proximal margin (*arrow*) of stent was protruded from LAD to left main (LM) coronary artery. LCX left circumflex artery

alone, with a greater fluoroscopy time [13]. In addition, the volume of contrast medium and the dose of radiation delivered were greater in patients receiving OCT guidance [13].

14.5 Summary

During PCI, the OCT evaluation has several advantages (Table 14.5). OCT can evaluate the severity and morphology of coronary atherosclerotic plaques and guide the proper treatment of coronary artery diseases. Various OCT findings affect the decision-making process of the physicians, leading to a change in interventional strategy. Compared to angiography guidance, OCT-guided PCI has a better stent expansion and FFR at the end of PCI and shows a lower percentage of uncovered or malapposed struts at follow-up. In complex lesions, detailed information derived from OCT helps to improve procedural outcomes. These beneficial findings were parallel to better outcomes of clinical end points in patients treated by OCT guidance. However, additional prospective studies are required to establish OCT guidance as standard use in patients with coronary artery diseases.

Table 14.5 Considerations for the usage of optical coherence tomography during percutaneous coronary intervention

Advantage
Good image resolution
Identification of morphometric characteristics of coronary atherosclerosis
Excellent visualization of stent apposition, dissection, and tissue prolapse
Evaluation of bioabsorbable stent
Disadvantage
Poor tissue penetration
Additional usage of contrast
Lesional limitations such as large-sized vessel (left main disease) or ostial stenosis
Insufficient clinical outcome data compared to intravascular ultrasound

References

1. Windecker S, Kolh P, Alfonso F, Collet JP, Cremer J, Falk V, et al. 2014 ESC/EACTS Guidelines on myocardial revascularization. Eur Heart J. 2014;35:2541–619.
2. Levine GN, Bates ER, Blankenship JC, Bailey SR, Bittl JA, Cercek B, et al. 2011 ACCF/AHA/SCAI guideline for percutaneous coronary intervention. J Am Coll Cardiol. 2011;58:44–122.
3. Gonzalo N, Serruys PW, García García HM, van Soest G, Okamura T, Ligthart J, et al. Quantitative ex vivo and in vivo comparison of lumen dimensions measured by optical coherence tomography and intravascular ultrasound in human coronary arteries. Rev Esp Cardiol. 2009;62:615–24.
4. Gonzalo N, Escaned J, Alfonso F, Nolte C, Rodriguez V, Jimenez-Quevedo P, et al. Morphometric assessment of coronary stenosis relevance with optical coherence tomography: a comparison with fractional flow reserve and intravascular ultrasound. J Am Coll Cardiol. 2012;59:1080–9.
5. Shiono Y, Kitabata H, Kubo T, Masuno T, Ohta S, Ozaki Y, et al. Optical coherence tomography-derived anatomical criteria for functionally significant coronary stenosis assessed by fractional flow reserve. Circ J. 2012;76:2218–25.
6. Pawlowski T, Prati F, Kulawik T, Ficarra E, Bil J, Gil R. Optical coherence tomography criteria for defining functional severity of intermediate lesions: a comparative study with FFR. Int J Cardiovasc Imaging. 2013;29:1685–91.
7. Reith S, Battermann S, Jaskolka A, Lehmacher W, Hoffmann R, Marx N, et al. Relationship between optical coherence tomography derived intraluminal and intramural criteria and haemodynamic relevance as determined by fractional flow reserve in intermediate coronary stenoses of patients with type 2 diabetes. Heart. 2013;99:700–7.
8. Ha J, Kim JS, Lim J, Kim G, Lee S, Lee JS, et al. Assessing computational fractional flow reserve from optical coherence tomography in patients with intermediate coronary stenosis in the left anterior descending artery. Circ Cardiovasc Interv. 2016;9:e003613.
9. Jang I-K, Bouma BE, Kang D-H, Park S-J, Park S-W, Seung K-B, et al. Visualization of coronary atherosclerotic plaques in patients using optical coherence tomography: comparison with intravascular ultrasound. J Am Coll Cardiol. 2002;39:604–9.
10. Yabushita H, Bouma BE, Houser SL, Aretz HT, Jang I, Schlendorf KH, et al. Characterization of human atherosclerosis by optical coherence tomography. Circulation. 2002;106:1640–5.
11. Lee T, Yonetsu T, Koura K, Hishikari K, Murai T, Iwai T, et al. Impact of coronary plaque morphology assessed by optical coherence tomography on

cardiac troponin elevation in patients with elective stent implantation. Circ Cardiovasc Interv. 2011;4:378–86.
12. Prati F, Di Vito L, Biondi-Zoccai G, Occhipinti M, La Manna A, Tamburino C, et al. Angiography alone versus angiography plus optical coherence tomography to guide decision-making during percutaneous coronary intervention: the centro per la lotta contro l'infarto-optimisation of percutaneous coronary intervention (CLI-OPCI) study. EuroIntervention. 2012;8:823–9.
13. Meneveau N, Souteyrand G, Motreff P, Caussin C, Amabile N, Ohlmann P, et al. Optical coherence tomography to optimize results of percutaneous coronary intervention in patients with non-ST-elevation acute coronary syndrome: results of the multicenter, randomized DOCTORS (does optical coherence tomography optimize results of stenting) study. Circulation. 2016;134:906. doi:10.1161/CIRCULATIONAHA.116.024393.
14. Wijns W, Shite J, Jones MR, Lee SW, Price MJ, Fabbiocchi F, et al. Optical coherence tomography imaging during percutaneous coronary intervention impacts physician decision-making: ILUMIEN I study. Eur Heart J. 2015;36:3346–55.
15. Sheth TN, Kajander OA, Lavi S, Bhindi R, Cantor WJ, Cheema AN, et al. Optical coherence tomography-guided percutaneous coronary intervention in ST-segment-elevation myocardial infarction: a prospective propensity-matched cohort of the thrombectomy versus percutaneous coronary intervention alone trial. Circ Cardiovasc Interv. 2016;9:e003414.
16. Maehara A, Ben-Yehuda O, Ali Z, Wijns W, Bezerra HG, Shite J, et al. Comparison of stent expansion guided by optical coherence tomography versus intravascular ultrasound: the ILUMIEN II study (observational study of optical coherence tomography [OCT] in patients undergoing fractional flow reserve [FFR] and percutaneous coronary intervention). J Am Coll Cardiol Intv. 2015;8:1704–14.
17. Ali ZA, Maehara A, Genereux P, Shlofmitz RA, Fabbiocchi F, Nazif TM, et al. Optical coherence tomography compared with intravascular ultrasound and with angiography to guide coronary stent implantation (ILUMIEN III: OPTIMIZE PCI): a randomised controlled trial. Lancet. 2016;388:2618. doi:10.1016/S0140-6736(16)31922-5.
18. Kim JS, Shin DH, Kim BK, Ko YG, Choi D, Jang Y, et al. Randomized comparison of stent strut coverage following angiography- or optical coherence tomography-guided percutaneous coronary intervention. Rev Esp Cardiol. 2015;68:190–7.
19. Finn AV, Joner M, Nakazawa G, Kolodgie F, Newell J, John MC, et al. Pathological correlates of late drug-eluting stent thrombosis: strut coverage as a marker of endothelialization. Circulation. 2007;115:2435–41.
20. Pijls NHJ, Klauss V, Siebert U, Powers E, Takazawa K, Fearon WF, et al. Coronary pressure measurement after stenting predicts adverse events at follow-up. Circulation. 2002;105:2950–4.
21. Burzotta F, Dato I, Trani C, Pirozzolo G, De Maria GL, Porto I, et al. Frequency domain optical coherence tomography to assess non-ostial left main coronary artery. EuroIntervention. 2015;10:1–8.
22. Fujino Y, Bezerra HG, Attizzani GF, Wang W, Yamamoto H, Chamie D, et al. Frequency-domain optical coherence tomography assessment of unprotected left main coronary artery disease-a comparison with intravascular ultrasound. Catheter Cardiovasc Interv. 2013;82:173–83.
23. Cho S, Kim JS, Ha J, Shin DH, Kim BK, Ko YG, et al. Three-dimensional optical coherence tomographic analysis of eccentric morphology of the jailed side-branch ostium in coronary bifurcation lesions. Can J Cardiol. 2016;32:234–9.
24. Ha J, Kim JS, Mintz GS, Kim BK, Shin DH, Ko YG, et al. 3D OCT versus FFR for jailed side-branch ostial stenoses. J Am Coll Cardiol Img. 2014;7:204–5.
25. Burzotta F, Talarico GP, Trani C, De Maria GL, Pirozzolo G, Niccoli G, et al. Frequency-domain optical coherence tomography findings in patients with bifurcated lesions undergoing provisional stenting. Eur Heart J Cardiovasc Imaging. 2014;15:547–55.
26. Alegria-Barrero E, Foin N, Chan PH, Syrseloudis D, Lindsay AC, Dimopolous K, et al. Optical coherence tomography for guidance of distal cell recrossing in bifurcation stenting: choosing the right cell matters. EuroIntervention. 2012;8:205–13.
27. Barlis P, Gonzalo N, Di Mario C, Prati F, Buellesfeld L, Rieber J, et al. A multicentre evaluation of the safety of intracoronary optical coherence tomography. EuroIntervention. 2009;5:90–5.
28. Imola F, Mallus MT, Ramazzotti V, Manzoli A, Pappalardo A, Di Giorgio A, et al. Safety and feasibility of frequency domain optical coherence tomography to guide decision making in percutaneous coronary intervention. EuroIntervention. 2010;6:575–81.

15
Pre-interventional Lesion Assessment

Hyuck-Jun Yoon

15.1 Introduction

Coronary angiography (CAG) is an established method to diagnose coronary artery disease, and it regards as gold standard method for guidance of percutaneous coronary intervention (PCI). However, it cannot directly assess atherosclerotic change of vessel wall because CAG can detect only the luminal contrast agent filling. Therefore, need for a compensated method to evaluate cross-sectional images of the lumen and vessel wall had been increased [1–3].

Recently, optical coherence tomography (OCT) was adopted in increasing numbers of catheterization laboratory. Because of its high-resolution power, OCT provides precise luminal narrowing as well as more detailed information about microscopic atherosclerotic change, the presence of high-risk plaque, tiny calcium and thrombus, plaque rupture (PR), and plaque erosion (PE) [4–7]. In this chapter, we will discuss about the pre-procedural OCT imaging.

H.-J. Yoon
Keimyung University Dongsan Medical Center,
Keimyung University, Daegu, South Korea
e-mail: hippsons@dsmc.or.kr

15.2 How to Obtain Good Pre-interventional OCT Image

Contrary to intravascular ultrasound (IVUS), OCT imaging requires blood in the coronary artery to be cleared. The infrared rays used in OCT are scattered by red blood cell in blood which will lead to severe signal attenuation, so it is essential to remove blood from the coronary arteries to obtain a good-quality image. However, in the currently used frequency-domain (FD)-OCT, image acquisition is done by a simultaneous process which consisted of the filling with injected contrast agents that replaced blood in vessel and automatic rapid pullback. The introduction of FD-OCT significantly reduces blood flow interruption time, and it makes simple the process of preparing the examination [8].

Imaging catheters are designed for rapid exchange over a 0.014-inch guide-wire delivery, have a crossing profile of 2.4–2.8 Fr, and are compatible with 6 Fr or larger guiding catheters. In case of severe stenotic lesion, predilatation with small-caliber balloon can be needed because the imaging catheter itself cannot pass the lesion or if it passes the contrast agent cannot be delivered to the distal portion and therefore the distal portion of severe stenosis cannot be visualized (Fig. 15.1).

Several checkpoints were summarized in Table 15.1 to obtain high-quality OCT images.

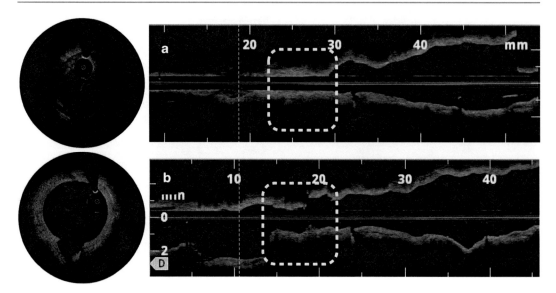

Fig. 15.1 Representative images of pre-procedural OCT images with poor preparation. Visualization of the distal portion of the MLA was not achieved because contrast agent could not pass the MLA site (indicated by a *rectangle* composed of *white dashed lines*). After balloon dilatation (2.0 mm balloon), visualization of the distal portion was much improved. Distal reference area (cross-sectional image) became clear after balloon dilatation. (**a**) Baseline OCT image and (**b**) post-balloon OCT image. *OCT* optical coherence tomography, *MLA* minimal lumen area

Table 15.1 Checklist for obtaining good pre-interventional OCT image

Proper guiding catheter position	The guiding catheter must be deep seated in coaxially maintained position
Adequate vessel preparation	1. Intracoronary nitrate must be used prior of OCT examination to avoid coronary spasm and to obtain accurate vessel size 2. In case of severe stenosis, small-caliber balloon can be needed to visualize the distal portion
Imaging catheter position	Be sure to include the area of interest because the indicator is located within proximal and distal part of imaging catheter
Imaging catheter status	Make sure that there is no blood or air in the imaging catheter before pullback
Adequate synchronization of OCT pullback after flushing	It is important to minimize the use of contrast by keeping in mind the synchronization of OCT pullback after flushing. Both the operator and the assistants must match their feet

15.3 Artifacts

As with other intravascular imaging methods, OCT also shows several types of artifacts requiring interpretation. Here are some typical examples of artifact (Fig. 15.2).

The most common avoidable artifact is the signal attenuation caused by the suboptimal purge of blood in the imaging catheter (Fig. 15.2a). Even if enough preparation was done prior to examination, during the passage of the guiding catheter and lesion, some blood may enter the imaging catheter. So, rechecking of standby imaging before pullback is necessary. This artifact can be prevented with additional purge.

Incomplete blood clearing also frequently causes artifacts (Fig. 15.2b). Incomplete guiding catheter engagement, too large vessel size, significantly angulated vessel, and inadequate contrast agent filling can affect this artifact. The stagnation of blood can be confused with thrombus as seen from a single cut.

15 Pre-interventional Lesion Assessment

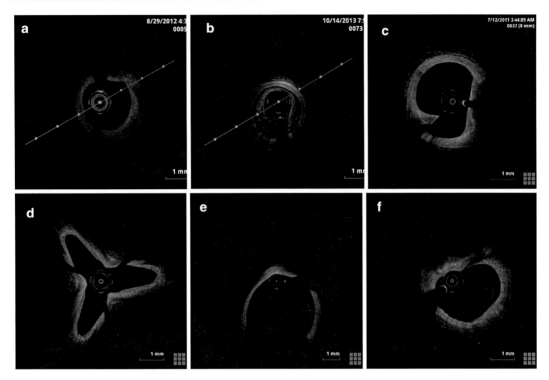

Fig. 15.2 Frequently observed OCT artifact images. (**a**) Incomplete imaging catheter preparation makes signal attenuation. (**b**) Incomplete blood clearance of coronary artery makes attenuated image. (**c**) Seam-line artifact is caused by rapid artery or wire movement during single-frame formation. (**d**) Deformation of imaging catheter can make mirror artifact. (**e**) Fold-over artifact. This artifact is the result of "phase wrapping" or "aliasing" along the Fourier transform when the structural signal is selected outside the system's field of view. (**f**) Tangential signal drop can be confused with plaque disruption or thin-cap fibroatheroma

The "sew-up artifact (seam-line artifact)" is misalignment on the lumen surface due to fast imaging wire movement during the construction of a single frame (Fig. 15.2c).

Nonuniform rotational distortion can lead to shape distortion and mirror artifact (Fig. 15.2d).

"Fold-over artifact" is caused by the inherent property of the FD-OCT. This artifact is the result of "phase wrapping" or "aliasing" along the Fourier transform when the structural signal is selected outside the system's field of view (Fig. 15.2e).

Tangential signal drop due to the location of the eccentric catheter adjacent to the lumen wall may result in misinterpretation of the lesion as thin-cap fibroatheroma (TCFA) or cap disruption (Fig. 15.2f).

Most of artifacts can be distinguished from true lesions by observing continuous changes in the serial cuts.

15.4 The Role of OCT in Pre-procedural Assessment

OCT provides detailed information on vessel walls and microstructure due to its superior resolution compared to CAG and IVUS. Especially, OCT may be more helpful in the following cases.

15.4.1 Role of OCT in Ambiguous Lesions

OCT provides a lot of information about plaque extension and characteristics that have not been fully evaluated in conventional CAG due to its excellent resolution.

This allows accurate diagnosis of suspicious findings that could not be confirmed and quantitative analysis of intracoronary thrombus.

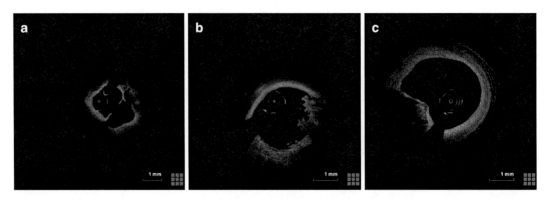

Fig. 15.3 Culprit lesion OCT findings of acute coronary syndrome. (**a**) Plaque rupture, (**b**) plaque erosion, (**c**) calcified nodule

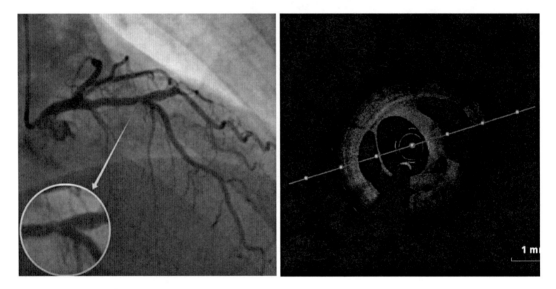

Fig. 15.4 Representative case of unstable angina without evident coronary disease. Coronary angiogram only revealed minimal stenosis and OCT revealed presence of recanalized thrombus at proximal left anterior descending artery

Ambiguous angiographic visualization of lesion is not infrequent in real practice, and OCT provides us the correct answer, especially when mixed with intermediate lesions, short lesions, thrombus, or calcification [9].

Kubo and colleagues conducted a comparison study using OCT, IVUS, and angioscopy in 30 consecutive patients with acute myocardial infarction (AMI) to assess the ability of each imaging method to detect the specific characteristics of culprit lesion. OCT was superior in detecting plaque rupture, plaque erosion, and thrombus, respectively [5].

Recent OCT studies have revealed three major mechanisms in acute coronary syndrome (ACS): PR, PE, and calcified nodule [10, 11] (Fig. 15.3).

In case of haziness on CAG without significant stenosis, there are cases of thrombus, dissection, heavy calcification, and ruptured plaque when examined through OCT [9].

If no evident lesion was seen in ACS presenting vasospastic angina, plaque disruption or thrombus was identified in OCT on a significant number of cases [12, 13] (Figs. 15.4, 15.5 and 15.6).

The diagnosis of spontaneous coronary artery dissection (SCAD) is not always apparent on

15 Pre-interventional Lesion Assessment

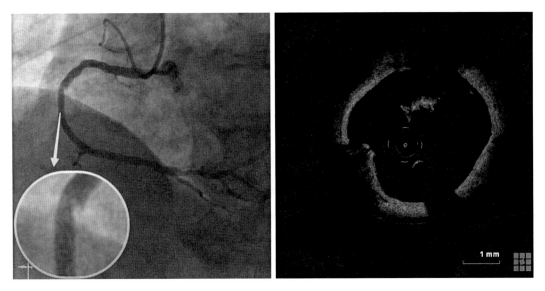

Fig. 15.5 Representative case of ambiguous coronary lesion which confirmed by OCT. After stent deployment at mid-RCA, linear slit-like lesion was observed. OCT confirmed clear image of edge dissection with thrombus

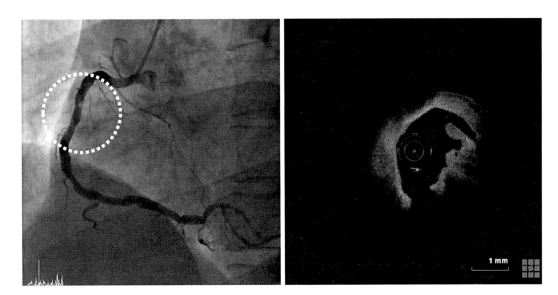

Fig. 15.6 Another case with ambiguous coronary lesion which confirmed by OCT. In coronary angiography, there was round filling defect on right coronary artery (in *circle* with *white dashed line*). OCT clearly revealed presence of thrombus without evidence of plaque disruption, lipid plaque, or calcified nodule. This is a representative image of probable plaque erosion

coronary angiography, and OCT has been greatly helpful to the diagnosis of this unfamilial disease entity (Fig. 15.7).

Abovementioned findings were not previously identified as CAG alone, which led to breakthroughs in the diagnosis by introduction of OCT.

15.4.2 Role of OCT in Lesion Severity Assessment

In addition, OCT can be helpful to determine functionally significant lesion in intermediate stenosis lesion.

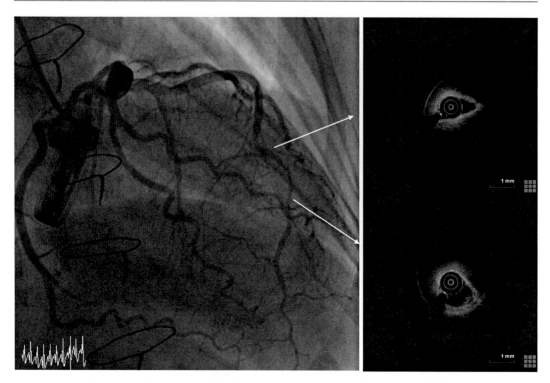

Fig. 15.7 Representative case of spontaneous coronary artery dissection. The 60-year-old woman received mitral valve replacement 10 years before and was on anticoagulation. Clinical presentation was acute ST-segment elevation myocardial infarction. The angiography showed intermediate stenosis in distal portion of left anterior descending artery. OCT revealed presence of hematoma in coronary artery without evidence of atherosclerosis. This patient was conservatively treated due to patent coronary perfusion, and she was well recovered

Although the gold standard method to identify functional significance of the coronary lesion is fractional flow reserve (FFR) in intermediate angiographic stenotic lesion, minimal lumen area (MLA) can be used as surrogate marker for functional significance. In comparison with IVUS, OCT showed slightly superior in identifying hemodynamically severe coronary stenosis (especially in vessel diameter less than 3 mm) [14].

Recently, a dedicated, semiautomated contour detection system (OPTIS™, St. Jude, MN, USA) is used for measurements. A contour detection algorithm that automatically traces lumen boundaries of the longitudinal (L)-mode view was implemented, and it allowed us to automatically detect MLA position within few seconds (Fig. 15.6). In a comparison study regarding automatically detected and manually detected lumen analysis, there was excellent correlation with two methods [15].

Through this method, we can easily identify the location and severity of the minimal lumen area (Figs. 15.8 and 15.9).

15.4.3 Role of OCT in Determining Vessel Sizing

OCT allows clear delineation between the lumen and vessel wall, although there may be a limit in the detail of the whole vessel structure visualized due to shallow penetration as compared with IVUS imaging.

OCT has good correlation in lumen measurement with IVUS. Moreover, OCT showed good inter-study correlation for FD-OCT in evaluating both stented and native coronary arteries undergoing PCI ($R^2 = 0.99$ and $P < 0.001$ for MLA) [16].

In a phantom model comparison, the mean lumen area per FD-OCT was equal to the actual

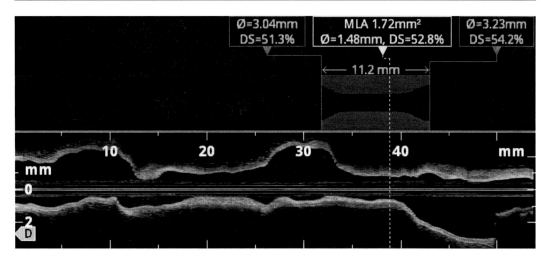

Fig. 15.8 Longitudinal view and automated detected lumen measurement (*upper panel*). Within few seconds, reconstruction of longitudinal view and automated contour detection were finished. After then operator can easily identify the location and severity of lesion

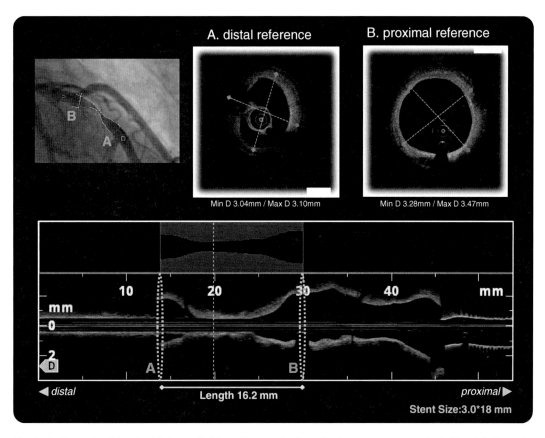

Fig. 15.9 Example of OCT-guided vessel sizing. Cross-sectional analysis of the distal reference identified EEL measurements of 3.04 and 3.10 mm. Cross-sectional analysis of the proximal reference area identified EEL visualization allowing measurement of a 3.28 and 3.47 mm in diameter. The distance from proximal to distal reference was 16.2 mm, and thus a 3.0 mm diameter by 18 mm stent was chosen

lumen area, with low standard deviation; IVUS overestimated the lumen area and was less reproducible than FD-OCT [17, 18]. OCT accurately measured the MLA compared with the actual phantom, whereas IVUS significantly overestimated the MLA in 10% of the patients ($P < 0.001$ vs. OCT), and it was less reproducible.

In early comparison of FD-OCT and IVUS to guide PCI (both groups finally confirmed by FD-OCT), OCT guidance showed smaller stent expansion compared with IVUS [19]. However, recently a conducted larger-sized randomized study (ILUMIEN III) showed OCT guidance had similar final MSA compared with IVUS guidance [20]. In this trial, OCT guidance group used external elastic lamina (EEL) as determinant to vessel size. In the reference area, EEL can be easily detected in OCT with three clearly separated layers. However, in lipid-rich plaque, vessel circular arcs could not be identified due to OCT signal attenuation in lipid plaque. In that case, clear discrimination of EEL might be only available at some portion. Kubo and colleague suggested approximation algorithm that can be applied in lipid plaque for assumption of OCT invisible circular arc. In comparison with IVUS, the assumption of the vessel area was well correlated with the actual EEM measured by IVUS [21].

If tapering vessel is severe in L-mode view, it will be able to determine whether to apply post-dilatation or not.

15.4.4 Role of OCT in Plaque Characterization

Pre-procedural OCT can reliably visualize the plaque extent, characteristics, and lumen dimensions.

Compared to CAG, the greatest advantage of intracoronary imaging such as OCT or IVUS is to show not only the degree of stenosis but also precise changes from the early stage of atherosclerosis to the advanced stage.

Rupture of atherosclerotic plaque is responsible for most thrombotic coronary event. So, detection of rupture-prone plaque (vulnerable plaque) might have clinical relevance to prevent thrombotic complication [5, 22]. Histologically, the vulnerable plaque was defined as a large lipid core, a TCFA (fibrous cap thickness <65 μm), and an accumulation of macrophages localized at the subsurface of the fibrous cap. Because fibrous cap thickness in most ruptured plaques is less than 65 μm, the resolution required for imaging of risky plaques should be at the level of 50 μm or better. OCT is the only available imaging technique with enough resolution to detect TCFAs [4].

With this information, the operator can identify the exact location of the culprit lesion, determine the adequate stent size and length, and avoid landing on a vulnerable plaque.

For determining optimal stent landing site, it is better to avoid lipid-rich plaque, TCFA with high plaque burden [23–25].

15.4.5 Role of OCT in Determining Treatment Strategy

OCT finding of the pre-procedural culprit lesion may have a significant impact on the decision of the therapeutic direction.

In the presence of large lipid pools or necrotic core contents, a no-reflow phenomenon can be predicted, and adjunctive medical therapy or distal filter device protection may also be considered.

In severely calcified lesions, OCT finding can be helpful to determine whether scoring balloon, wire-cutting, or rotablation is needed. Kubo and colleagues showed that the thickness of calcified plaque by OCT can predict the possibility of calcium plate break in encircled calcified lesion [26]. It can offer useful information about the lesions which need further plaque modification prior to stenting especially in the severe calcified lesion (Fig. 15.10).

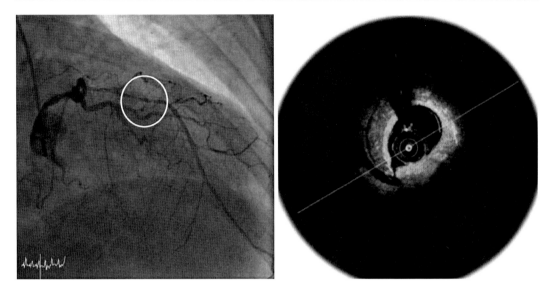

Fig. 15.10 Representative case of heavy calcified lesion. After balloon dilatation, OCT clearly revealed broken calcified plate

15.4.6 The Role of OCT in Acute Coronary Syndrome

OCT imaging technique enables detailed evaluation of plaque morphology in patients with ACS and helps to understand the underlying mechanisms.

Plaque disruption and subsequent thrombus formation is the main mechanism for the onset of acute coronary syndrome. Current guideline regards catheter-based reperfusion of the infarct-related artery with stent placement as preferential treatment of acute myocardial infarction (AMI) and thrombus aspiration as supplementary treatment option. Although advanced atherosclerosis with obstructive stenosis is often accompanied in ACS, previous postmortem studies and recent clinical study using optical coherence tomography (OCT) revealed that about one third of ACS cases showed plaque disruption accompanying thrombus formation without severe stenosis [27, 28]. After removal or reduction of thrombus by aspiration, OCT can provide assessment of the underlying plaque pathology including PR or PE. Because plaque erosion has an intact fibrous cap and represents less luminal narrowing after thrombus removal compared to PR, medical treatment without stent implantation has been challenged as an alternative strategy for the treatment of culprit lesion in ACS.

A study by Prati et al. reported clinical follow-up (median 753 days) results of two treatment strategies including medical therapy with and without percutaneous revascularization in 31 STEMI patients with plaque erosion by OCT after thrombectomy. In their study, 40% of patients who had intact fibrous cap with nonobstructive lesions were treated with dual antiplatelet therapy alone and showed comparable outcomes without symptoms compared to the remaining 60% of patients who underwent stent implantation [29]. Similarly, favorable clinical outcomes were reported in young patients with STEMI after thrombus aspiration without stent implantation [29].

IK Jang and his colleague conducted prospective study for this approach. Among the 405 ACS patients, 103 patients showed plaque erosion as culprit morphology. Sixty plaque erosion patients were treated with dual antiplatelet therapy (aspirin and ticagrelor), without stent deployment. Significant number of patients (78.3%) showed >50% reduction of thrombus volume at 1-month follow-up [30].

Therefore, OCT evaluation of culprit lesion will identify ACS patients who can defer coronary stenting.

15.5 Summary

Pre-procedural OCT can provide us much more accurate morphometric information and plaque characters compared with CAG or IVUS. This may be helpful in selecting the optimal treatment method and avoiding unnecessary stents in coronary artery disease.

References

1. Nissen SE, Yock P. Intravascular ultrasound: novel pathophysiological insights and current clinical applications. Circulation. 2001;103(4):604–16.
2. Nissen S, Gurley J, Grines C, Booth D, McClure R, Berk M, et al. Intravascular ultrasound assessment of lumen size and wall morphology in normal subjects and patients with coronary artery disease. Circulation. 1991;84(3):1087–99.
3. Sones FM Jr. Cine-coronary arteriography. Ohio State Med J. 1962;58:1018–9.
4. Jang IK, Bouma BE, Kang DH, Park SJ, Park SW, Seung KB, et al. Visualization of coronary atherosclerotic plaques in patients using optical coherence tomography: comparison with intravascular ultrasound. J Am Coll Cardiol. 2002;39(4):604–9.
5. Kubo T, Imanishi T, Takarada S, Kuroi A, Ueno S, Yamano T, et al. Assessment of culprit lesion morphology in acute myocardial infarction. J Am Coll Cardiol. 2007;50(10):933–9.
6. Kume T, Akasaka T, Kawamoto T, Ogasawara Y, Watanabe N, Toyota E, et al. Assessment of coronary arterial thrombus by optical coherence tomography. Am J Cardiol. 2006a;97(12):1713–7.
7. Kume T, Akasaka T, Kawamoto T, Okura H, Watanabe N, Toyota E, et al. Measurement of the thickness of the fibrous cap by optical coherence tomography. Am Heart J. 2006b;152(4):755.e1–4.
8. Stefano GT, Bezerra HG, Mehanna E, Yamamoto H, Fujino Y, Wang W, et al. Unrestricted utilization of frequency domain optical coherence tomography in coronary interventions. Int J Cardiovasc Imaging. 2012;29(4):741–52.
9. Yoon H-J, Cho Y-K, Nam C-W, Kim K-B, Hur S-H. Angiographically minimal but functionally significant coronary lesion confirmed by optical coherence tomography. Korean J Intern Med. 2016;31(4):807–8.
10. Jia H, Abtahian F, Aguirre AD, Lee S, Chia S, Lowe H, et al. In vivo diagnosis of plaque erosion and calcified nodule in patients with acute coronary syndrome by intravascular optical coherence tomography. J Am Coll Cardiol. 2013;62(19):1748–58.
11. Tian J, Ren X, Vergallo R, Xing L, Yu H, Jia H, et al. Distinct morphological features of ruptured culprit plaque for acute coronary events compared to those with silent rupture and thin-cap fibroatheroma. J Am Coll Cardiol. 2014;63(21):2209–16.
12. Shin E-S, Ann SH, Singh GB, Lim KH, Yoon H-J, Hur S-H, et al. OCT-defined morphological characteristics of coronary artery spasm sites in vasospastic angina. JACC Cardiovasc Imaging. 2015;8(9):1059–67.
13. Shin ES, Her AY, Ann SH, Balbir Singh G, Cho H, Jung EC, et al. Thrombus and plaque erosion characterized by optical coherence tomography in patients with vasospastic angina. Rev Esp Cardiol (Engl Ed). 2016;70(6):459–66.
14. Gonzalo N, Escaned J, Alfonso F, Nolte C, Rodriguez V, Jimenez-Quevedo P, et al. Morphometric assessment of coronary stenosis relevance with optical coherence tomography. J Am Coll Cardiol. 2012;59(12):1080–9.
15. Sihan K, Botha C, Post F, de Winter S, Gonzalo N, Regar E, et al. Fully automatic three-dimensional quantitative analysis of intracoronary optical coherence tomography: method and validation. Catheter Cardiovasc Interv. 2009;74(7):1058–65.
16. Jamil Z, Tearney G, Bruining N, Sihan K, van Soest G, Ligthart J, et al. Interstudy reproducibility of the second generation, Fourier domain optical coherence tomography in patients with coronary artery disease and comparison with intravascular ultrasound: a study applying automated contour detection. Int J Cardiovasc Imaging. 2013;29(1):39–51.
17. Kim I-C, Nam C-W, Cho Y-K, Park H-S, Yoon H-J, Kim H, et al. Discrepancy between frequency domain optical coherence tomography and intravascular ultrasound in human coronary arteries and in a phantom in vitro coronary model. Int J Cardiol. 2016;221:860–6.
18. Kubo T, Akasaka T, Shite J, Suzuki T, Uemura S, Yu B, et al. OCT compared with IVUS in a coronary lesion assessment. JACC Cardiovasc Imaging. 2013;6(10):1095–104.
19. Habara M, Nasu K, Terashima M, Kaneda H, Yokota D, Ko E, et al. Impact of frequency-domain optical coherence tomography guidance for optimal coronary stent implantation in comparison with intravascular ultrasound guidance. Circ Cardiovasc Interv. 2012;5(2):193–201.
20. Ali ZA, Maehara A, Genereux P, Shlofmitz RA, Fabbiocchi F, Nazif TM, et al. Optical coherence tomography compared with intravascular ultrasound and with angiography to guide coronary stent implantation (ILUMIEN III: OPTIMIZE PCI): a randomised controlled trial. Lancet. 2016;388(10060):2618–28.

21. Kubo T, Yamano T, Liu Y, Ino Y, Shiono Y, Orii M, et al. Feasibility of optical coronary tomography in quantitative measurement of coronary arteries with lipid-rich plaque. Circ J. 2015a;79(3):600–6.
22. Tanaka A, Imanishi T, Kitabata H, Kubo T, Takarada S, Tanimoto T, et al. Morphology of exertion-triggered plaque rupture in patients with acute coronary syndrome: an optical coherence tomography study. Circulation. 2008;118(23):2368–73.
23. Gonzalo N, Serruys PW, Okamura T, Shen ZJ, Garcia-Garcia HM, Onuma Y, et al. Relation between plaque type and dissections at the edges after stent implantation: an optical coherence tomography study. Int J Cardiol. 2011;150(2):151–5.
24. Imola F, Occhipinti M, Biondi-Zoccai G, Di Vito L, Ramazzotti V, Manzoli A, et al. Association between proximal stent edge positioning on atherosclerotic plaques containing lipid pools and postprocedural myocardial infarction (from the CLI-POOL study). Am J Cardiol. 2013;111(4):526–31.
25. Mudra H, Regar E, Klauss V, Werner F, Henneke KH, Sbarouni E, et al. Serial follow-up after optimized ultrasound-guided deployment of Palmaz-Schatz stents. In-stent neointimal proliferation without significant reference segment response. Circulation. 1997;95(2):363–70.
26. Kubo T, Shimamura K, Ino Y, Yamaguchi T, Matsuo Y, Shiono Y, et al. Superficial calcium fracture after PCI as assessed by OCT. JACC Cardiovasc Imaging. 2015b;8(10):1228–9.
27. Burke AP, Farb A, Malcom GT, Liang YH, Smialek J, Virmani R. Coronary risk factors and plaque morphology in men with coronary disease who died suddenly. N Engl J Med. 1997;336(18):1276–82.
28. Falk E, Nakano M, Bentzon JF, Finn AV, Virmani R. Update on acute coronary syndromes: the pathologists' view. Eur Heart J. 2013;34(10):719–28.
29. Prati F, Uemura S, Souteyrand G, Virmani R, Motreff P, Di Vito L, et al. OCT-based diagnosis and management of STEMI associated with intact fibrous cap. JACC Cardiovasc Imaging. 2013;6(3):283–7.
30. Jia H, Dai J, Hou J, Xing L, Ma L, Liu H, et al. Effective anti-thrombotic therapy without stenting: intravascular optical coherence tomography-based management in plaque erosion (the EROSION study). Eur Heart J. 2016;38(11):792–800.

# Immediate Post-Stent Evaluation with Optical Coherence Tomography

16

Seung-Yul Lee, Yangsoo Jang, and Myeong-Ki Hong

Compared to intravascular ultrasound, optical coherence tomography (OCT) enables detailed evaluations regarding immediate post-stent deployment. Contents of this chapter are limited to metallic coronary drug-eluting stents. Absorbable stents with polymeric scaffolds will be reviewed in the following chapter.

16.1 Stent Expansion

Adequate stent expansion has been considered as a key component of successful percutaneous coronary intervention (PCI). In ILUMIEN II study (Observational Study of OCT in Patients Undergoing Fractional Flow Reserve and Percutaneous Coronary Intervention) that analyzed the matched cohort with intravascular ultrasound-guided PCI, the degree of optimal stent expansion was not different between OCT and intravascular ultrasound guidance (median [first, third quartiles] = 72.8% [63.3, 81.3] versus 70.6% [62.3, 78.8], respectively, $p = 0.29$) [1]. Based on previous intravascular ultrasound study [2], optimal stent expansion was defined by the CLI-OPCI (Centro per la Lotta contro l'Infarto-Optimisation of Percutaneous Coronary Intervention) study as in-stent minimal lumen area $\geq 90\%$ of average reference lumen area or $\geq 100\%$ of lumen area of reference segment with lowest lumen area [3]. Also, in the CLI-OPCI II study that retrospectively compared the OCT findings with clinical outcomes [4], in-stent minimum lumen area < 4.5 mm^2 was an independent predictor of major adverse cardiac events. Although most of these findings resulted from observational studies, stent expansion can be reliably evaluated by OCT [4, 5]. Figure 16.1 represents causative mechanisms of in-stent restenosis, as assessed by OCT.

16.2 Stent Apposition

The position of individual struts from the vessel wall can be assessed qualitatively (apposed or malapposed) or quantitatively (distance or area) with OCT. Immediately after stent implantation, individual stent struts are in touch with the vessel wall (apposed) or not (malapposed). Apposed struts are further classified into embedded or protruding. Struts are defined as embedded if more than half of the strut thickness is below the level of luminal surface [6] and as protruding if the adluminal strut surface is just above the vessel

S.-Y. Lee (✉)
Department of internal Medicine, Sanbon Hospital, Wonkwang University College of Medicine, Gunpo, South Korea
e-mail: seungyul79@gmail.com

Y. Jang · M.-K. Hong
Division of Cardiology, Severance Cardiovascular Hospital, Yonsei University College of Medicine, Seoul, South Korea

© Springer Nature Singapore Pte Ltd. 2018
M.-K. Hong (ed.), *Coronary Imaging and Physiology*,
https://doi.org/10.1007/978-981-10-2787-1_16

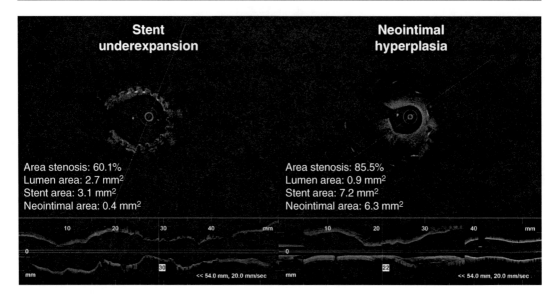

Fig. 16.1 Optical coherence tomographic images of in-stent restenosis 1 year after stent implantation. On *left panel*, the percentage of area stenosis was 60.1%, showing in-stent restenosis. The cross-sectional area of stent and neointima was, respectively, 3.1 mm² and 0.4 mm², suggesting that stent underexpansion was a causative mechanism for in-stent restenosis. However, on *right panel*, neointimal hyperplasia was a restenotic mechanism. The cross-sectional area of stent and neointima was 7.2 mm² and 6.3 mm², respectively

Table 16.1 Definitions of malapposed strut according to types of drug-eluting stents [8, 9]

Trade name	Company	Distance between vessel wall and strut
Cypher	Cordis	≥160 µm
Taxus	Boston scientific	≥130 µm
Endeavor resolute	Medtronic	≥110 µm
Resolute integrity	Medtronic	≥110 µm
Xience	Abbott vascular	≥100 µm
Nobori	Terumo	≥130 µm
Biomatrix	Biosensors	≥130 µm

wall [6]. Considering the thickness of stent strut or abluminal polymer, struts that are detached from the vessel wall are defined as malapposed [7]. Accordingly, the distances for defining malapposed struts are different among types of drug-eluting stents (Table 16.1). Representative OCT images of apposed and malapposed struts are shown in Fig. 16.2.

The malapposed struts are commonly observed immediately after stent implantation.

The incidence of acute stent malapposition was approximately 40–60% (Table 16.2), and about 70% of acute stent malapposition spontaneously disappeared at follow-up of 1 year [8, 10]. Determinants of the spontaneous resolution were malapposed distance or area, indicating that *tiny* malapposition can be resolved at follow-up [8, 10, 11]. The coverage of malapposed struts is delayed compared with apposed struts [12]. Recent two registries investigating mechanisms of stent thrombosis showed that stent malapposition was the most frequent finding (about one third of all cases) that caused stent thrombosis [13, 14]. In these studies, a significant or observed maximal distance of malapposition was >200–300 µm (Fig. 16.3) [13, 14]. Accordingly, it may be important to interpret malapposition as a quantitative, rather than binary phenomenon (present or absent), and to define the threshold of malapposition detachment that may benefit from optimization during stent implantation [15]. However, simple presence of acute stent malapposition >200 µm itself was not associated with worse outcomes [4], and results from prospective trials have not been

16 Immediate Post-Stent Evaluation with Optical Coherence Tomography

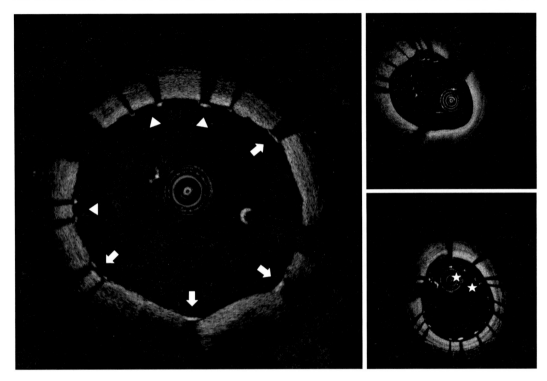

Fig. 16.2 Representative optical coherence tomographic images of apposed (*left panel*) and malapposed struts (*right panel*) immediately after implantation. *Arrows* indicate embedded struts and arrowheads protruding struts. All struts were malapposed on top right panel, whereas partial struts (*asterisks*) were malapposed on bottom *right panel*

Table 16.2 Frequency of acute malapposition, detected by optical coherence tomography

Authors	No. of study population	Stent type	Frequency of acute malapposition
Kawamori et al. [10]	40	DES	65%
Im et al. [8]	356	DES	62%
Soeda et al. [5]	1001	BMS and DES	39%
Prati et al. [4]	1002	BMS, DES, and BVS	49%

BMS bare metal stent; *BVS* bioresorbable vascular scaffold; *DES* drug-eluting stent

contour plot analysis [17], the associations between strut apposition immediately after intervention and strut coverage at follow-up can be identified at a glance. On the contour plot, x-axis represented circumferential arc length of individual stent strut, and y-axis represented stent length. In (x, y) format, the locations of stent struts were delineated by their pixel coordinates [17]. In addition, individual stent struts could be marked regarding the status of apposition or coverage. Figure 16.4 is an example of serial contour plot analysis at post-intervention and follow-up.

addressed yet. On the contrary to malapposed struts, embedded struts at post-intervention were highly covered at follow-up of 6 months (median percentage of uncovered struts, 0% in embedded and 26.8% in malapposed, $p < 0.001$) [16]. Using

16.3 Dissection

During PCI, dissections can occur within stented segment or at stent edge. Intra-stent dissection is defined as a disruption of the luminal vessel surface in the stent segment [18]. It can appear in two forms:

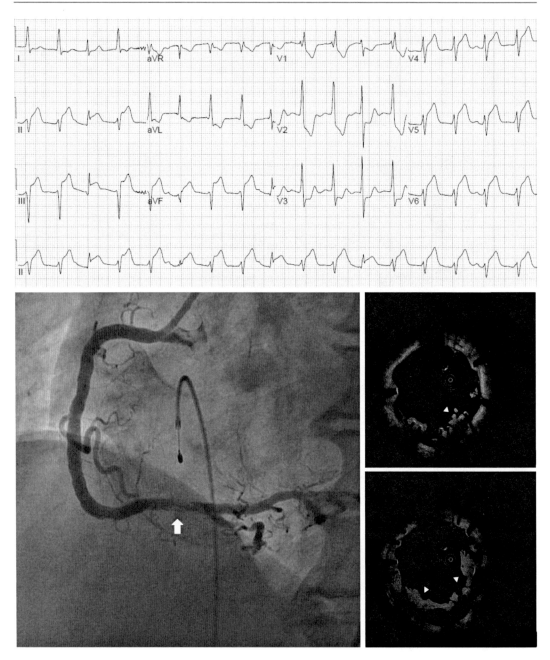

Fig. 16.3 A case of very late stent thrombosis related to malapposed struts. A 69-year-old man visited the emergency department for severe angina. He underwent biolimus-eluting stent (3.5 mm × 18 mm) implantation 24 months ago. Electrocardiogram showed ST-segment elevation on inferior leads, suggesting acute myocardial infarction. Emergent angiography showed intraluminal haziness (*arrow*) within stent of right coronary artery. After thrombosuction, optical coherence tomography revealed that irregular-shaped thrombi (*arrowheads*) were attached to malapposed struts. The maximal distance between malapposed struts and vessel wall was 560 μm

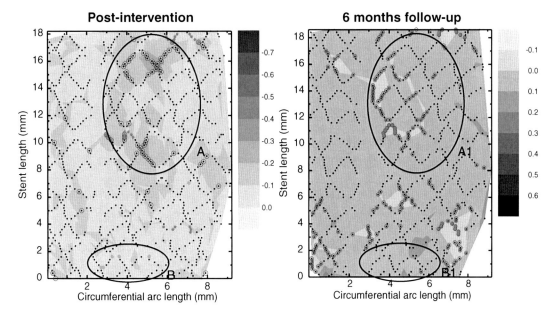

Fig. 16.4 Serial contour plot analyses at post-intervention and 6-month follow-up. Compared to embedded struts (*green circle* in Circle B), malapposed struts (*red circle* in Circle A) at post-intervention were more uncovered (*blue circle* in Circle A1) at follow-up of 6 months. *Orange* circle indicated strut above the ostium of side-branch vessel

(1) dissection: the vessel surface is disrupted, and a dissection flap is visible; (2) Cavity: the vessel surface is disrupted, and an empty cavity can be seen [18].

Edge dissection is defined as a disruption of the luminal vessel surface in the edge segments (within 5 mm proximal and distal to the stent, no struts are visible) [18]. Figures 16.5 and 16.6 show OCT images of these dissections. Intra-stent dissection was more frequently observed compared to edge dissection, and the incidence of intra-stent dissection was 86.6% by Gonzalo et al. [18]. The incidence of edge dissection was approximately 25–40% after stent implantation, and most was not apparent on coronary angiography [18–21]. At 1 year of follow-up, edge dissections that were small and non-flow limiting were completely healed, and consequently there were no associated major adverse cardiac events [20, 21]. However, large dissection (>200 to 300 μm) at the distal stent edge increased the risk of major adverse cardiac events during follow-up [4, 19].

16.4 Tissue Prolapse

Tissue prolapse is found in more than 90% of stented segments immediately after successful PCI [18]. On OCT, tissue prolapse is defined as follows: convex-shaped protrusion of tissue between adjacent stent struts toward the lumen, without disruption of the continuity of the luminal vessel surface (Fig. 16.7) [18]. Tissue prolapse (>500 μm of protrusion) was not associated with worse clinical outcomes in the CLI-OPCI II study [4], but one OCT study suggested that irregular shape of tissue protrusion was an independent predictor of adverse cardiac events at 1 year of follow-up [5]. Accordingly, clinical significance regarding tissue prolapse has not been established yet.

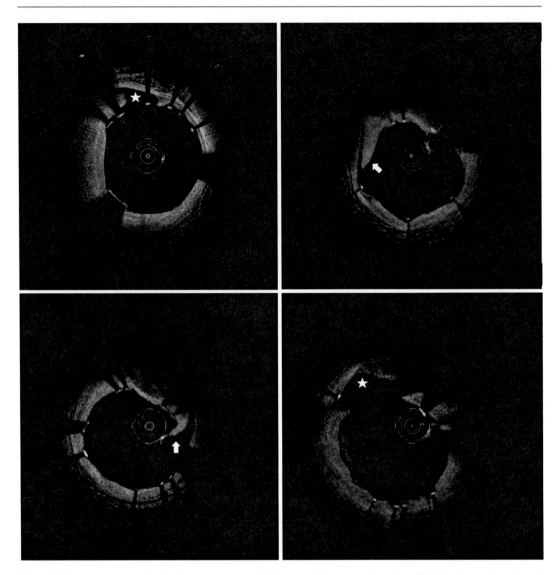

Fig. 16.5 Representative optical coherence tomographic images of intra-stent dissection. With the disruption of vessel surface, dissection flap (*arrows*) or empty cavity (*asterisks*) was seen

16 Immediate Post-Stent Evaluation with Optical Coherence Tomography 161

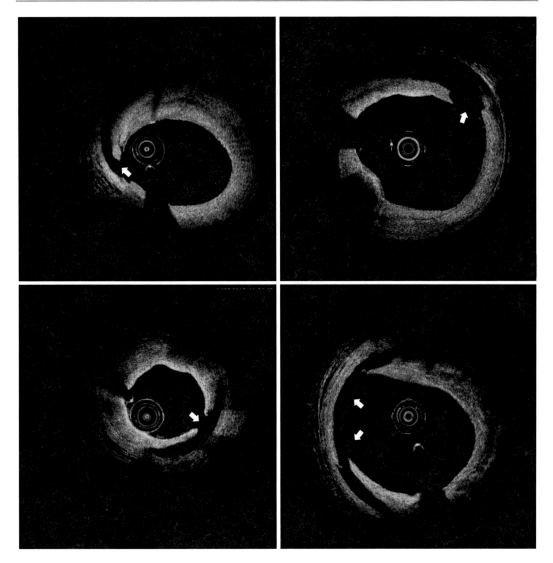

Fig. 16.6 Representative optical coherence tomographic images of edge dissection. The luminal vessel surface was disrupted in the edge segments (*arrows*)

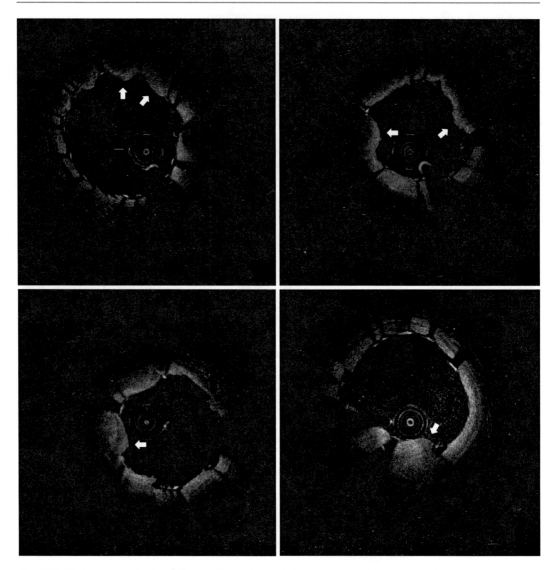

Fig. 16.7 Various morphologies of tissue prolapse. *Arrows* indicated prolapsed tissues

16.5 Summary

Based on previous studies, suggestions regarding post-stenting OCT findings are summarized in Table 16.3. Stent underexpansion and large-sized dissection with flow limitation should be considered for further treatment if these are shown at post-intervention. Although large malapposition and tissue prolapse may be associated with worse outcomes, clinical impacts of these OCT findings need further investigations.

Table 16.3 Suggestions regarding post-stenting optical coherence tomographic findings

Findings	Comments
Adequate stent expansion	Minimum lumen area or minimum stent area > 4–5 mm^2 or 80% of reference lumen area
Stent malapposition	May consider further treatment if significant malapposition (>200 μm) was identified
Edge dissection	Intervene if coronary flow is limited
Tissue prolapse/ thrombus	Presence of minor findings is possibly insignificant

References

1. Maehara A, Ben-Yehuda O, Ali Z, Wijns W, Bezerra HG, Shite J, et al. Comparison of stent expansion guided by optical coherence tomography versus intravascular ultrasound: the ILUMIEN II study (observational study of optical coherence tomography [OCT] in patients undergoing fractional flow reserve [FFR] and percutaneous coronary intervention). J Am Coll Cardiol Interv. 2015;8:1704–14.
2. de Jaegere P, Mudra H, Figulla H, Almagor Y, Doucet S, Penn I, et al. Intravascular ultrasound-guided optimized stent deployment. Immediate and 6 months clinical and angiographic results from the multicenter ultrasound stenting in coronaries study (MUSIC study). Eur Heart J. 1998;19:1214–23.
3. Prati F, Di Vito L, Biondi-Zoccai G, Occhipinti M, La Manna A, Tamburino C, et al. Angiography alone versus angiography plus optical coherence tomography to guide decision-making during percutaneous coronary intervention: the centro per la lotta contro l'infarto-optimisation of percutaneous coronary intervention (CLI-OPCI) study. EuroIntervention. 2012;8:823–9.
4. Prati F, Romagnoli E, Burzotta F, Limbruno U, Gatto L, La Manna A, et al. Clinical impact of OCT findings during PCI: the CLI-OPCI II study. J Am Coll Cardiol Img. 2015;8:1297–305.
5. Soeda T, Uemura S, Park SJ, Jang Y, Lee S, Cho JM, et al. Incidence and clinical significance of poststent optical coherence tomography findings: one-year follow-up study from a multicenter registry. Circulation. 2015;132:1020–9.
6. Tanigawa J, Barlis P, Di Mario C. Intravascular optical coherence tomography: optimisation of image acquisition and quantitative assessment of stent strut apposition. EuroIntervention. 2007;3:128–36.
7. Tanigawa J, Barlis P, Dimopoulos K, Dalby M, Moore P, Di Mario C. The influence of strut thickness and cell design on immediate apposition of drug-eluting stents assessed by optical coherence tomography. Int J Cardiol. 2009;134:180–8.
8. Im E, Kim BK, Ko YG, Shin DH, Kim JS, Choi D, et al. Incidences, predictors, and clinical outcomes of acute and late stent malapposition detected by optical coherence tomography after drug-eluting stent implantation. Circ Cardiovasc Interv. 2014;7:88–96.
9. Lee SY, Hong MK. Stent evaluation with optical coherence tomography. Yonsei Med J. 2013;54:1075–83.
10. Kawamori H, Shite J, Shinke T, Otake H, Matsumoto D, Nakagawa M, et al. Natural consequence of post-intervention stent malapposition, thrombus, tissue prolapse, and dissection assessed by optical coherence tomography at mid-term follow-up. Eur Heart J Cardiovasc Imaging. 2013;14:865–75.
11. Gutiérrez-Chico JL, Wykrzykowska J, Nüesch E, van Geuns RJ, Koch KT, Koolen J, et al. Vascular tissue reaction to acute malapposition in human coronary arteries: sequential assessment with optical coherence tomography. Circ Cardiovasc Interv. 2012;5:20–9.
12. Gutiérrez-Chico JL, Regar E, Nüesch E, Okamura T, Wykrzykowska J, di Mario C, et al. Delayed coverage in malapposed and side-branch struts with respect to well-apposed struts in drug-eluting stents: in vivo assessment with optical coherence tomography. Circulation. 2011;124:612–23.
13. Souteyrand G, Amabile N, Mangin L, Chabin X, Meneveau N, Cayla G, et al. Mechanisms of stent thrombosis analysed by optical coherence tomography: insights from the national PESTO French registry. Eur Heart J. 2016;37:1208–16.
14. Taniwaki M, Radu MD, Zaugg S, Amabile N, Garcia-Garcia HM, Yamaji K, et al. Mechanisms of very late drug-eluting stent thrombosis assessed by optical coherence tomography. Circulation. 2016;133:650–60.
15. Foin N, Gutiérrez-Chico JL, Nakatani S, Torii R, Bourantas CV, Sen S, et al. Incomplete stent apposition causes high shear flow disturbances and delay in neointimal coverage as a function of strut to wall detachment distance: implications for the management of incomplete stent apposition. Circ Cardiovasc Interv. 2014;7:180–9.
16. Kim JS, Ha J, Kim BK, Shin DH, Ko YG, Choi D, et al. The relationship between post-stent strut apposition and follow-up strut coverage assessed by a contour plot optical coherence tomography analysis. J Am Coll Cardiol Interv. 2014;7:641–51.
17. Ha J, Kim B, Kim J, Shin D, Ko Y, Choi D, et al. Assessing neointimal coverage after DES implantation by 3D OCT. J Am Coll Cardiol Img. 2012;5:852–3.
18. Gonzalo N, Serruys PW, Okamura T, Shen ZJ, Onuma Y, Garcia-Garcia HM, et al. Optical coherence tomography assessment of the acute effects of stent implantation on the vessel wall: a systematic quantitative approach. Heart. 2009;95:1913–9.
19. Bouki KP, Sakkali E, Toutouzas K, Vlad D, Barmperis D, Phychari S, et al. Impact of coronary artery stent edge dissections on long-term clinical outcome in patients with acute coronary syndrome: an optical coherence tomography study. Catheter Cardiovasc Interv. 2015;86:237–46.
20. Chamie D, Bezerra HG, Attizzani GF, Yamamoto H, Kanaya T, Stefano GT, et al. Incidence, predictors, morphological characteristics, and clinical outcomes of stent edge dissections detected by optical coherence tomography. J Am Coll Cardiol Interv. 2013;6:800–13.
21. Radu MD, Raber L, Heo J, Gogas BD, Jorgensen E, Kelbaek H, et al. Natural history of optical coherence tomography-detected non-flow-limiting edge dissections following drug-eluting stent implantation. EuroIntervention. 2014;9:1085–94.

Late Stent Evaluation (Neoatherosclerosis)

Jung-Hee Lee, Yangsoo Jang, and Jung-Sun Kim

Using light instead of ultrasound, optical coherence tomography (OCT) can provide high-resolution in vivo images of the coronary artery and evaluate the stent status and neointimal tissue after coronary stent implantation more accurately than intravascular ultrasound (IVUS). Intravascular OCT assessment is useful for the detection of strut coverage, malapposition, and the characterization of neointimal tissue during stent follow-up due to high resolution. Furthermore, OCT enables detailed assessment of the morphological characteristics of late stent failure, including neoatherosclerosis. In this chapter, late stent-based changes evaluated by OCT will be reviewed.

17.1 Stent Strut Coverage

Percutaneous coronary intervention is currently the standard treatment for symptomatic coronary artery disease [1], and drug-eluting stent (DES) has minimized the limitation of bare metal stent (BMS) [2, 3]. However, delayed stent strut coverage following DES implantation is considered as one of the crucial pathological mechanism of late stent thrombosis [4]. Neointimal formation is a common feature of healing response after stent implantation that occurs in vascular tissue [5]. It is well recognized that neointimal coverage after BMS implantation requires around 1 month and excessive neointimal formation results in in-stent restenosis (ISR) [6]. Several OCT studies showed that almost complete stent strut coverage was observed in BMS treated lesion in both the early and late periods [7–10]. DES remarkably reduced the rate of ISR and subsequent target lesion revascularization (TLR) by releasing drugs that can inhibit neointimal growth [2, 3]. However, excessive inhibition of neointimal formation and the vascular inflammatory response caused delayed vascular healing process with incomplete endothelialization, which has been associated with an increased risk of late stent thrombosis (LST) [11, 12]. Representative OCT images of late stent thrombosis within incomplete endothelialization of struts are shown in Fig. 17.1. There are many observational OCT studies about the rate of stent strut coverage of various DES and BMS in each period after stent implantation (Table 17.1) [7–10, 13–23]. Some studies evaluated neointimal coverage according to initial clinical presentation, such as acute coronary syndrome (ACS) and non-ACS, and suggested that strut coverage might be associated with initial clinical presentation (Table 17.2) [7, 9, 22, 24–27]

J.-H. Lee
Division of Cardiology, Yeungnam University Medical Center, Yeungnam University College of Medicine, Daegu, South Korea

Y. Jang • J.-S. Kim (✉)
Division of Cardiology, Severance Cardiovascular Hospital, Yonsei University College of Medicine, Seoul, South Korea
e-mail: kjs1218@yuhs.ac

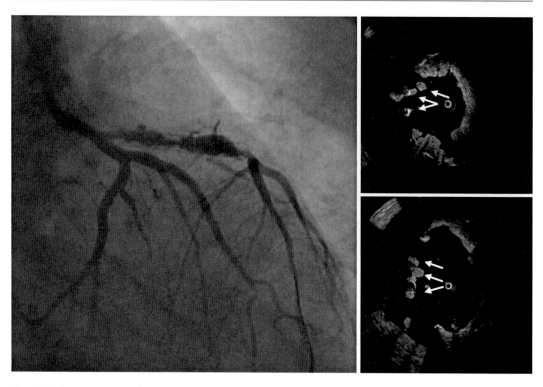

Fig. 17.1 Representative optical coherence tomography (OCT) images of late stent thrombosis within incomplete stent struts coverage

Table 17.1 Proportions of uncovered stent struts observed by optical coherence tomography

Stent type	Stent position	Stent duration			
		1 month	3 months	6–9 months	≥12 months
BMS	Single layered		0.1%	0.3–2.0%	0.3–1.1%
	Overlapped			3.4%	
SES	Single layered		13–18%	12.3%	3.2–11.6%
	Overlapped			9.6%	
PES	Single layered		3.8%	4.9%	0.9%
	Overlapped			16.5%	
ZES-P	Single layered		0.1%	0.02–1.2%	
	Overlapped			0.37%	
EES	Single layered	26.7%	4.7%	1.6–2.3%	1.9–5.8%
	Overlapped	51.6%			
	Side branch	89.4%		35.7%	
ZES-R	Single layered		6.2%	4.4%	
	Side branch			35.7%	
BES	Single layered		21.3%	15.9–21.8%	4.1%
	Side branch			35.7%	
BP-EES	Single layered		3%	1.8%	

BMS bare metal stent, *SES* sirolimus-eluting stent, *PES* paclitaxel-eluting stent, *ZES-P* sprint zotarolimus-eluting stent, *EES* everolimus-eluting stent, *ZES-R* resolute zotarolimus-eluting stent, *BES* biolimus-eluting stent, *OLP* overlapped, *BP* bioabsorbable
Other values are for single layered stents.
The values, which were derived from Bayesian hierarchical models, were excluded in this table.

The application of these findings to real clinical practice is the important task. A pathological study showed that uncovered stent strut after DES implantation was the best morphometric predictor of late stent thrombosis; the odds ratio for stent thrombosis in a stent with a ratio of uncovered to total stent struts per section >30% was 9.0 (95% confidence interval, 3.5–22) [28]. In a case-control study to evaluate uncovered stent strut on stent thrombosis with OCT, the length of an uncovered stent strut segment was one of the independent predictor of late stent thrombosis [29]. Another OCT study revealed that a greater percentage of uncovered struts (the cutoff value of ≥5.9% uncovered struts), as assessed by OCT at the 6–18-month follow-up, in asymptomatic DES-treated patients might predict increases in major adverse cardiac events which is very relevant to stent safety in the future [30]. Based on these studies, strut coverage assessed by OCT is an important marker for predicting serious adverse cardiovascular events in daily clinical practice. Typical representative examples of follow-up strut coverage by using cross-sectional OCT are shown in Fig. 17.2.

Table 17.2 Proportions of uncovered stent struts and malapposed struts according to initial clinical presentation

Clinical presentation	Authors	Stent type	Stent duration	Uncovered stent struts	Malapposed stent struts
ACS	Takano et al. [22]	SES	3 months	18%	8%
	Kim et al. [24]	ZES	3 months	0.1%	0.4%
	Guagliumi et al. [7]	BMS	6 months	1.98%	0.15%
	Guagliumi et al. [7]	ZES	6 months	0.00%	0.00%
	Davlouros et al. [25]	PES	6 months	8.6%	2.2%
	Kim et al. [26]	SES, PES, ZES	9 months	8.9%	2.2%
	Guagliumi et al. [9]	BMS	13 months	1.1%	0.1%
	Guagliumi et al. [9]	PES	13 months	5.7%	0.9%
	Räber et al. [27]	SES, PES	5 years	1.7%	0.5%
Non-ACS	Takano et al. [22]	SES	3 months	13%	5%
	Kim et al. [24]	ZES	3 months	0.1%	0.02%
	Kim et al. [26]	SES, PES, ZES	9 months	2.9%	0.5%
	Räber et al. [27]	SES, PES	5 years	0.7%	0.13%

ACS acute coronary syndrome, *BMS* bare metal stent, *SES* sirolimus-eluting stent, *PES* paclitaxel-eluting stent, *ZES* zotarolimus-eluting stent

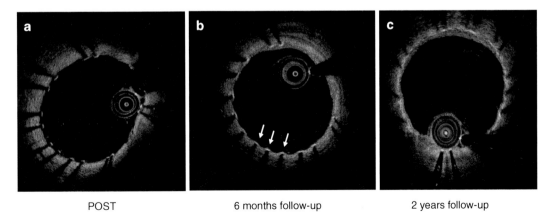

Fig. 17.2 Typical representative examples of follow-up strut coverage by using cross-sectional OCT. (**a**) OCT images of immediate after stent implantation revealed well-apposed stent struts, (**b**) OCT images of 6-month follow-up showed that uncovered stent strut at 6 o'clock (*white arrow*), (**c**) OCT images of 2-year follow-up showed well-covered strut

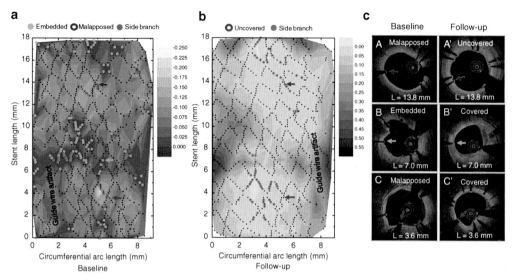

25. Kim JS, Ha J, Kim BK, Shin DH, Ko YG, Choi D, et al. The relationship between post-stent strut apposition and follow-up strut coverage assessed by a contour plot optical coherence tomography analysis. JACC Cardiovasc Interv. 2014;7(6):641-51.

Fig. 17.3 Representative contour plot images at post-stenting and 12-month follow-up. (**a**) Baseline plot of artery-strut spacing post-intervention. Malapposed and embedded struts are indicated with *red* and *green circles*, respectively. Grayscale indicates the artery-strut distance post-intervention (range 0.0–0.7 mm). (**b**) The neointimal coverage at follow-up as a function of circumferential arc length and stent length in a 3.0 × 18 mm biolimus-eluting stent; covered struts and struts crossing over the side branches are indicated with *blue* and *orange circles*, respectively. Grayscale indicates a stent strut coverage thickness range of −0.1 to 0.6 mm. (**c**) At a stent length of 13.6 mm from the distal stent margin, a malapposed strut at post-intervention turns into an uncovered strut at follow-up without malapposition (*red arrows* on contour plots and cross sections) in A and A'. At a stent length of 6.0 mm from the distal stent margin, an embedded strut becomes a covered strut (*green arrows*) in B and B'. At a stent length of 4.6 mm from the distal stent margin, a malapposed strut post-intervention becomes a covered strut without malapposition at follow-up (*blue arrows*) in C and C'. Adapted with permission from Kim et al. [32]

Serial analysis of the malapposed and uncovered struts at the strut level by current OCT analysis with conventional methods might be challenging during serial follow-up at different time points. Recent study suggested that a contour plot OCT analysis could be a possible method of assessing individual stent struts at the strut level practically; this comprehensive monitoring of stent strut status at different time points would then provide useful information regarding vascular healing status after DES implantation [31, 32]. Representative contour plot images at post-stenting and 12-month follow-up are shown in Fig. 17.3 [32].

17.2 Neointimal Characteristics

Pathological studies have demonstrated that neointima in stented coronary artery is consisted of various tissue components including proteoglycan, collagen, smooth muscle, fibrin, or thrombus [33, 34]. By using previous imaging modalities, such as conventional angiography or intravascular ultrasound, there are several limitations for detecting distinct neointimal characteristics due to their low resolution. However, intravascular OCT has higher resolution and is useful for both the qualitative as well as quantitative evaluation of neointimal tissue [35, 36]. The neointima within a stent

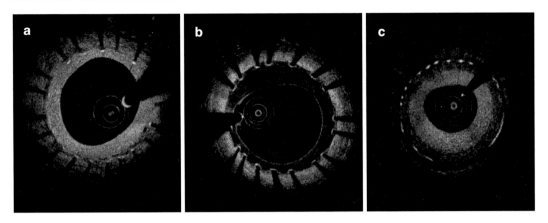

Fig. 17.4 Representative OCT images of neointimal tissue. (**a**) Homogeneous, (**b**) heterogeneous, (**c**) layered neointimal tissue

could be assessed qualitatively to characterize the neointimal tissue as (1) homogeneous neointima, a uniform signal-rich band without focal variation or attenuation; (2) heterogeneous neointima, focally changing optical properties and various backscattering patterns; and (3) layered neointima, layers with different optical properties, namely, an adluminal high scattering layer and an abluminal low scattering layer [35–37]. Pathological studies have reported differential morphological characteristics of neointimal tissue, which was well correlated with histological findings [37, 38]. Representative OCT images of neointimal tissue are shown in Fig. 17.4. Comparing different OCT morphological characteristics with different in-stent neointimal tissue types analyzed by histology with swine in-stent restenosis models, the optical characteristics of neointimal formation seen in OCT were consistent with the histological studies on stent healing [37]. Fibrous connective tissue deposition was more frequently present in the homogeneous pattern (71.6%, $P < 0.001$), whereas significant fibrin deposits were more commonly seen in the heterogeneous pattern (56.9%, $P = 0.007$). Peristrut inflammation was less frequently found in the homogeneous pattern (19.8%, $P < 0.001$) in comparison with the layered (73.9%) or heterogeneous patterns (43.1%). The presence of external elastic lamina (EEL) rupture was also more commonly seen in layered (73.9%) and heterogeneous (46.6%) patterns than in the homogeneous pattern (22.4%, $P < 0.001$) [37]. A recent histopathological OCT studies investigated 22 autopsy cases with a total of 36 lesions and 42 implanted stents (17 BMS, 11 first generation DES, and 14 second generation DES) [39] In this study, stented segments neointimal histologic characteristics revealed great variability of tissue components, which were not consistent with characteristics OCT features, except in the case of restenotic tissue (Fig. 17.5) [39]. This study suggested that it required more attention to interpret OCT imaging in non-restenotic tissues.

A recent study determined the detailed relationship between different OCT-based neointimal tissues regardless of neoatherosclerosis and clinical outcomes [40]. Heterogeneous neointima was frequently detected in 21.7% of DES-treated lesions and significantly associated with both old age and initial clinical presentation of ACS [40]. Major adverse cardiac events (MACE), a composite of cardiac death, nonfatal myocardial infarction, or target lesion revascularization, were more frequently in patients with heterogeneous neointima over a median 31-month follow-up period after OCT examination (13.7% vs. 2.9% in homogeneous vs. 7.3% in layered, $p = 0.001$) [40]. In that study, the heterogeneous neointima (hazard ratio: 3.925, 95%

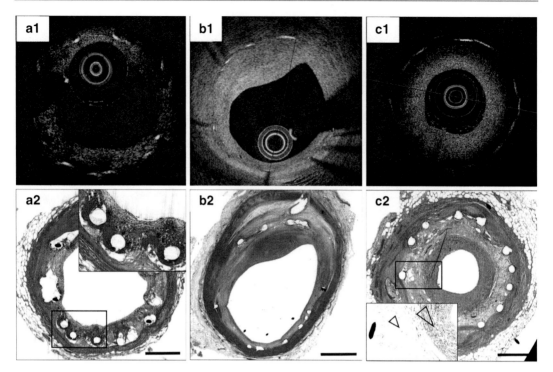

Fig. 17.5 Neointimal pattern and histologic findings. (**a**) Drug-eluting stent(s) (DES) (resolute) in the left anterior descending coronary artery, 238 days after implantation in the setting of stable coronary artery disease. (*A1*) Optical coherence tomographic (OCT) image shows heterogeneous backscattering. (*A2*) Corresponding histological cross section (hematoxylin and eosin [H&E]) shows an intense inflammatory reaction and focal fibrin deposits in the peri-strut regions. Higher magnification shows massive leukocyte infiltration and fibrin accumulation (scale bar = 1000 mm). (**b**) Bare metal stent(s) (BMS) (vision) 3 years after revascularization. (*B1*) Optical frequency domain image shows a homogeneous appearance. (*B2*) H&E-stained histopathological cross section showing smooth muscle cell-rich neointimal tissue coverage above all struts (scale bar = 1000 mm). (**c**) DES (endeavor) in the right coronary artery, 2 years after implantation in the setting of stable CAD. (*C1*) OCT image shows a layered pattern. (*C2*) Corresponding histological cross section, stained with hematoxylin and eosin, shows a layer of loose neointimal tissue with neovascularization and inflammation close to stent struts (*small arrowhead; black bar* represents strut) and a smooth muscle cell (SMC)-rich neointimal layer toward the lumen (*large arrowhead*) (scale bar = 1000 mm). Immunohistochemical staining (identification of SMCs by a-actin). Adapted with permission from Lutter et al. [39]

CI: 1.445–10.662, $P = 0.007$) was one of the independent risk factor for MACE. This data suggested that the neointimal tissue pattern in the heterogeneous pattern might be associated with future adverse clinical events in undergoing follow-up OCT surveillance after DES implantation. It means that OCT surveillance for neointimal characteristics might be useful in future clinical practice. A representative case of heterogeneous neointima who should be required repeat revascularization at 2-year follow-up is shown in Fig. 17.6.

17.3 Neoatherosclerosis

In-stent neoatherosclerosis, defined as an accumulation of lipid-laden foamy macrophage with or without necrotic core formation and/or calcification within the neointima, is an important mechanism of late DES failure [41, 42]. A pathological study showed that the incidence of neoatherosclerosis was higher in DES than BMS lesions (31% vs. 16%, $P < 0.001$) and the median stent duration with neoatherosclerosis was shorter in DES than BMS (420 days [IQR,

17 Late Stent Evaluation (Neoatherosclerosis)

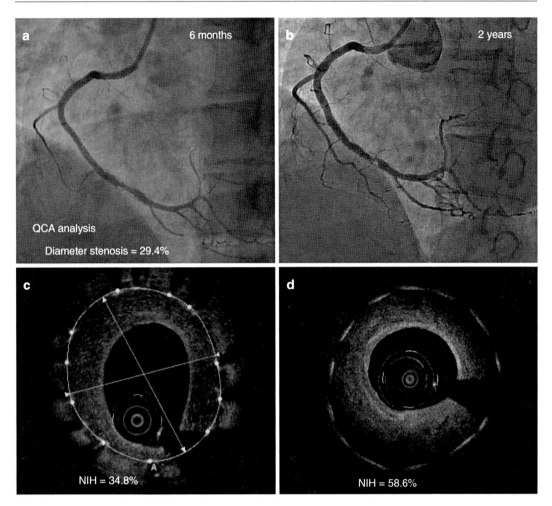

Fig. 17.6 A representative case of heterogeneous neointima who should be required repeat revascularization at 2-year follow-up. (**a**) 59-year-old female was performed follow-up coronary angiography without clinical symptoms at 6 months after sirolimus-eluting stent implantation at distal right coronary artery, and it revealed mild neointimal hyperplasia. (**b**) After 2 years, she was rehospitalized for ongoing chest pain, and follow-up angiography showed in-stent restenosis of previous drug-eluting stent. (**c**) OCT images of neointimal characteristics showed typical heterogeneous pattern at 6 months after sirolimus-eluting stent implantation. (**d**) At 2 years after stent implantation, OCT revealed more progressed neointimal hyperplasia and lipid laden neointima

361–683 days] vs. 2160 days [IQR, 1800–2880], $P < 0.001$) [41]. Representative OCT images of in-stent neoatherosclerosis are shown in Fig. 17.7. OCT study about neointimal characteristics investigated BMS segments according to stent follow duration: early phase (<6 months) and late phase (≥5 years) [43]. Lipid-laden neointima, intimal disruption, and thrombus were more frequently observed in late phase in comparison with the early phase [43]. Another OCT study compared neointimal characteristics within BMS between early ISR (≤1 year) and very late ISR (>5 years) and found that heterogeneous intima, similar to atherosclerotic plaque, was more frequently observed in very late ISR [44]. Several OCT surveillance studies have investigated the prevalence and characteristics of neoatherosclerosis in patients with in-stent restenosis. The

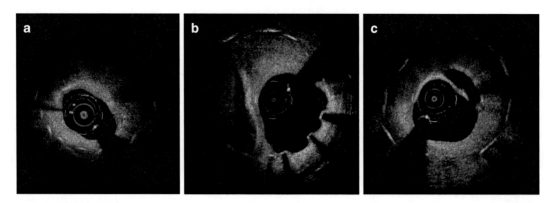

Fig. 17.7 Representative OCT images of in-stent neoatherosclerosis (**a**) Lipid laden neointima, (**b**) Calcified neointima, (**c**) Thin-cap fibroatheroma

Table 17.3 Frequency of in-stent neoatherosclerosis by optical coherence tomography

Authors	Definition of neoatherosclerosis	DES type	Stent duration	Frequency
Kim et al. [36]	Lipid-laden neointima	1st and 2nd generation	9 months	14.5%
Kim et al. [36]	Lipid-laden neointima	1st and 2nd generation	24 months	27.6%
Lee et al. [45]	Lipid-laden neointima, calcification, or TCFA	DES and BMS	70.7 months	35.5%
Yonetsu et al. [46]	Lipid-laden neointima	1st and 2nd generation	<9 months	37%
Lee et al. [47]	Lipid-laden neointima, calcification, or TCFA	1st and 2nd generation	12.4 vs. 55.4 months (2nd vs. 1st generation)	27.4%
Nakamura et al. [48]	Lipid-laden noeintima or calcified neointima	DES and BMS	57.9 months	49.2%

DES drug-eluting stent

reported prevalence of neoatherosclerosis in the OCT studies is summarized in Table 17.3 [36, 45–48]. An OCT study about in-stent neoatherosclerosis after DES found that 90% of lesion had lipid-containing neointima, 52% had TCFA-containing neointima, and 58% had at least one in-stent neointimal rupture during 32.2 months' follow-up time [49]. A retrospective OCT study about predictors for neoatherosclerosis which assessed 179 stents demonstrated that longer duration of implant (≥48 months), DES usage, current smoking, chronic kidney disease, and an absence of angiotensin-converting enzyme inhibitors or angiotensin II receptor blockade usage were independent determinants of OCT-detected in-stent neoatherosclerosis [50].

17.4 Summary

Stent evaluations using OCT have been well addressed in previous investigations and provide important information for clinical decisions at late stent evaluation. OCT-derived assessments of stent strut coverage, malapposition, or neointimal characteristics may also be very useful for predicting and preventing future adverse cardiac events. Moreover, advances in OCT technology and developments in its applications will likely provide further information on coronary stents and improve their practical usefulness in future clinical practice.

References

1. Serruys PW, de Jaegere P, Kiemeneij F, Macaya C, Rutsch W, Heyndrickx G, et al. A comparison of balloon-expandable-stent implantation with balloon angioplasty in patients with coronary artery disease: Benestent Study Group. N Engl J Med. 1994;331:489–95.
2. Moses JW, Leon MB, Popma JJ, Fitzgerald PJ, Holmes DR, O'Shaughnessy C, et al. Sirolimus-eluting stents versus standard stents in patients with stenosis in a native coronary artery. N Engl J Med. 2003;349(14):1315–23.
3. Stone GW, Ellis SG, Cox DA, Hermiller J, O'Shaughnessy C, Mann JT, et al. A polymer-based, paclitaxel-eluting stent in patients with coronary artery disease. N Engl J Med. 2004;350(3):221–31.
4. Farb A, Burke AP, Kolodgie FD, Virmani R. Pathological mechanisms of fatal late coronary stent thrombosis in humans. Circulation. 2003;108(14):1701–6.
5. Forrester JS, Fishbein M, Helfant R, Fagin J. A paradigm for restenosis based on cell biology: clues for the development of new preventive therapies. J Am Coll Cardiol. 1991;17(3):758–69.
6. Virmani R, Farb A. Pathology of in-stent restenosis. Curr Opin Lipidol. 1999;10(6):499–506.
7. Guagliumi G, Sirbu V, Bezerra H, Biondi-Zoccai G, Fiocca L, Musumeci G, et al. Strut coverage and vessel wall response to zotarolimus-eluting and bare-metal stents implanted in patients with ST-segment elevation myocardial infarction: the OCTAMI (optical coherence tomography in acute myocardial infarction) Study. JACC Cardiovasc Interv. 2010;3(6):680–7.
8. Guagliumi G, Musumeci G, Sirbu V, Bezerra HG, Suzuki N, Fiocca L, et al. Optical coherence tomography assessment of in vivo vascular response after implantation of overlapping bare-metal and drug-eluting stents. JACC Cardiovasc Interv. 2010;3(5):531–9.
9. Guagliumi G, Costa MA, Sirbu V, Musumeci G, Bezerra HG, Suzuki N, et al. Strut coverage and late malapposition with paclitaxel-eluting stents compared with bare metal stents in acute myocardial infarction: optical coherence tomography substudy of the harmonizing outcomes with revascularization and stents in acute myocardial infarction (HORIZONS-AMI) Trial. Circulation. 2011;123(3):274–81.
10. Xie Y, Takano M, Murakami D, Yamamoto M, Okamatsu K, Inami S, et al. Comparison of neointimal coverage by optical coherence tomography of a sirolimus-eluting stent versus a bare-metal stent three months after implantation. Am J Cardiol. 2008;102(1):27–31.
11. Daemen J, Wenaweser P, Tsuchida K, Abrecht L, Vaina S, Morger C, et al. Early and late coronary stent thrombosis of sirolimus-eluting and paclitaxel-eluting stents in routine clinical practice: data from a large two-institutional cohort study. Lancet. 2007;369(9562):667–78.
12. Jeremias A, Sylvia B, Bridges J, Kirtane AJ, Bigelow B, Pinto DS, et al. Stent thrombosis after successful sirolimus-eluting stent implantation. Circulation. 2004;109(16):1930–2.
13. de la Torre Hernández JM, Tejedor P, Camarero TG, Duran JM, Lee DH, Monedero J, et al. Early healing assessment with optical coherence tomography of everolimus-eluting stents with bioabsorbable polymer (synergy™) at 3 and 6 months after implantation. Catheter Cardiovasc Interv. 2016;88(3):E67–73.
14. Kim JS, Jang IK, Kim TH, Takano M, Kume T, Hur NW, et al. Optical coherence tomography evaluation of zotarolimus-eluting stents at 9-month follow-up: comparison with sirolimus-eluting stents. Heart. 2009;95(23):1907–12.
15. Kim JS, Kim TH, Fan C, Lee JM, Kim W, Ko YG, et al. Comparison of neointimal coverage of sirolimus-eluting stents and paclitaxel-eluting stents using optical coherence tomography at 9 months after implantation. Circ J. 2010;74(2):320–6.
16. Kim BK, Kim JS, Park J, Ko YG, Choi D, Jang Y, et al. Comparison of optical coherence tomographic assessment between first- and second-generation drug-eluting stents. Yonsei Med J. 2012;53(3):524–9.
17. Kim JS, Kim BK, Jang IK, Shin DH, Ko YG, Choi D, et al. comparison of neointimal coverage between zotarolimus-eluting stent and everolimus-eluting stent using optical coherence tomography (COVER OCT). Am Heart J. 2012;163(4):601–7.
18. Kim S, Kim JS, Shin DH, Kim BK, Ko YG, Choi D, et al. Comparison of early strut coverage between zotarolimus- and everolimus-eluting stents using optical coherence tomography. Am J Cardiol. 2013;111(1):1–5.
19. Kim BK, Hong MK, Shin DH, Kim JS, Ko YG, Choi D, et al. Optical coherence tomography analysis of strut coverage in biolimus- and sirolimus-eluting stents: 3-month and 12-month serial follow-up. Int J Cardiol. 2013;168(5):4617–23.
20. Kim BK, Ha J, Mintz GS, Kim JS, Shin DH, Ko YG, et al. Randomised comparison of strut coverage between Nobori biolimus-eluting and sirolimus-eluting stents: an optical coherence tomography analysis. EuroIntervention. 2014;9(12):1389–97.
21. Takahara M, Kitahara H, Nishi T, Miura K, Miyayama T, Sugimoto K, et al. Very early tissue coverage after drug-eluting stent implantation: an optical coherence tomography study. Int J Cardiovasc Imaging. 2016; Sep 6. [Epub ahead of print]
22. Takano M, Inami S, Jang IK, Yamamoto M, Murakami D, Seimiya K, et al. Evaluation by optical coherence

tomography of neointimal coverage of sirolimus-eluting stent three months after implantation. Am J Cardiol. 2007;99(8):1033–8.
23. Watanabe M, Uemura S, Kita Y, Sugawara Y, Goryo Y, Ueda T, et al. Impact of branching angle on neointimal coverage of drug-eluting stents implanted in bifurcation lesions. Coron Artery Dis. 2016;27(8):682–9.
24. Kim JS, Jang IK, Fan C, Kim TH, Kim JS, Park SM, et al. Evaluation in 3 months duration of neointimal coverage after zotarolimus-eluting stentimplantation by optical coherence tomography: the ENDEAVOR OCT trial. JACC Cardiovasc Interv. 2009;2:1240–7.
25. Davlouros PA, Nikokiris G, Karantalis V, Mavronasiou E, Xanthopoulou I, Damelou A, et al. Neointimal coverage and stent strut apposition six months after implantation of a paclitaxel eluting stent in acute coronary syndromes: an optical coherence tomography study. Int J Cardiol. 2011;151(2):155–9.
26. Kim JS, Fan C, Choi D, Jang IK, Lee JM, Kim TH, et al. Different patterns of neointimal coverage between acute coronary syndrome and stable angina after various types of drug-eluting stents implantation; 9-month follow-up optical coherence tomography study. Int J Cardiol. 2011;146(3):341–6.
27. Räber L, Zanchin T, Baumgartner S, Taniwaki M, Kalesan B, Moschovitis A, et al. Differential healing response attributed to culprit lesions of patients with acute coronary syndromes and stable coronary artery after implantation of drug-eluting stents: an optical coherence tomography study. Int J Cardiol. 2014;173(2):259–67.
28. Finn AV, Joner M, Nakazawa G, Kolodgie F, Newell J, John MC, et al. Pathological correlates of late drug-eluting stent thrombosis: strut coverage as a marker of endothelialization. Circulation. 2007;115(18):2435–41.
29. Guagliumi G, Sirbu V, Musumeci G, Gerber R, Biondi-Zoccai G, Ikejima H, et al. Examination of the in vivo mechanisms of late drug-eluting stent thrombosis: findings from optical coherence tomography and intravascular ultrasound imaging. JACC Cardiovasc Interv. 2012;5(1):12–20.
30. Won H, Shin DH, Kim BK, Mintz GS, Kim JS, Ko YG, et al. Optical coherence tomography derived cut-off value of uncovered stent struts to predict adverse clinical outcomes after drug-eluting stent implantation. Int J Cardiovasc Imaging. 2013;29(6):1255–63.
31. Ha J, Kim BK, Kim JS, Shin DH, Ko YG, Choi D, et al. Assessing neointimal coverage after DES implantation by 3D OCT. JACC Cardiovasc Imaging. 2012;5(8):852–3.
32. Kim JS, Ha J, Kim BK, Shin DH, Ko YG, Choi D, et al. The relationship between post-stent strut apposition and follow-up strut coverage assessed by a contour plot optical coherence tomography analysis. JACC Cardiovasc Interv. 2014;7(6):641–51.
33. Farb A, Kolodgie FD, Hwang JY, Burke AP, Tefera K, Weber DK, et al. Extracellular matrix changes in stented human coronary arteries. Circulation. 2004;110(8):940–7.
34. Nakano M, Vorpahl M, Otsuka F, Taniwaki M, Yazdani SK, Finn AV, et al. Ex vivo assessment of vascular response to coronary stents by optical frequency domain imaging. JACC Cardiovasc Imaging. 2012;5(1):71–82.
35. Gonzalo N, Serruys PW, Okamura T, van Beusekom HM, Garcia-Garcia HM, van Soest G, et al. Optical coherence tomography patterns of stent restenosis. Am Heart J. 2009;158(2):284–93.
36. Kim JS, Hong MK, Shin DH, Kim BK, Ko YG, Choi D, et al. Quantitative and qualitative changes in DES-related neointimal tissue based on serial OCT. JACC Cardiovasc Imaging. 2012;5(11):1147–55.
37. Kim JS, Afari ME, Ha J, Tellez A, Milewski K, Conditt G, et al. Neointimal patterns obtained by optical coherence tomography correlate with specific histological components and neointimal proliferation in a swine model of restenosis. Eur Heart J Cardiovasc Imaging. 2014;15(3):292–8.
38. Malle C, Tada T, Steigerwald K, Ughi GJ, Schuster T, Nakano M, et al. Tissue characterization after drug-eluting stent implantation using optical coherence tomography. Arterioscler Thromb Vasc Biol. 2013;33(6):1376–83.
39. Lutter C, Mori H, Yahagi K, Ladich E, Joner M, Kutys R, et al. Histopathological differential diagnosis of optical coherence tomographic image interpretation after stenting. JACC Cardiovasc Imaging. 2016;9:2511.
40. Kim JS, Lee JH, Shin DH, Kim BK, Ko YG, Choi D, et al. Long-term outcomes of neointimal hyperplasia without neoatherosclerosis after drug-eluting stent implantation. JACC Cardiovasc Imaging. 2014;7(8):788–95.
41. Nakazawa G, Otsuka F, Nakano M, Vorpahl M, Yazdani SK, Ladich E, et al. The pathology of neoatherosclerosis in human coronary implants: bare-metal and drug-eluting stents. J Am Coll Cardiol. 2011;57:1314–22.
42. Otsuka F, Byrne RA, Yahagi K, Mori H, Ladich E, Fowler DR, et al. Neoatherosclerosis: overview of histopathologic findings and implications for intravascular imaging assessment. Eur Heart J. 2015;36:2147–59.
43. Takano M, Yamamoto M, Inami S, Murakami D, Ohba T, Seino Y, et al. Appearance of lipid-laden intima and neovascularization after implantation of bare-metal stents extended late-phase observation by intracoronary optical coherence tomography. J Am Coll Cardiol. 2009;55:26–32.
44. Habara M, Terashima M, Nasu K, Kaneda H, Inoue K, Ito T, et al. Difference of tissue characteristics between early and very late restenosis lesions after bare-metal stent implantation: an optical coherence tomography study. Circ Cardiovasc Interv. 2011;4:232–8.
45. Lee SY, Shin DH, Mintz GS, Kim JS, Kim BK, Ko YG, et al. Optical coherence tomography-based eval-

uation of in-stent neoatherosclerosis in lesions with more than 50% neointimal cross-sectional area stenosis. EuroIntervention. 2013;9:945–51.
46. Yonetsu T, Kim JS, Kato K, et al. Comparison of incidence and time course of neoatherosclerosis between bare metal stents and drug-eluting stents using optical coherence tomography. Am J Cardiol. 2012;110:933–9.
47. Lee SY, Hur SH, Lee SG, Kim SW, Shin DH, Kim JS, et al. Optical coherence tomographic observation of in-stent neoatherosclerosis in lesions with more than 50% neointimal area stenosis after second-generation drug-eluting stent implantation. Circ Cardiovasc Interv. 2015;8:e001878.
48. Nakamura D, Attizzani GF, Toma C, Sheth T, Wang W, Soud M, et al. Failure mechanisms and neoatherosclerosis patterns in very late drug-eluting and bare-metal stent thrombosis. Circ Cardiovasc Interv. 2016;9:e003785.
49. Kang SJ, Mintz GS, Akasaka T, Park DW, Lee JY, Kim WJ, et al. Optical coherence tomographic analysis of in-stent neoatherosclerosis after drug-eluting stent implantation. Circulation. 2011;123:2954–63.
50. Yonetsu T, Kato K, Kim SJ, Xing L, Jia H, McNulty I, et al. Predictors for neoatherosclerosis: a retrospective observational study from the optical coherence tomography registry. Circ Cardiovasc Imaging. 2012;5:660–6.

Bioresorbable Vascular Scaffold Evaluation by Optical Coherence Tomography

18

Soo-Joong Kim

18.1 Introduction

Bioresorbable vascular scaffold (BVS) has been introduced as the latest revolution in the field of percutaneous coronary intervention (PCI), which could overcome the long-term limitations of the permanent stent implantation [1]. This device is designed to provide the temporary scaffolding of the vessel before being resorbed completely within the vessel, leaving nothing behind. It makes BVS offer a potential solution to the weakness of drug-eluting stents, which include endothelial dysfunction and hypersensitivity reactions, leading to late stent failure, and disturbance of future surgical revascularization at the same lesion [2–4].

Optical coherence tomography (OCT) has been introduced for in vivo vascular study as a high-resolution imaging modality [5–7]. It allows good visualization of the surface of vessel lumen and fine structures including stent struts, with a tenfold higher axial resolution (10–15 μm) compared to IVUS. Therefore, it is widely accepted that OCT be an in vivo "gold standard" imaging modality for the detection of stent malapposition, dissections, tissue protrusion, and thrombus, which could be very useful in guiding BVS implantation. In this chapter, feasibility and advantages of OCT application for BVS evaluation will be discussed.

18.2 OCT as Intravascular Imaging Tool in BVS

Although OCT has an excellent resolution, it is of intrinsic limitations to metal stent which is powerful light reflectors and induces posterior shadowing and blooming artifacts on the surface and edges (Fig. 18.1). However, in case of BVS, polymeric struts are transparent to the light so that BVS permits the assessment of the vessel wall behind the scaffold without any shadowing of metallic struts (Fig. 18.2) [8]. Scaffold integrity, apposition to the underlying vessel wall, luminal dimensions, and presence of thrombus or tissue prolapse on the scaffold surface can be evaluated by OCT immediately after BVS implantation (Fig. 18.3). Also it is possible to assess the tissue coverage of scaffold, the changes in the scaffold properties with resorption, and the response of the vessel wall over time [9, 10]. Thanks to the high resolution of OCT and optically transparent characteristics of BVS, OCT has been used for BVS implantation in most of the available studies, which

S.-J. Kim
Division of Cardiology, Department of Internal Medicine, Kyung Hee University Hospital, Kyung Hee University College of Medicine,
Seoul, South Korea
e-mail: soojoong@dreamwiz.com

Fig. 18.1 Representative optical coherence tomography (OCT) image of metal stent. OCT has a limitation to be unable to show the vessel behind the metal stent which is powerful light reflector and induces posterior shadowing and blooming artifact on the surface and edges

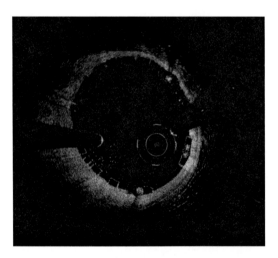

Fig. 18.2 Representative optical coherence tomography (OCT) image of bioresorbable vascular scaffold. With the property of polymeric struts transparent to the light, OCT can evaluate the vessel wall behind the scaffold without any shadowing of metallic struts

contributed to figure out the findings of BVS and interaction of vessel wall [8, 11]. OCT is now considered the gold standard for the evaluation of immediate and follow-up results of BVS implantation.

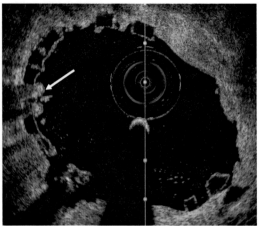

Fig. 18.3 Tissue prolapse on the scaffold surface after bioresorbable vascular scaffold (BVS) implantation. Optical coherence tomography can demonstrate tissue prolapse (*white arrow*) on the scaffold surface immediately after BVS implantation

18.2.1 OCT for Evaluation of Healing Process with Resorption

Stent deployment in coronary artery produces a series of physiological responses, which sequentially lead to platelet and fibrin deposition, inflammatory cell recruitment, smooth muscle cell hyperplasia, deposition of cellular matrix, and re-endothelialization in the segment treated by stent [12]. Unfortunately, the persistence of metal and/or durable polymers in the vessel induces chronic inflammation and hypersensitivity reaction, which can cause complications including neoatherosclerosis and late or very late stent thrombosis [13–17]. BVS can offer potential benefits over metallic stents for these problems with the process of "bioresorption" of scaffold. Intracoronary imaging techniques such as intravascular ultrasound (IVUS) and virtual histology intravascular ultrasound (VH-IVUS) have been used to analyze the process of bioresorption of BVS [10]. Polymeric strut is recognized as hyperechogenic tissue in IVUS and as areas of apparent dense calcium surrounded by necrotic core due to the strong backscattering properties of the polymer in VH-IVUS, and resorption process can be assessed by the reduction in the percentage hyperechogenicity and by

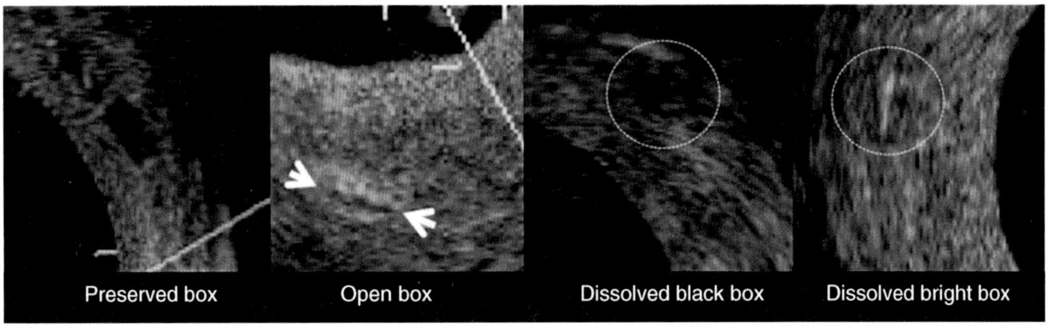

Fig. 18.4 Classification of scaffold appearances assessed with optical coherence tomography in porcine coronary arteries. *Preserved box* is defined as a box appearance with sharply defined borders with bright reflection; strut body shows low reflection. *Open box* is characterized by luminal and abluminal "long-axis" borders thickened with bright reflection and short-axis borders that are no longer visible. *Black* and *bright* dissolved boxes are defined as *black* spot with poorly defined contours, often confluent but with no box-shaped appearance and partially visible *bright spot* with poorly defined contours and no box-shaped appearance, respectively

Table 18.1 Proportion and sequential changes of optical coherence tomography findings

Strut appearance %	Immediately after implantation	At 28 days	At 2 years	At 3 years	At 4 years
Preserved box	100	82	80.4	5.4	0
Open box	0	18	2.4	16.1	0
Dissolved bright box	0	0	0	34.8	51.2
Dissolved black box	0	0	17.2	43.7	48.8

change in quantitative analyses of these areas, respectively.

OCT also provided crucial information for the BVS resorption process. Thorax center investigators have proposed the terminology to describe OCT findings associated with various stages of BVS strut resorption in the vessel wall (Fig. 18.4) [18]. An intact scaffold strut footprint is denominated as a "preserved box," which is defined as a box appearance with sharply defined borders with bright reflection, and the strut body shows low reflection. The first OCT changes in the strut footprint are named as "open box" which is characterized by luminal and abluminal "long-axis" borders thickened with bright reflection and short-axis borders that are no longer visible at follow-up. The last change on OCT in the process of resorption is "black" and "bright" "dissolved boxes," which are defined as black spot with poorly defined contours, often confluent but with no box-shaped appearance and partially visible bright spot with poorly defined contours and no box-shaped appearance, respectively [18]. This serial change of OCT findings reflecting resorption process of BVS was firstly evaluated with histology in porcine coronary artery model [19]. In this study, BVS was serially assessed immediately, at 1 month and 2, 3, and 4 years after implantation. The proportion and sequential changes of OCT findings over time are summarized in Table 18.1. Immediately after implantation, all struts had a preserved box appearance. However, the proportion of box appearance decreased over time and only dissolved boxes were seen at 4 years. The preserved box in OCT corresponded well (86.4%) with 2-year histology in which the struts were first covered by a thin, fibromuscular neointima and then replaced by proteoglycan-rich matrix gradually over time, whereas the dissolved bright and black boxes corresponded well (88.0 and 90.7%, respectively) to 3-year histology showing inspissations of the provisional matrix and connective tissue infiltration in the region of the pre-existing struts. Struts indiscernible by OCT corresponded to the integrated strut footprints seen at 4 years (100%) [19].

OCT also demonstrated that BVS implantation led to the formation of a symmetrical neointima with a mean thickness of 220 µm during 6–12 months [18], which nearly completed the healing process without further increase of neointima over time [20, 21]. This formation of a circumferential neointimal layer, with resorption of polymeric struts, creates a "de novo" cap, which may help to seal a thin-cap fibroatheroma [20].

18.2.2 OCT for Evaluation of Strut Coverage and Malapposition

OCT is the gold standard for the evaluation of metallic stent strut tissue coverage with its high resolution [22, 23]. It is important to assess the tissue coverage of strut after stent implantation because this coverage is generally considered a marker of endothelialization [24]. BVS has translucent polymeric struts which enable OCT to image the abluminal surface of scaffold. Gutiérrez-Chico et al. demonstrated that most of the malapposed and side-branch struts were covered by neointimal tissues on both the abluminal and adluminal side 6 months after BVS implantation, with thicker neointimal coverage on the abluminal side (101 vs. 71 µm; 95% confidence interval [CI] of the difference: 20–40 µm) (Fig. 18.5) [8]. This OCT finding for BVS strut coverage may provide the understanding of the mechanism by which acute stent malapposition could be spontaneously corrected over time. Long-term follow-up data of BVS showed that all incomplete appositions (incomplete, persistent, and late-acquired incomplete stent apposition) were resolved over 2 years [10]. ABSORB JAPAN trial using OCT also demonstrated that the incidence of malapposed struts decreased from 4.9% immediately after BVS implantation to 0.12% at 2-year follow-up with 0.6% of uncovered struts [25].

OCT was able to reveal the advantage of BVS for early vascular healing with optimal strut coverage [26]. In this study, overall 99% of BVS struts were covered at mean 47.6 ± 6.3 days, in the setting of acute coronary syndrome and stable angina. ABSORB-STEMI TROFI II, which enrolled the ST-segment elevation myocardial infarction patients undergoing primary PCI with BVS or everolimus-eluting metal stent (EES), evaluated the 6-month OCT healing score (HS) based on the presence of uncovered and/or malapposed stent struts and intraluminal filling defects. BVS showed a nearly complete arterial healing with lower HS when compared with EES arm [1.74 (2.39) vs. 2.80 (4.44); difference (90% CI) −1.06 (−1.96, −0.16); $P_{\text{non-inferiority}} < 0.001$] [27]. Accordingly, OCT is considered the gold standard for strut coverage evaluation of BVS at follow-up. Indeed, OCT made it possible to clearly identify the fibrotic de novo cap (a neointimal layer covering the scaffold struts) manifested by signal-rich low-attenuating tissue layer, even when struts are no more identifiable (Fig. 18.6) [11].

18.2.3 OCT for Evaluation of BVS Optimization and Late Lumen Gain

BVS has the potential for greater scaffold underexpansion and malapposition due to its intrinsic differences in recoil characteristics and its less distensibility as compared with metallic stents [28]. Therefore, it is very important during BVS implantation to get the accurate measurement of vascular lumen, to select the appropriate size and length of BVS, and to achieve optimal apposition after deployment. OCT can allow more accurate detection of luminal border at both lesion and reference segments, which enables to select the optimal size of BVS, and quantification of scaffold malapposition and underexpansion with its high resolution, as compared with conventional intravascular imaging modalities (Fig. 18.7) [29]. However, the limitation of clinical data

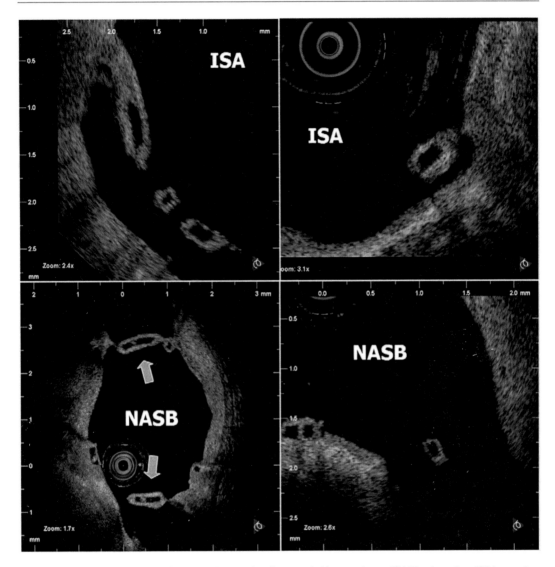

Fig. 18.5 Neointimal coverage of strut. Malapposed and side-branch struts were covered by neointimal tissues on both the abluminal and adluminal side 6 months after bioresorbable vascular scaffold implantation. *ISA* incomplete strut apposition, *NASB* non-apposed side branch

and the lack of standardized criteria for OCT measurements are still problematic in its clinical use, although a comprehensive consensus document has been issued from international working group for OCT standardization and validation [24]. Also several studies demonstrated that lumen dimensions measured by OCT were smaller than those measured by IVUS [30, 31]. Despite these limitations, OCT is now considered a useful intravascular imaging modality for the evaluation of BVS, thanks to the characteristics of BVS to allow the assessment of the vessel wall behind the struts without any metal shadowing [8]. Recently, one study revealed that further optimization after BVS implantation was required in over a quarter of lesions on the basis of OCT findings, despite angiographic success [32].

Another important advantage of OCT in the evaluation of BVS treatment is that it can prove

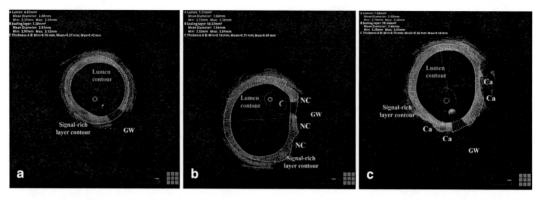

Fig. 18.6 Quantification of fibrotic de novo cap by optical coherence tomography (OCT) after bioresorbable vascular scaffold resorption. OCT can detect a signal-rich layer, which consisted of the neointimal layer, resorbed struts, and preexisting fibrous tissue, even when struts are no more identifiable. In the absence of attenuating intimal regions, the contour is traced at the internal elastic lamina (**a**). In plaques with necrotic core, the abluminal contour is traced at the attenuating region boundary (**b**). In plaques with calcifications, the signal-rich layer is segmented at the calcification edge (**c**). *Ca* calcium, *GW* guidewire, *NC* necrotic core

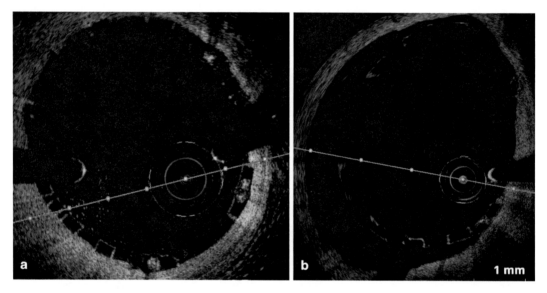

Fig. 18.7 Apposition of bioresorbable vascular scaffold (BVS) detected by optical coherence tomography (OCT). After BVS implantation, OCT can evaluate the status of apposition. (**a**) represents the well-apposed strut. (**b**) represents malapposition, which is defined as a discontinuity between the backscattering frame of the translucent strut and the vessel wall, appearing as a contrast-filled gap between these two structures

the potential benefit of BVS to get late lumen gain [11], which is a common phenomenon between 6 months and up to 5 years after successful balloon angioplasty with myointimal regression at the lesion site. Two-year follow-up data of ABSORB with OCT demonstrated that there was an increase in minimal and mean luminal area with a significant decrease in plaque volume without change in vessel size between 6 months and 2 years [10]. On the other hand, there was a decrease in lumen area between the immediate post-procedural and

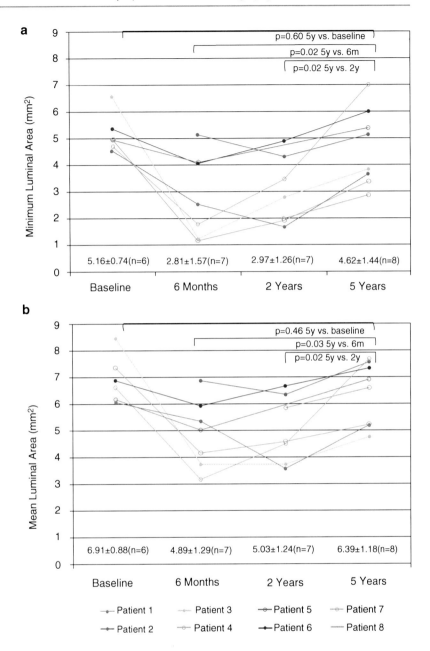

Fig. 18.8 Serial luminal measurements after bioresorbable vascular scaffold implantation. Despite an initial decrease in lumen dimensions from baseline to 6 months, minimal (**a**) and mean (**b**) lumen areas were increased at 5 years compared with previous follow-ups and were not significantly different from baseline

6-month follow-up measurements. There was no significant vascular remodeling over 2 years. Long-term follow-up study using OCT showed that both minimum and mean luminal area increased from 2 to 5 years (Figs. 18.8 and 18.9) [11]. Therefore, OCT can provide the information for the late lumen enlargement and vascular remodeling during follow-up after BVS implantation, although late lumen enlargement is a phenomenon that needs to be confirmed.

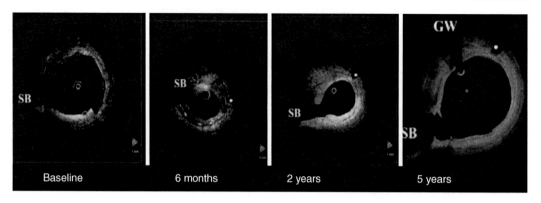

Fig. 18.9 Representative case of luminal enlargement over 5 years. Optical coherence tomography can evaluate the serial changes in luminal dimension, which initially decreased during 6 months after bioresorbable vascular scaffold implantation and then increased over 5 years continuously. *GW* guidewire, *SB* side branch

18.3 OCT for Guidance of BVS Implantation

Adequate lesion preparation and post-implantation optimization should be done in BVS implantation to improve the efficacy and to reduce the risk of post-procedural complications including scaffold thrombosis. For this reason, intravascular imaging modalities, especially OCT, have been recommended to guide the BVS implantation. OCT can provide the informations about plaque characteristics which are very important for lesion preparation, lumen dimensions, the size and length of BVS, and scaffold expansion after deployment. It also allows reliable qualitative and quantitative scaffold analysis and follow-up evaluations. Brown et al. recently reported excellent acute results after OCT-guided BVS implantation, in which 1:1 balloon/vessel predilation improves scaffold expansion (82.81% of predicted scaffold cross-sectional area in 1:1 predilation vs. 78.65% in no 1:1 predilation, $P < 0.0001$) [33]. OCT could help the selection of optimal BVS with accurate measurement of lumen dimensions [34]. Due to higher strut thickness and conformability of BVS, the need for postdilation to achieve better scaffold expansion is increasing, especially in complex lesions. In fact, the final minimal scaffold area (MLA) is a strong predictor of restenosis and stent thrombosis [35, 36]. OCT can accurately detect the luminal and scaffold border and easily measure the minimal diameter or area, which make postdilation with the detection of underexpansion of scaffold done. Consistently, the studies reporting high postdilation rates (over 90%) with OCT have shown similar post-procedural results of area stenosis and minimal lumen area and target lesion revascularization (3.3% vs. 5.4%, $P = 0.41$) and major adverse cardiac events (3.3% vs. 7.6%, $P = 0.19$) with lower rates of stent thrombosis at 6-month follow-up, as compared with second-generation drug-eluting stent [37, 38]. This suggests that improvements in the implantation technique with OCT could favorably affect the BVS performance, especially in special conditions such as complex lesions, bifurcations, and CTOs [39–41].

Another advantage of OCT in BVS guidance is the ability to detect incomplete stent apposition, tissue or plaque protrusions, dissections, thrombus, and number of struts. On the basis of OCT evaluation in BVS implantation, further optimization was done in over a quarter of lesions in spite of successful angiographic findings [32].

In summary, OCT can guide BVS deployment in three steps: (1) preintervention step, when OCT provides the informations for lesion characteristics and lumen dimensions at both lesion and reference segments and guides the treatment options for lesion preparation including cutting balloon, noncompliant balloon, or rotational atherectomy with plaque characterization; (2) prestenting step, when OCT can check the lesion preparation and guide to select the most appropriate length and size of BVS; and (3) poststenting step, when OCT checks the need of postdilation with usually noncompliant balloon, assesses lesion coverage and scaffold apposition, and finally guides additional postdilation in case of underexpansion or malapposition of BVS.

18.4 Summary

With the advent of BVS, intravascular imaging modality should be used for the optimization of procedure and the best outcomes. OCT is both safe and feasible for guidance of BVS coronary intervention. It provides many informations about luminal dimensions, plaque characteristics, treatment choice for lesion preparation, and the size and length of BVS before implantation. Also it is of incomparable value for evaluating optimal BVS deployment with scaffold analysis and guiding further optimization with postdilation. Clear visualization of border between vascular lumen and wall with high resolution and optically transparent scaffold enables OCT to measure lumen dimensions accurately and to identify malapposed struts. Taking it into account that both stent underexpansion and strut malapposition may contribute to the development of scaffold thrombosis and restenosis, especially in ACS, the remedy of these conditions with OCT is essential for BVS intervention in coronary artery disease. In addition, operators might be encouraged to use OCT guidance in challenging and complex lesions treated with BVS to get both high procedural success rates and favorable clinical outcomes.

References

1. Lu C, Filion KB, Eisenberg MJ. The safety and efficacy of absorb bioresorbable vascular scaffold: a systematic review. Clin Cardiol. 2016;39:48–55.
2. Hofma SH, van der Giessen WJ, van Dalen BM, Lemos PA, McFadden EP, Sianos G, et al. Indication of long-term endothelial dysfunction after sirolimus-eluting stent implantation. Eur Heart J. 2006;27:166–70.
3. Kay IP, Wardeh AJ, Kozuma K, Foley DP, Knook AH, Thury A, et al. Radioactive stents delay but do not prevent in-stent neointimal hyperplasia. Circulation. 2001;103:14–7.
4. McFadden EP, Stabile E, Regar E, Cheneau E, Ong AT, Kinnaird T, et al. Late thrombosis in drug-eluting coronary stents after discontinuation of antiplatelet therapy. Lancet. 2004;364:1519–21.
5. Fujimoto JG, Boppart SA, Tearney GJ, Bouma BE, Pitris C, Brezinski ME. High resolution in vivo intra-arterial imaging with optical coherence tomography. Heart. 1999;82:128–33.
6. Jang IK, Bouma BE, Kang DH, Park SJ, Park SW, Seung KB, et al. Visualization of coronary atherosclerotic plaques in patients using optical coherence tomography: comparison with intravascular ultrasound. J Am Coll Cardiol. 2002;39:604–9.
7. Yabushita H, Bouma BE, Houser SL, Aretz HT, Jang IK, Schlendorf KH, et al. Characterization of human atherosclerosis by optical coherence tomography. Circulation. 2002;106:1640–5.
8. Gutiérrez-Chico JL, Gijsen F, Regar E, Wentzel J, de Bruyne B, Thuesen L, et al. Differences in neointimal thickness between the adluminal and the abluminal sides of malapposed and side-branch struts in a polylactide bioresorbable scaffold: evidence in vivo about the abluminal healing process. JACC Cardiovasc Interv. 2012;5:428–35.
9. Farooq V, Serruys PW, Heo JH, Gogas BD, Onuma Y, Perkins LE, et al. Intracoronary optical coherence tomography and histology of overlapping everolimus eluting bioresorbable vascular scaffolds in a porcine coronary artery model: the potential implications for clinical practice. JACC Cardiovasc Interv. 2013;6:523–32.
10. Serruys PW, Ormiston JA, Onuma Y, Regar E, Gonzalo N, Garcia-Garcia HM, et al. A bioabsorbable everolimus-eluting coronary stent system (ABSORB): 2-year outcomes and results from multiple imaging methods. Lancet. 2009;373:897–910.
11. Karanasos A, Simsek C, Gnanadesigan M, van Ditzhuijzen NS, Freire R, Dijkstra J, et al. OCT assessment of the long-term vascular healing response 5 years after everolimus-eluting bioresorbable vascular scaffold. J Am Coll Cardiol. 2014;64:2343–56.
12. Oberhauser JP, Hossainy S, Rapoza RJ. Design principles and performance of bioresorbable polymeric vascular scaffolds. EuroIntervention. 2009;5:F15–22.
13. Otsuka F, Byrne RA, Yahagi K, Mori H, Ladich E, Fowler DR, et al. Neoatherosclerosis: overview of histopathologic findings and implications for intravascular imaging assessment. Eur Heart J. 2015;36:2147–59.
14. Palmerini T, Biondi-Zoccai G, Della Riva D, Mariani A, Sabaté M, Smits PC, et al. Clinical outcomes with bioabsorbable polymer- versus durable polymer-based drug-eluting and bare-metal stents: evidence from a comprehensive network meta-analysis. J Am Coll Cardiol. 2014;63:299–307.
15. Serruys PW, Farooq V, Kalesan B, de Vries T, Buszman P, Linke A, et al. Improved safety and reduction in stent thrombosis associated with biodegradable

polymer-based biolimus-eluting stents versus durable polymer-based sirolimus eluting stents in patients with coronary artery disease: final 5-year report of the LEADERS (Limus Eluted from a Durable versus Erodable Stent coating) randomized, noninferiority trial. JACC Cardiovasc Interv. 2013;6:777–89.
16. Stefanini GG, Byrne RA, Serruys PW, de Waha A, Meier B, Massberg S, et al. Biodegradable polymer drug-eluting stents reduce the risk of stent thrombosis at 4 years in patients undergoing percutaneous coronary intervention: a pooled analysis of individual patient data from the ISAR-TEST 3, ISAR-TEST 4, and LEADERS randomized trials. Eur Heart J. 2012;33:1214–22.
17. Yoneda S, Abe S, Kanaya T, Oda K, Nishino S, Kageyama M, et al. Late-phase inflammatory response as a feature of in-stent restenosis after drug-eluting stent implantation. Coron Artery Dis. 2013;24:368–73.
18. Ormiston JA, Serruys PW, Regar E, Dudek D, Thuesen L, Webster MW, et al. A bioabsorbable everolimus-eluting coronary stent system for patients with single de-novo coronary artery lesions (ABSORB): a prospective open-label trial. Lancet. 2008;371:899–907.
19. Onuma Y, Serruys PW, Perkins LE, Okamura T, Gonzalo N, García-García HM, et al. Intracoronary optical coherence tomography and histology at 1 month and 2, 3, and 4 years after implantation of everolimus-eluting bioresorbable vascular scaffolds in a porcine coronary artery model: an attempt to decipher the human optical coherence tomography images in the ABSORB trial. Circulation. 2010;122:2288–300.
20. Brugaletta S, Radu MD, Garcia-Garcia HM, Heo JH, Farooq V, Girasis C, et al. Circumferential evaluation of the neointima by optical coherence tomography after ABSORB bioresorbable vascular scaffold implantation: can the scaffold cap the plaque? Atherosclerosis. 2012;221:106–12.
21. Otsuka F, Pacheco E, Perkins LE, Lane JP, Wang Q, Kamberi M, et al. Long-term safety of an everolimus-eluting bioresorbable vascular scaffold and the cobalt–chromium XIENCE V stent in a porcine coronary artery model. Circ Cardiovasc Interv. 2014;7:330–42.
22. Finn AV, Joner M, Nakazawa G, Kolodgie F, Newell J, John MC, et al. Pathological correlates of late drug eluting stent thrombosis: strut coverage as a marker of endothelialization. Circulation. 2007;115:2435–41.
23. Maehara A, Mintz GS, Weissman NJ. Advances in intravascular imaging. Circ Cardiovasc Interv. 2009;2:482–90.
24. Tearney GJ, Regar E, Akasaka T, Adriaenssens T, Barlis P, Bezerra HG, et al. Consensus standards for acquisition, measurement, and reporting of intravascular optical coherence tomography studies: a report from the international working Group for Intravascular Optical Coherence Tomography Standardization and Validation. J Am Coll Cardiol. 2012;59:1058–72.
25. Onuma Y, Sotomi Y, Shiomi H, Ozaki Y, Namiki A, Yasuda S, et al. Two-year clinical, angiographic, and serial optical coherence tomographic follow-up after implantation of an everolimus-eluting bioresorbable scaffold and an everolimus-eluting metallic stent: insights from the randomised ABSORB Japan trial. EuroIntervention. 2016;12:1090–101.
26. Baquet M, Brenner C, Wenzler M, Eickhoff M, David J, Brunner S, et al. Impact of clinical presentation on early vascular healing after bioresorbable vascular scaffold implantation. J Interv Cardiol. 2016; doi:10.1111/joic.12359.
27. Sabaté M, Windecker S, Iñiguez A, Okkels-Jensen L, Cequier A, Brugaletta S, et al. Everolimus-eluting bioresorbable stent vs. durable polymer everolimus-eluting metallic stent in patients with ST-segment elevation myocardial infarction: results of the randomized ABSORB ST-segment elevation myocardial infarction-TROFI II trial. Eur Heart J. 2016;37:229–40.
28. Tanimoto S, Serruys PW, Thuesen L, Dudek D, de Bruyne B, Chevalier B, et al. Comparison of in vivo acute stent recoil between the bioabsorbable everolimus-eluting coronary stent and the everolimus-eluting cobalt chromium coronary stent: insights from the ABSORB and SPIRIT trials. Catheter Cardiovasc Interv. 2007;70:515–23.
29. Takarada S, Imanishi T, Liu Y, Ikejima H, Tsujioka H, Kuroi A, et al. Advantage of next-generation frequency-domain optical coherence tomography compared with conventional time-domain system in the assessment of coronary lesion. Cardiovasc Interv. 2010;75:202–6.
30. Gonzalo N, Serruys PW, García-García HM, van Soest G, Okamura T, Ligthart J, et al. Quantitative ex vivo and in vivo comparison of lumen dimensions measured by optical coherence tomography and intravascular ultrasound in human coronary arteries. Rev Esp Cardiol. 2009;62:615–24.
31. Yamaguchi T, Terashima M, Akasaka T, Hayashi T, Mizuno K, Muramatsu T, et al. Safety and feasibility of an intravascular optical coherence tomography image wire system in the clinical setting. Am J Cardiol. 2008;101:562–7.
32. Allahwala UK, Cockburn JA, Shaw E, Figtree GA, Hansen PS, Bhindi R. Clinical utility of optical coherence tomography (OCT) in the optimisation of absorb bioresorbable vascular scaffold deployment during percutaneous coronary intervention. EuroIntervention. 2015;10:1154–9.
33. Brown AJ, LM MC, Braganza DM, Bennett MR, Hoole SP, West NE. Expansion and malapposition characteristics after bioresorbable vascular scaffold implantation. Catheter Cardiovasc Interv. 2014;84:37–45.
34. Foin N, Alegria E, Sen S, Petraco R, Nijjer S, Di Mario C, et al. Importance of knowing stent design threshold diameters and post-dilatation capacities to optimize stent selection and prevent stent overexpansion/incomplete apposition during PCI. Int J Cardiol. 2013;166:755–8.
35. Doi H, Maehara A, Mintz GS, Yu A, Wang H, Mandinov L, et al. Impact of post-intervention minimal stent area on 9-month follow-up patency of

paclitaxel-eluting stents: an integrated intravascular ultrasound analysis from the TAXUS IV, V, and VI and TAXUS ATLAS workhorse, long lesion, and direct stent trials. JACC Cardiovasc Interv. 2009;2:1269–75.
36. Fujii K, Carlier SG, Mintz GS, Yang YM, Moussa I, Weisz G, et al. Stent underexpansion and residual reference segment stenosis are related to stent thrombosis after sirolimus-eluting stent implantation: an intravascular ultrasound study. J Am Coll Cardiol. 2005;45:995–8.
37. Costopoulos C, Latib A, Naganuma T, Miyazaki T, Sato K, Figini F, et al. Comparison of early clinical outcomes between absorb bioresorbable vascular scaffold and everolimus-eluting stent implantation in a real-world population. Catheter Cardiovasc Interv. 2015;85:E10–5.
38. Mattesini A, Secco GG, Dall'Ara G, Ghione M, Rama-Merchan JC, Lupi A, et al. ABSORB biodegradable stents versus second-generation metal stents: a comparison study of 100 complex lesions treated under OCT guidance. JACC Cardiovasc Interv. 2014;7:741–50.
39. Alegría-Barrero E, Foin N, Chan PH, Syrseloudis D, Lindsay AC, Dimopolous K, et al. Optical coherence tomography for guidance of distal cell recrossing in bifurcation stenting: choosing the right cell matters. EuroIntervention. 2012;8:205–13.
40. Foin N, Ghione M, Mattesini A, Davies JE, Di Mario C. Bioabsorbable scaffold optimization in provisional stenting: insight from optical coherence tomography. Eur Heart J Cardiovasc Imaging. 2013;14:1149.
41. Vaquerizo B, Barros A, Pujadas S, Bajo E, Estrada D, Miranda-Guardiola F, et al. Bioresorbable everolimus-eluting vascular scaffold for the treatment of chronic total occlusions: CTO-ABSORB pilot study. EuroIntervention. 2015;11:555–63.

Novel Application of OCT in Clinical Practice

Sunwon Kim and Jin Won Kim

19.1 Usefulness of Three-Dimensional Optical Coherence Tomography in Current Interventional Cardiology

Optical coherence tomography (OCT) offers cross-sectional images of coronary structure at microscopic level. Advanced three-dimensional OCT (3D OCT), first described by Tearney et al., offers more intuitive and comprehensive appreciation of the complex three-dimensional (3D) structure of coronary arteries by providing a large volume of tomographic data in a single stacked composite image [1]. Here, we will discuss how 3D OCT can be applied for clinical practice in interventional cardiology and aid the management of coronary artery disease.

S. Kim
Department of Cardiology, Korea University Ansan Hospital, Ansan, South Korea

Multimodal Imaging and Theranostic Lab, Korea University Medical School, Seoul, South Korea

J.W. Kim (✉)
Multimodal Imaging and Theranostic Lab, Korea University Medical School, Seoul, South Korea

Cardiovascular Center, Korea University Guro Hospital, Seoul, South Korea
e-mail: kjwmm@korea.ac.kr

19.1.1 Assessment of Ambiguous Angiographic Lesions

Coronary angiography has been widely performed as a diagnostic tool, providing a unique overview of the coronary tree. This conventional luminography is, however, a relatively poor representation of coronary structure, as it is highly dependent on the projection angle, and thus the accurate estimation of angiographic "hazy" lesions is limited [2]. The use of 3D OCT is known to be useful in many forms of ambiguous lesions such as intraluminal thrombus [3], spontaneous coronary artery dissection [4, 5], recanalized thrombus [6], and coronary evagination [7]. 3D OCT provides not only an accurate luminal visualization but also a more comprehensive understanding of whole lesion structure, which guides correct diagnosis and helps to optimize percutaneous coronary intervention (PCI) (Fig. 19.1) [5].

19.1.2 Coronary Bifurcation and Jailed Side Branch Evaluation

Despite remarkable advances in procedural techniques during the past decades, coronary bifurcation lesions, which account for approximately 10–20% of all PCIs, remain a challenge [8]. The application of 3D OCT rendering has allowed visualization of bifurcation lesions in

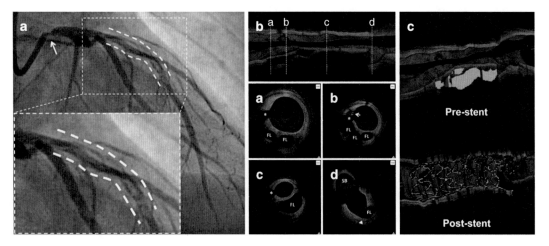

Fig. 19.1 (a) Coronary angiography of a patient presenting with unstable angina shows a significant left main ostial stenosis (*arrow*) with concomitant ambiguous angiographic lesion in proximal left anterior descending coronary artery (*dotted line*). (b) 3D-rendering OCT image clearly identifies the presence of dual lumen with thick intimal membrane, confirmative of spontaneous coronary artery dissection. (c) 3D OCT provides an accurate imaging guidance to ensure appropriate wire positioning and complete lesion coverage. *FL* false lumen. Reprinted from JACC Cardiovasc Interv. 2014;7(6):e57–9, by Lee S et al., with permission from Elsevier

detail not achieved by any imaging diagnostic modalities including 2D OCT. Recently, through 3D OCT image analysis of human bifurcation lesions, Farooq et al. reported that there is a variability of carina structure according to takeoff angle of side branch (SB): perpendicular (e.g., septal, mid-distal diagonal, and obtuse marginal branch) vs. parallel takeoff (e.g., proximal diagonal, right ventricular branch). They suggested that stenting across SBs with parallel takeoff is more susceptible to the carina shift rather than SBs with perpendicular takeoff [9]. This study highlights the potential role of 3D OCT in enhancing our understanding of the complex coronary anatomy and the effect of PCI on adjacent structure.

Accurate sizing of SB is crucial to circumvent SB injury during PCI with final kissing ballooning. It is well known that angiographic appearance of SB ostium after stent crossover is inconclusive [10]. According to a recent report using 3D cut-plane analysis, it is feasible to determine an accurate SB ostial diameter in a single OCT imaging of the main branch, by correcting the misalignment errors between pullback direction (main branch) and SB centerline [11].

This method could be utilizable in catheterization laboratory because it reduces the need for burdensome SB rewiring and additional pullback. Meanwhile, during the provisional stenting with kissing balloon inflation, it is recommended to rewire SB via a distal cell (Fig. 19.2a, a*, b, b*) because, otherwise, there remain large unopposed struts at the carina (Fig. 19.2c, c*, d, d*), which potentially cause disturbance in shear flow, delay in re-endothelialization, and thrombosis [8, 12]. 3D OCT is expected to provide an intuitive and accurate imaging guidance to ensure distal cell recrossing (Fig. 19.2).

As clinical trials failed to demonstrate the benefits of routine kissing ballooning in bifurcation lesions [13, 14], jailed SB is usually left untreated unless indicated (e.g., SB flow compromise, SB dissection, etc.). However, significant alterations of SB ostium morphology during strut coverage still warrant further investigation regarding the natural course of jailed SB [15, 16]. 3D OCT could offer unique opportunity for visualization of anatomical modifications occurring at the SB ostium (Fig. 19.3, left panel). Indeed, serial 3D demonstration of jailed SB has shown that overhanging struts may serve as a focus for

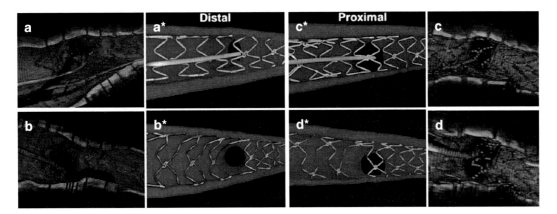

Fig. 19.2 3D OCT as an imaging guidance to optimize bifurcation PCI. Representative 3D OCT images and corresponding illustrations of bifurcation bench model (*) highlight the significance of distal cell rewiring (**a**, **a***) resulting in optimal reopening of a jailed SB (**b**, **b***), while kissing ballooning after proximal cell recrossing (**c**, **c***) ends up with large residual unopposed struts at carina (**d**, **d***). Reprinted from EuroIntervention. 2012;8(2):205–13, by Alegría-Barrero E et al., with permission from Europa Digital & Publishing

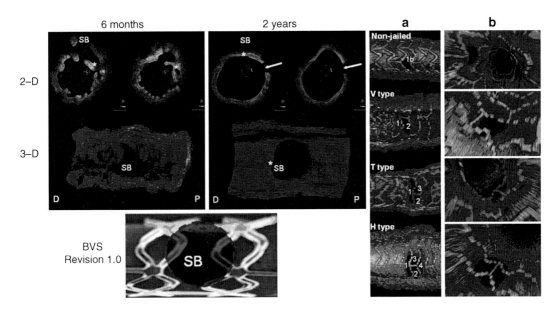

Fig. 19.3 *Left panel*, serial 3D OCT imaging of jailed SB after bioresorbable vascular scaffold implantation. At 2 years, neointimal tissue at the distal border of the side branch orifice extended to form a thick membranous structure at the carina (neocarina*) whereas overhanging strut at the proximal border was fully degraded. *Right panel*, jailed SB classification based on 3D morphology of the overhanging struts (Types V, T, and H) and the number of compartment outlined by the struts. Reprinted from JACC Cardiovasc Interv. 2010;3(8):836–44, by Okamura T et al., with permission from Elsevier

excessive neointima formation and thrombosis, suggesting the potential mechanism regarding delayed SB compromise [15]. Theoretically, the use of a bioresorbable vascular scaffold could be a solution to this issue because restoration of normal bifurcation anatomy can be expected after full biodegradation [17]. In this regard, there is an attempt to categorize jailed SB according to

3D morphology to elucidate the fate of scaffold during bioresorption (Fig. 19.3, right panel) [18]. Application of 3D OCT will help to clarify the roles of stent design, strut-tissue interaction, and optimized PCI on long-term patency of jailed SB.

19.1.3 Assessment of Coronary Stent Configuration

Coronary stent fracture is an important cause of late stent failure associated with major adverse cardiovascular outcome [19]. However, even with the use of current Fourier domain OCT, it is challenging to accurately identify fracture sites in a small mesh-like structure. In particular, newer-generation open-cell design stent exhibits non-uniform strut allocation on cross-sectional images (Fig. 19.4, left lower panel), and conventional 2D OCT criteria for stent fracture (e.g., lack of circumferential stent strut) appear to be inconclusive [20, 21]. Volumetric 3D OCT offers significant advantages over other imaging modalities in terms of accurate delineation of 3D configuration (Fig. 19.4).

The quality and spatial accuracy of 3D-rendered images are hampered by cardiac motion and under-sampling (still only 12% of the lumen is sampled with current Fourier domain OCT, Fig. 19.5b, d) [22, 23]. Recent progress of ultra-high-speed OCT, a novel method that achieves a 5–10 times faster imaging speed, enables more accurate assessment of stent configuration by sampling larger data during a short period of diastole, an optimal phase for coronary imaging [24]. As ultrahigh-speed 3D OCT enables high-fidelity, motion-free imaging, it seems to be promising for more precise evaluation of stent integrity (Fig. 19.5a, c).

With the introduction of the first commercial 3D-rendering technology, 3D OCT is now finding its way into interventional practice. Further studies are warranted to determine whether the beneficial advantages outlined above will translate into improved clinical outcomes.

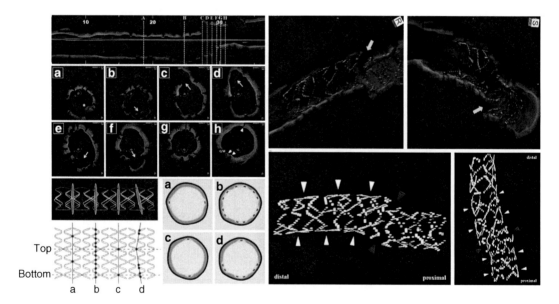

Fig. 19.4 A case of stent fracture (SF) diagnosed by 3D OCT. Conventional 2D OCT imaging (*left upper panel*) reveals several pathological findings: thrombus (**a**), malapposition (**b**), peristrut ulcer (**b, e**), aneurysmal deformations (**c–g**), and also a cross section with a lack of circumferential stent struts (**h**, only three struts, *arrowheads*) suggestive of SF. However, 2D OCT findings are inconclusive for SF because newer-generation open-cell stents exhibit various strut patterns on cross-sectional images (*left lower panel*). The volume-rendered 3D OCT strut mapping clearly identifies the breakage of interconnecting links (*right panels, yellow arrow*, and *red arrowheads*). Reprinted from Circulation. 2014;129:24–7, by Kim S et al., with permission from Wolters Kluwer Health, Inc

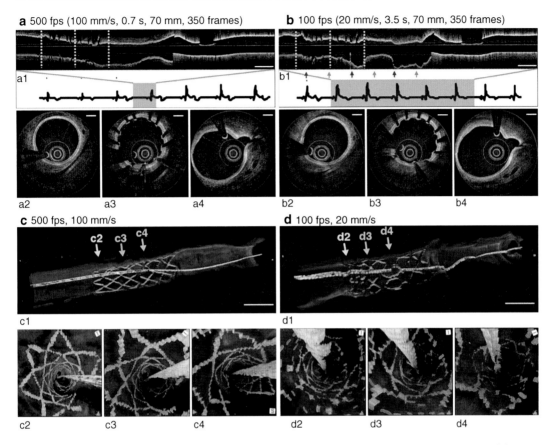

Fig. 19.5 In conjunction with electrocardiography-triggering module, ultrahigh-speed OCT (UHS OCT) enables rapid imaging acquisition during a brief period of diastole where the cardiac motion could be minimized (**a**). Unlike conventional OCT influenced by ventricular contraction (**b**, *b1*, *d1*) and under-sampling (**d**, *d2*, *d3*, *d4*, note the "grainy" appearance of strut), UHS OCT provides images with smooth, uninterrupted vascular contour (**a**, *a1*). 3D reconstruction could provide high-fidelity images (**c**, *c1*, *c2*, *c3*, *c4*). Reprinted from JACC Cardiovasc Imaging. 2016;9(5):623–5, by Jang SJ et al., with permission from Elsevier

19.2 Near-Future Technologies: Multimodal Intravascular Biological Imaging Integrated with OCT

Coronary plaque rupture is a dynamic biological process driven by chronic maladaptive immune response against subendothelial lipoproteins, which involves growth of lipid-enriched necrotic core, increases of inflammation and protease activities, and thinning of fibrous cap by gradual loss of collagen and smooth muscle cell [25, 26]. Despite the clinical need to predict future coronary events, current structural imaging alone does not estimate rupture risk enough to guide clinical decisions [27]. This concise overview will address recent advances in biological cardiovascular imaging for the assessment of plaque vulnerability, focusing specifically on multimodal integrative imaging approaches combined with OCT (Table 19.1).

19.2.1 Integrated Optical Coherence Tomography and Near-Infrared Fluorescence Molecular Imaging

With the favorable optical properties of near-infrared bandwidth to detect fluorescence signals through blood, near-infrared fluorescence

Table 19.1 Comparison of the multimodal biological imaging combined with intravascular optical coherence tomography

	Standalone OCT	Spectroscopic OCT	PS-OCT	OCT + NIRF	OCT + NIRAF
Detection	N/A	Attenuation coefficient	Polarization status	Target-specific NIRF	Autofluorescence
Additional equipment	N/A	None	Polarization modulator	NIRF console, DCF hybrid rotary junction, DCF catheter, exogenous NIRF imaging agent	NIRF console, DCF hybrid rotary junction, DCF catheter
Identifiable plaque characteristics					
Cap thickness	+++	+++	+++	+++	+++
Collagen and SMC	−	+	+++	−	−
Inflammation	+	+	+	+++	+
Protease	−	−	−	+++	−
Lipid	++	+++	++	++	++
Necrotic core	−	−	−	−	+++
Calcium	++	++	++	++	++
Thrombus	++	++	+++	++	++

OCT optical coherence tomography, *PS-OCT* polarization-sensitive OCT, *NIRF* near-infrared fluorescence, *NIRAF* near-infrared autofluorescence, *DCF* double-clad fiber, *SMC* smooth muscle cell

(NIRF) imaging, in combination with target-specific imaging agents, provides in vivo readout regarding key markers of vulnerable plaque such as protease and macrophage activity [28–30]. After the first feasibility report in 2008 [31], intravascular NIRF imaging has shown a remarkable progress. One of the major breakthroughs is the fabrication of a fully integrated dual-modal OCT-NIRF system based on double-cladding fiber probe, which simultaneously provides distance-calibrated quantitative NIRF imaging with co-registered OCT structural information [30, 32]. Furthermore, the use of indocyanine green, a FDA-approved NIRF agent, has made it the most promising strategy for translational molecular cardiovascular imaging [32]. Its capability to quantitate molecular activities contributing to plaque vulnerability (Fig. 19.6), in synergy with high-resolution structural imaging by OCT, could provide an incremental value in risk stratification of coronary plaque.

19.2.2 Spectroscopic Optical Coherence Tomography

One major drawback of current OCT is the lack of biochemical specificity, which may significantly impede the qualitative differentiation between pathologic component and normal tissue with similar optical properties. Lipid is an important imaging target of vulnerable plaque because an atheroma enriched in lipid is prone to rupture. However, characterization of lipid-rich plaques using OCT is often challenging because signal attenuation could be influenced by a variety of sources such as macrophages, foam cells, thrombus, mixed lesion with calcification, and even intrinsic artifacts (e.g., shadowing, tangential signal dropout, and negative contrast) [33, 34]. By applying a predetermined attenuation coefficient, spectroscopic OCT facilitated robust and accurate detection of plaque lipids (Fig. 19.7) [35]. Also, this quantitative spectroscopic analysis can be applied to differentiate collagen-rich regions

19 Novel Application of OCT in Clinical Practice

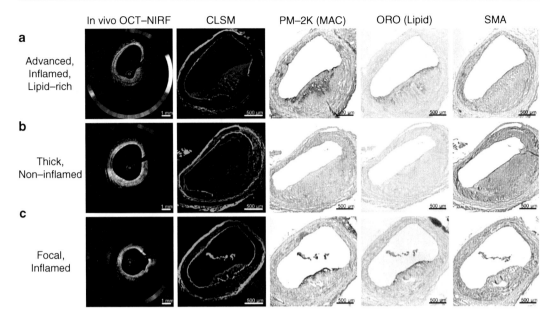

Fig. 19.6 Intracoronary dual-modal OCT-NIRF imaging using indocyanine green as NIRF signal enhancer. This imaging strategy enabling quantitative estimation of inflammatory activity has the potential to stratify the risk of individual plaque. Each row shows representative in vivo OCT-NIRF image and the corresponding histological sections validated by confocal laser scanning microscopy (CLSM) (ICG, *red*; autofluorescence, *green*) and immunohistochemistry: macrophage (PM-2K), lipids (ORO), and smooth muscle cells (SMA). (**a**) Advanced, high-risk plaque showing robust NIRF activity. Strong NIRF area (red) of CLSM colocalizes with macrophage- and lipid-positive areas. (**b**) Fibrotic, stable plaque on OCT-NIRF has abundant smooth muscle cells without overt inflammation. (**c**) Small plaque with focal inflammation shows less robust NIRF activity as compared to advanced plaque. Reprinted from Eur Heart J. 2016;37(37):2833–44, by Kim S et al., with permission from Oxford University Press

from lipid content [36]. As for a major strength of spectroscopic OCT, it requires no additional devices to acquire the compositional information and therefore can readily be implemented in cardiac catheterization laboratory.

19.2.3 Combined Optical Coherence Tomography and Near-Infrared Autofluorescence Imaging

Current OCT lacks the capability to discern necrotic cores from lipid pools because both lesions manifest as signal-poor regions with diffuse border [37]. Recent experimental research using human autopsied arteries has demonstrated that natural emission of light, autofluorescence in the NIR wavelength (NIRAF, excitation at 633 nm, emission at 675–950 nm), significantly increased in the plaques with necrotic cores, suggesting that high plaque NIRAF can be used as an indicator of vulnerability [38]. In the subsequent clinical study, the investigators also found a significant association between elevated NIRAF and OCT-defined high-risk plaque features (Fig. 19.8) [39]. Although considerable further work is required to elucidate the biological mechanism and potential sources of NIRAF production in atherosclerotic lesions, this first human study demonstrating feasibility and safety of the dual-modal imaging is an important step toward clinical NIRF molecular imaging in vivo.

19.2.4 Polarization-Sensitive Optical Coherence Tomography

Tissue birefringence can be utilized to assess plaque vulnerability because plaque-stabilizing components such as organized collagen and

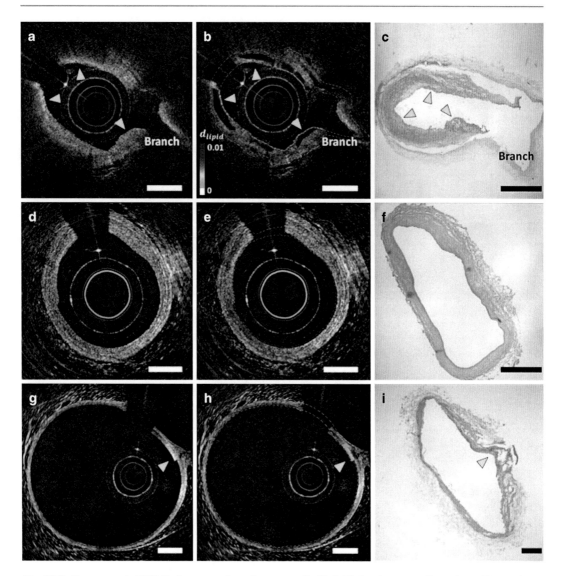

Fig. 19.7 Spectroscopic OCT. Each row consists of representative OCT image, spectroscopic mapping data overlaid on OCT image as a color contrast (*red*, high in lipid; *green*, low in lipid), and corresponding histology, respectively. *First row*, high spectroscopic contrast colocalizes well with lipid abundant areas. *Second row*, fibrous plaque yields low spectroscopic contrast. *Third row*, spectroscopic OCT is able to detect tiny lipid portion precisely. Reprinted from J Biomed Opt. 2016;21(7):75004, by Nam HS et al., with permission from SPIE

smooth muscle cell are highly birefringent. Polarization-sensitive OCT (PS-OCT), by analyzing polarization states of backscattered light from a sample, provides measurements of tissue birefringence as an additional image contrast (Fig. 19.9). With the development of catheter-based intracoronary PS-OCT system in 2008 [40], first human pilot study investigating its clinical feasibility and usefulness is now underway [41].

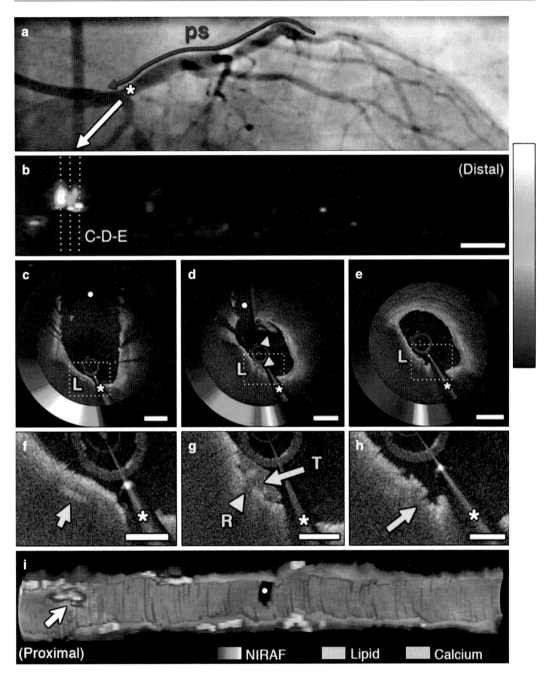

Fig. 19.8 An example of OCT-NIRAF imaging from a 66-year-old male who had an intermediate lesion in left anterior descending coronary artery (**a**, *asterisk*). Strong NIRAF signals are localized in OCT-defined thin cap fibroatheroma (**b**), which contains high-risk features such as cholesterol crystal (**c**, **f**, *arrow*), cap surface thrombus (**d**, **g**, *arrow*), and rupture of cap (**e**, **h**, *arrow*). 3D-rendering image shows focal NIRAF hot spot (arrow) within a large lipid pool (**i**). Reprinted from JACC Cardiovasc Imaging. 2016;9(11):1304–1314, by Ughi GJ et al., with permission from Elsevier

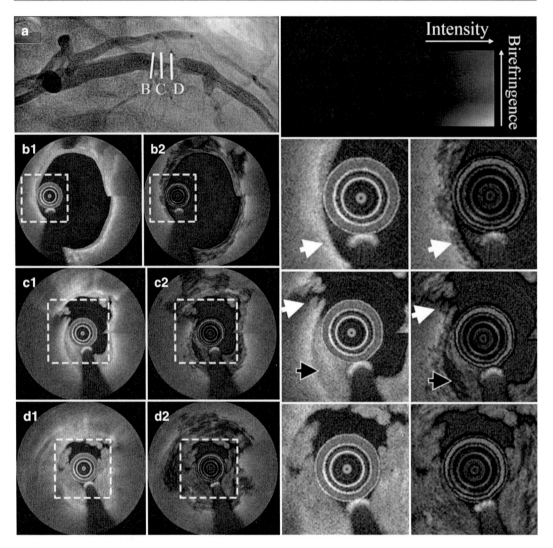

Fig. 19.9 PS-OCT imaging of a ruptured thin cap fibroatheroma. Low birefringence (*yellow color*) is noted near the rupture site, suggesting the loss of collagen within fibrous cap (*b1, b2, c1, c2, white arrow*). Intraluminal thrombus mostly displays low birefringence as it has unorganized architecture (*d1, d2*). Reprinted from Eur Heart J. 2016;37(24):1932, by van der Sijde JN et al., with permission from Oxford University Press

Conclusion

Cardiovascular imaging in recent years has focused on the biological aspects of coronary atherosclerosis, aiming to evaluate its natural course of the plaque and to predict future coronary events. Multimodal biological imaging combined with OCT, albeit still in its early stage, is extensively studied by leading research groups and rapidly evolving. This novel approach, as did intravascular ultrasound and OCT in the past decades, is expected to shed a new light in the study of coronary atherosclerosis.

References

1. Tearney GJ, Waxman S, Shishkov M, Vakoc BJ, Suter MJ, Freilich MI, et al. Three-dimensional coronary artery microscopy by intracoronary optical frequency domain imaging. JACC Cardiovasc Imaging. 2008;1(6):752–61. Epub 2009/04/10
2. Topol EJ, Nissen SE. Our preoccupation with coronary luminology. The dissociation between clinical and angiographic findings in ischemic-heart-disease. Circulation. 1995;92(8):2333–42.
3. Okamura T, Serruys PW, Regar E. Three-dimensional visualization of intracoronary thrombus during stent implantation using the second generation, Fourier domain optical coherence tomography. Eur Heart J. 2010;31(5):625. Epub 2009/12/08
4. Kim JB, Nam HS, Yoo H, Kim JW. A bi-directional assessment of spontaneous coronary artery dissection by three-dimensional flythrough rendering of optical coherence tomography images. Eur Heart J. 2015;36(17):1022. Epub 2015/01/04
5. Lee S, Kim CS, Oh DJ, Yoo H, Kim JW. Three-dimensional intravascular optical coherence tomography rendering assessment of spontaneous coronary artery dissection concomitant with left main ostial critical stenosis. JACC Cardiovasc Interv. 2014;7(6):E57–E9.
6. Khoueiry GM, Magnus P, Friedman BJ, Kaplan AV. Honeycomb-like appearance of hazy coronary lesions: OCT image report of a recanalized thrombus. Eur Heart J Cardiovasc Imaging. 2014;15(12):1427. Epub 2014/08/02
7. Radu MD, Raber L, Kalesan B, Muramatsu T, Kelbaek H, Heo J, et al. Coronary evaginations are associated with positive vessel remodelling and are nearly absent following implantation of newer-generation drug-eluting stents: an optical coherence tomography and intravascular ultrasound study. Eur Heart J. 2014;35(12):795–807. Epub 2013/10/18
8. Lassen JF, Holm NR, Stankovic G, Lefevre T, Chieffo A, Hildick-Smith D, et al. Percutaneous coronary intervention for coronary bifurcation disease: consensus from the first 10 years of the European Bifurcation Club meetings. EuroIntervention. 2014;10(5):545–60. Epub 2014/09/27
9. Farooq V, Serruys PW, Heo JH, Gogas BD, Okamura T, Gomez-Lara J, et al. New insights into the coronary artery bifurcation hypothesis-generating concepts utilizing 3-dimensional optical frequency domain imaging. JACC Cardiovasc Interv. 2011;4(8):921–31.
10. Koo BK, Kang HJ, Youn TJ, Chae IH, Choi DJ, Kim HS, et al. Physiologic assessment of jailed side branch lesions using fractional flow reserve. J Am Coll Cardiol. 2005;46(4):633–7.
11. Karanasos A, Tu S, van Ditzhuijzen NS, Ligthart JM, Witberg K, Van Mieghem N, et al. A novel method to assess coronary artery bifurcations by OCT: cut-plane analysis for side-branch ostial assessment from a main-vessel pullback. Eur Heart J Cardiovasc Imaging. 2015;16(2):177–89. Epub 2014/09/18
12. Alegria-Barrero E, Foin N, Chan PH, Syrseloudis D, Lindsay AC, Dimopolous K, et al. Optical coherence tomography for guidance of distal cell recrossing in bifurcation stenting: choosing the right cell matters. EuroIntervention. 2012;8(2):205–13.
13. Hildick-Smith D, de Belder AJ, Cooter N, Curzen NP, Clayton TC, Oldroyd KG, et al. Randomized trial of simple versus complex drug-eluting stenting for bifurcation lesions: the British bifurcation coronary study: old, new, and evolving strategies. Circulation. 2010;121(10):1235–43. Epub 2010/03/03
14. Niemela M, Kervinen K, Erglis A, Holm NR, Maeng M, Christiansen EH, et al. Randomized comparison of final kissing balloon dilatation versus no final kissing balloon dilatation in patients with coronary bifurcation lesions treated with main vessel stenting: the Nordic-Baltic bifurcation study III. Circulation. 2011;123(1):79–86. Epub 2010/12/22
15. Diletti R, Farooq V, Muramatsu T, Gogas BD, Garcia-Garcia HM, van Geuns RJ, et al. Serial 2- and 3-dimensional visualization of side branch jailing after metallic stent implantation: to kiss or not to kiss...? JACC Cardiovasc Interv. 2012;5(10):1089–90. Epub 2012/10/20
16. Foin N, Viceconte N, Chan PH, Lindsay AC, Krams R, Di Mario C. Jailed side branches: fate of unapposed struts studied with 3D frequency-domain optical coherence tomography. J Cardiovasc Med. (Hagerstown). 2011;12(8):581–2. Epub 2011/06/29
17. Iqbal J, Onuma Y, Ormiston J, Abizaid A, Waksman R, Serruys P. Bioresorbable scaffolds: rationale, current status, challenges, and future. Eur Heart J. 2014;35(12):765–76. Epub 2013/12/25
18. Okamura T, Onuma Y, Garcia-Garcia HM, Regar E, Wykrzykowska JJ, Koolen J, et al. 3-dimensional optical coherence tomography assessment of jailed side branches by bioresorbable vascular scaffolds: a proposal for classification. JACC Cardiovasc Interv. 2010;3(8):836–44. Epub 2010/08/21
19. Kuramitsu S, Iwabuchi M, Haraguchi T, Domei T, Nagae A, Hyodo M, et al. Incidence and clinical impact of stent fracture after everolimus-eluting stent implantation. Circ Cardiovasc Interv. 2012;5(5):663–71. Epub 2012/09/27
20. Francaviglia B, Capranzano P, Gargiulo G, Longo G, Tamburino CI, Ohno Y, et al. Usefulness of 3D OCT to diagnose a noncircumferential open-cell stent fracture. JACC Cardiovasc Imaging. 2016;9(2):210–1. Epub 2015/03/24
21. Kim S, Kim CS, Na JO, Choi CU, Lim HE, Kim EJ, et al. Coronary stent fracture complicated multiple aneurysms confirmed by 3-dimensional reconstruction of intravascular-optical coherence tomography in a patient treated with open-cell designed drug-eluting stent. Circulation. 2014;129(3):e24–7. Epub 2014/01/22

22. Farooq V, Gogas BD, Okamura T, Heo JH, Magro M, Gomez-Lara J, et al. Three-dimensional optical frequency domain imaging in conventional percutaneous coronary intervention: the potential for clinical application. Eur Heart J. 2013;34(12):875–85. Epub 2011/11/24
23. Wang T, Pfeiffer T, Regar E, Wieser W, van Beusekom H, Lancee CT, et al. Heartbeat OCT and motion-free 3D in vivo coronary artery microscopy. JACC Cardiovasc Imaging. 2016;9(5):622–3. Epub 2016/05/07
24. Jang SJ, Park HS, Song JW, Kim TS, Cho HS, Kim S, et al. ECG-triggered, single cardiac cycle, high-speed, 3D, intracoronary OCT. JACC Cardiovasc Imaging. 2016;9(5):623–5. Epub 2016/05/07
25. Bentzon JF, Otsuka F, Virmani R, Falk E. Mechanisms of plaque formation and rupture. Circ Res. 2014;114(12):1852–66. Epub 2014/06/07
26. Moore KJ, Tabas I. Macrophages in the pathogenesis of atherosclerosis. Cell. 2011;145(3):341–55. Epub 2011/05/03
27. Stone GW, Maehara A, Lansky AJ, de Bruyne B, Cristea E, Mintz GS, et al. A prospective natural-history study of coronary atherosclerosis. N Engl J Med. 2011;364(3):226–35. Epub 2011/01/21
28. Lee S, Lee MW, Cho HS, Song JW, Nam HS, Oh DJ, et al. Fully integrated high-speed intravascular optical coherence tomography/near-infrared fluorescence structural/molecular imaging in vivo using a clinically available near-infrared fluorescence-emitting indocyanine green to detect inflamed lipid-rich atheromata in coronary-sized vessels. Circ Cardiovasc Interv. 2014;7(4):560–9. Epub 2014/07/31
29. Press MC, Jaffer FA. Molecular intravascular imaging approaches for atherosclerosis. Curr Cardiovasc Imaging Rep. 2014;7(10):9293. Epub 2014/09/16
30. Yoo H, Kim JW, Shishkov M, Namati E, Morse T, Shubochkin R, et al. Intra-arterial catheter for simultaneous microstructural and molecular imaging in vivo. Nat Med. 2011;17(12):1680–4. Epub 2011/11/08
31. Jaffer FA, Vinegoni C, John MC, Aikawa E, Gold HK, Finn AV, et al. Real-time catheter molecular sensing of inflammation in proteolytically active atherosclerosis. Circulation. 2008;118(18):1802–9. Epub 2008/10/15
32. Kim S, Lee MW, Kim TS, Song JW, Nam HS, Cho HS, et al. Intracoronary dual-modal optical coherence tomography-near-infrared fluorescence structural-molecular imaging with a clinical dose of indocyanine green for the assessment of high-risk plaques and stent-associated inflammation in a beating coronary artery. Eur Heart J. 2016;37(37):2833–44. Epub 2016/01/21
33. Imanaka T, Hao H, Fujii K, Shibuya M, Fukunaga M, Miki K, et al. Analysis of atherosclerosis plaques by measuring attenuation coefficients in optical coherence tomography: thin-cap fibroatheroma or foam cells accumulation without necrotic core? Eur Heart J. 2013;34:1007–8.
34. van Soest G, Regar E, Goderie TPM, Gonzalo N, Koljenovic S, van Leenders GJLH, et al. Pitfalls in plaque characterization by OCT image artifacts in native coronary arteries. JACC Cardiovasc Imaging. 2011;4(7):810–3.
35. Nam HS, Song JW, Jang SJ, Lee JJ, Oh WY, Kim JW, et al. Characterization of lipid-rich plaques using spectroscopic optical coherence tomography. J Biomed Opt. 2016;21(7):75004.
36. Fleming CP, Eckert J, Halpern EF, Gardecki JA, Tearney GJ. Depth resolved detection of lipid using spectroscopic optical coherence tomography. Biomed Opt Express. 2013;4(8):1269–84.
37. Otsuka F, Joner M, Prati F, Virmani R, Narula J. Clinical classification of plaque morphology in coronary disease. Nat Rev Cardiol. 2014;11(7):379–89.
38. Wang H, Gardecki JA, Ughi GJ, Jacques PV, Hamidi E, Tearney GJ. Ex vivo catheter-based imaging of coronary atherosclerosis using multimodality OCT and NIRAF excited at 633 nm. Biomed Opt Express. 2015;6(4):1363–75.
39. Ughi GJ, Wang H, Gerbaud E, Gardecki JA, Fard AM, Hamidi E, et al. Clinical characterization of coronary atherosclerosis with dual-modality OCT and near-infrared autofluorescence imaging. JACC Cardiovasc Imaging. 2016;9(11):1304–14. Epub 2016/03/14
40. Oh WY, Yun SH, Vakoc BJ, Shishkov M, Desjardins AE, Park BH, et al. High-speed polarization sensitive optical frequency domain imaging with frequency multiplexing. Opt Express. 2008;16(2):1096–103.
41. van der Sijde JN, Karanasos A, Villiger M, Bouma BE, Regar E. First-in-man assessment of plaque rupture by polarization-sensitive optical frequency domain imaging in vivo. Eur Heart J. 2016;37(24):1932.

Part III

Physiology

Concept of Invasive Coronary Physiology: Focus on FFR

20

Bon-Kwon Koo and Joo Myung Lee

20.1 Functional Anatomy of Coronary Arterial Circulation

The coronary artery system has three components with different functions: conductive epicardial coronary arteries, arterioles, and capillaries although the borders of each compartment cannot be clearly defined anatomically (Fig. 20.1) [1]. The proximal compartment, conductive epicardial coronary arteries, has a capacitance function and possesses little resistance to coronary blood flow, and its diameter ranges from approximately 500 μg to 4 mm. The intermediate compartment is represented by prearterioles with measurable pressure drop along its length with diameter ranging from 100 to 500 μg. The most distal compartment is represented by intramural arterioles, which are characterized by a considerable pressure drop along its length. The arteriolar segment possesses diameter less than 100 μg. Prearterioles and arterioles cannot be clearly delineated by coronary angiography. These microvessels take most of the coronary vascular resistance and are called as resistance vessels. They can modulate the vascular tone and resistance under various physiologic and pharmacological conditions to control the myocardial blood flow. Prearteriolar vessels are responsive to flow and pressure changes, and their function is to maintain the pressure in a narrow range at the origin of arterioles when coronary perfusion flow or pressure changes. Intramural arterioles are the main part of the metabolic regulation of coronary blood flow. When oxygen consumption increases, arterioles are dilated and vascular resistance is reduced in response to myocardial metabolites. This induces the dilatation of other vessels by the increase in flow and shear stress (Fig. 20.2) [1, 2].

20.2 Physiologic Characteristics of Coronary Arterial Circulation

At resting status, coronary arterial blood flow is about 5% of total cardiac output, and flow across coronary arterial system largely depends on the pressure gradient between aortic root (the coronary driving pressure) and end-diastolic pressure of left and right ventricles. Therefore, coronary

B.-K. Koo (✉)
Department of Internal Medicine and Cardiovascular Center, Seoul National University Hospital, Seoul, South Korea

Institute of Aging, Seoul National University, Seoul, South Korea
e-mail: bkkoo@snu.ac.kr

J.M. Lee
Department of Internal Medicine and Cardiovascular Center, Samsung Medical Center, Sungkyunkwan University School of Medicine, Seoul, South Korea

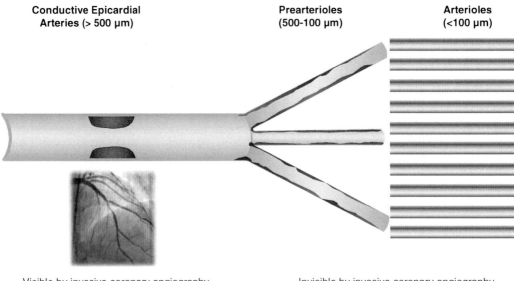

Fig. 20.1 Functional anatomy of coronary arterial system. The coronary artery system has three components with different functions: conductive epicardial coronary arteries, arterioles, and capillaries although the borders of each compartment cannot be clearly defined anatomically

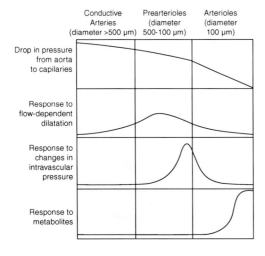
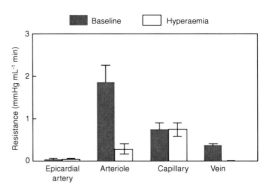

Fig. 20.2 Regulation of coronary flow and resistance. (**a**, **b**) Although prearterioles and arterioles cannot be clearly delineated by coronary angiography, these components mainly regulate coronary vascular resistance and myocardial blood flow. The *left panel* of figure was adapted from the review of Camici et al. NEJM 2007 under permission of the publisher. The *right panel* of figure was adapted and modified from the original article of Chilian et al. American Journal of Physiology 1989 under permission of the publisher

arterial blood flow occurs predominantly during diastole, and systolic component at hyperemia is less than 25% of total flow (Fig. 20.3). In case of right coronary artery, phasic blood flow in the right coronary artery proper occurs equally during systole and diastole; conversely, phasic blood flow in the posterior descending and posterolateral coronary arteries occurs predominantly in diastolic phase [3, 4]. In the absence of atherosclerotic narrowing in epicardial coronary artery, the diameter

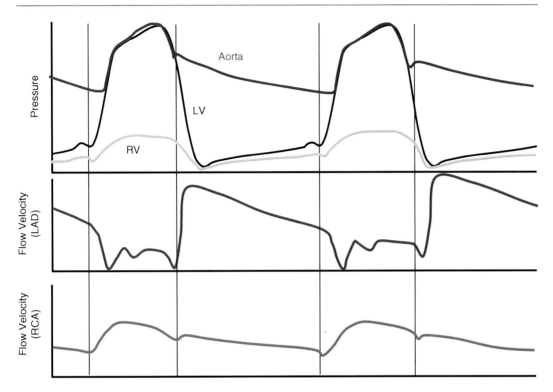

Fig. 20.3 Relationship between aortic pressure and coronary flow velocity. The coronary flow across coronary arterial system largely depends on the pressure gradient between aortic root (the coronary driving pressure) and end-diastolic pressure of left and right ventricles; therefore, coronary arterial blood flow occurs predominantly during diastole, and systolic component at hyperemia is less than 25% of total flow. Abbreviations: *LAD* left anterior descending artery, *LV* left ventricle, *RCA* right coronary artery, *RV* right ventricle

and cross-sectional area of epicardial coronary artery usually tapers from proximal to distal portion along with decreasing amount of regional myocardial mass, supplied by the coronary artery. Although body mass index or habitus can affect coronary arterial size and coronary flow, intracoronary pressure remains constantly as long as the absence of epicardial coronary stenosis (Fig. 20.4). As epicardial coronary arterial system has branching trees, the absolute amount of coronary flow and cross-sectional area (or diameter) decreases along with the course of epicardial coronary artery. However, as the total amount of flow in main vessel and side branch after bifurcation is same, the coronary flow velocity is not changed before and after coronary bifurcation (Fig. 20.5).

The myocardial oxygen demand (8–10 ml/min/100 g) is much higher than other organs (e.g., skeletal muscle 0.5 ml/min/100 g) even in the resting condition, and coronary capillary density is also higher to meet the high oxygen demand. Nevertheless, the oxygen extraction by the myocardium is much higher than the other organs and reaches near maximum. The oxygen saturation of coronary sinus venous blood is only about 20–30% (renal vein: 85%). According to Fick's principle, oxygen consumption is the product of blood flow and oxygen extraction. Therefore, coronary circulation can meet the increasing oxygen demand mainly by increasing the amount of coronary blood flow [5].

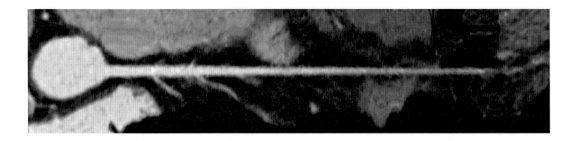

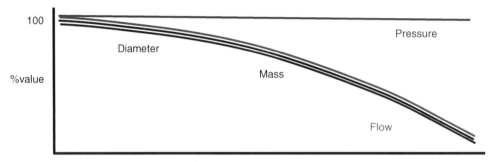

Fig. 20.4 The changes in vessel diameter, myocardial mass, coronary flow, and coronary pressure. In the absence of atherosclerotic narrowing in epicardial coronary artery, the diameter and cross-sectional area of epicardial coronary artery usually tapers from proximal to distal portion along with decreasing amount of regional myocardial mass, supplied by the coronary artery and coronary flow. However, intracoronary pressure remains constantly as long as the absence of epicardial coronary stenosis

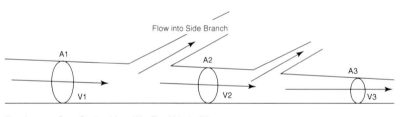

Fig. 20.5 The relationship among coronary flow, flow velocity, and cross-sectional area. As epicardial coronary arterial system has branching trees, the absolute amount of coronary flow and cross-sectional area (or diameter) decreases along with the course of epicardial coronary artery. However, as the total amount of flow in main vessel and side branch after bifurcation is same with that of proximal mother vessel before bifurcation, the coronary flow velocity is not changed before and after coronary bifurcation. Abbreviations: A area, F coronary flow, V coronary flow velocity

20.3 Coronary Autoregulation and Coronary Reserve

At resting status, coronary blood flow remains constant as coronary artery pressure is reduced below aortic pressure over wide range when the determinants of myocardial oxygen consumption are kept constant [6]. When coronary artery pressure falls below the range of autoregulation, coronary resistance arteries are maximally vasodilated to intrinsic stimuli and flow becomes pressure dependent. Resting coronary blood flow under normal hemodynamic conditions averages 0.7–1.0 ml/min/g and can increase up to three- to fivefold during

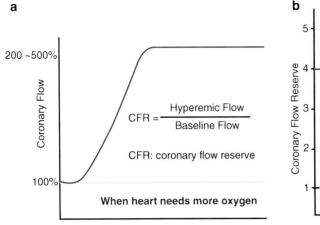
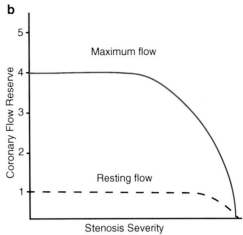

Fig. 20.6 Coronary autoregulation and concept of coronary flow reserve. (**a**) Resting coronary blood flow under normal hemodynamic conditions averages 0.7–1.0 ml/min/g and can increase up to three- to fivefold during vasodilation. The ability to increase flow above resting values in response to pharmacologic vasodilation is termed coronary flow reserve. (**b**) Due to coronary autoregulation, coronary flow remains constant as stenosis severity of epicardial coronary artery increases; therefore, assessment of resting perfusion cannot identify hemodynamically significant stenoses. This figure was adapted and modified from the original article of Gould LK et al. Am J Cardiol 1974 under permission of the publisher. Abbreviations: *CFR* coronary flow reserve

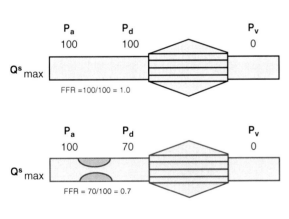
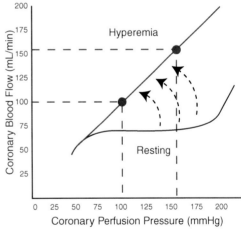

Fig. 20.7 The concept of maximal perfusion. When maximal vasodilation of resistance arteries occurs, coronary flow is mainly dependent on coronary artery pressure, and this maximally vasodilated pressure-flow relationship is much more sensitive for detecting increases in stenosis severity. In this maximum vasodilatory condition, for example, 30% decrease in distal coronary pressure linearly correlates 30% decrease in coronary flow. Abbreviations: *FFR* fractional flow reserve, P_a aortic pressure, P_d distal coronary pressure, P_v venous pressure

vasodilation [7]. The ability to increase flow above resting values in response to pharmacologic vasodilation is termed coronary flow reserve (Fig. 20.6a) [8]. Due to coronary autoregulation, coronary flow remains constant as stenosis severity of epicardial coronary artery increases; therefore, assessment of resting perfusion cannot identify hemodynamically significant stenoses (Fig. 20.6b). When maximal vasodilation of resistance arteries occurs, coronary flow is mainly dependent on coronary artery pressure, and this maximally vasodilated pressure-flow relationship is much more sensitive for detecting increases in stenosis severity. In this maximum vasodilatory condition, for example, 30% decrease in distal coronary pressure linearly correlates 30% decrease in coronary flow (Fig. 20.7). When stenosis

Table 20.1 Coronary flow and coronary flow reserve in normal controls

Doppler wire ($n = 301$)	Status	Flow velocity (cm/s)	CFR
	Resting	17.8 ± 6.9	2.64 ± 0.76
	Hyperemic	44.9 ± 16.0	
PET ($n = 3484$)	Status	Absolute flow (ml/min/g)	CFR
	Resting	0.82 ± 0.06	3.55 ± 1.36
	Stress	2.86 ± 1.29	

Adapted from Nijjer et al. EHJ 2016;37:2069–2080 and Gould et al. J Am Coll Cardiol. 2013;62:1639–53

severity exceeds over 40–60% diameter reduction, stenosis resistance begins to increase, distal coronary pressure decreases, and maximal vasodilatory flow decreases. In this condition, coronary flow reserve (CFR) can reflect the functional significance of epicardial coronary stenosis, unless the stenosis accompanied with diffuse atherosclerosis, LV hypertrophy, or disease causing microcirculation impairment. As absolute coronary flow cannot be easily measured in human, CFR can be quantified using Doppler wire-measured coronary flow velocity, thermodilution flow measurement, or absolute tissue perfusion-based method using positron emission tomography (PET). Table 20.1 summarizes Doppler wire-measured coronary flow velocity and CFR or PET-based measurement of absolute flow and CFR for insignificant coronary artery disease or normal controls. Clinically important reductions in maximum flow correlating with stress-induced ischemia on SPECT are generally associated with CFR value below 2 [9].

20.4 Stenosis Pressure and Flow Relationship

The normal conductive artery without atherosclerotic involvement is normally able to accommodate large increases in coronary flow without producing any significant pressure drop. Therefore, the epicardial coronary artery serves a conduit function to the resistance arteries. However, when atherosclerotic plaque deposits in epicardial coronary artery, resistance across a stenosis can become a dominant factor in coronary flow and limits maximal myocardial perfusion. The coronary flow distal to the stenosis can vary according to pressure drop across a stenosis and microcirculatory function. The total pressure drop across a stenosis is governed by three hydrodynamic factors—friction losses, separation losses, and turbulence. The single most important determinant of stenosis resistance for any given level of flow is the minimal lesion cross-sectional area within the stenosis [10]. As resistance is inversely proportionated to the square of the cross-sectional area, small changes in cross-sectional area lead to major changes in stenosis pressure-flow relationship and reduce maximal perfusion during vasodilation (Fig. 20.8). Separation losses are mainly influenced by flow rates distal to the stenosis and determine curve linearity or steepness of the stenosis pressure-flow relationship. As stenosis resistance increases exponentially as minimum lesion cross-sectional area decreases, it is also flow dependent and varies with the square of flow or flow velocity (Fig. 20.9).

20.5 The Concept of Maximal Perfusion and Fractional Flow Reserve

In maximal vasodilation induced by pharmacologic agents, distal coronary pressure is directly proportional to maximum vasodilated perfusion and coronary flow (Fig. 20.10). Fractional flow reserve (FFR) is an indirect index determined by measuring driving pressure for microcirculatory flow distal to the stenosis (distal coronary pressure—coronary venous pressure), which is linearly proportional to stenotic coronary flow (Q^S_{max}), relative to the coronary driving pressure assumed in the absence of a stenosis (mean aor-

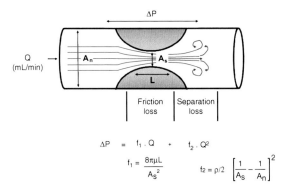

Fig. 20.8 Hydrodynamic factors for pressure drop across a stenosis. The coronary flow distal to the stenosis can vary according to pressure drop across a stenosis and microcirculatory function. The total pressure drop across a stenosis is governed by three hydrodynamic factors—friction losses, separation losses, and turbulence. Separation losses are mainly influenced by flow rates distal to the stenosis and determine curve linearity or steepness of the stenosis pressure-flow relationship. Abbreviations: A_s cross-sectional area of the stenosis segment, A_n cross-sectional area of the reference segment, ΔP pressure gradient across stenosis, f_1 friction coefficient, f_2 separation coefficient, ρ blood density, L stenosis length, μ absolute blood viscosity, Q coronary flow

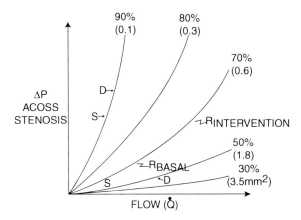

Fig. 20.9 Stenosis pressure and flow relationship. When exploring the relationship between the pressure gradient and absolute coronary flow resting to hyperemic status according to the stenosis severity, a unique curve-linear relationship between the pressure gradient and absolute coronary flow is shown across the different range of stenosis severity. This figure was adapted and modified from the original article of Klocke et al. JACC 1983 under permission of the publisher

tic pressure—coronary venous pressure), which is linearly proportional to assumed normal coronary flow (Q^N_{max}) [11, 12]. As venous pressures are generally negligible compared to arterial pressure, this results in the simplified clinical index of mean distal coronary pressure/mean aortic pressure (P_d/P_a) (Fig. 20.11) [13–15]. FFR of 0.80 means that diseased coronary artery with epicardial stenosis supplies 80% of the assumed normal maximal flow without epicardial stenosis. The strength of FFR is that FFR can assess the degree and presence of epicardial lesion-specific inducible myocardial ischemia, not only in cases with negative or ambiguous results of noninvasive functional tests but also in the presence of multivessel disease. FFR also has excellent reproducibility, regardless of changes in hemodynamics or myocardial contractility [16–19].

As previously mentioned, the fundamental assumption in the concept of FFR is a linear relationship between coronary flow and pressure. As

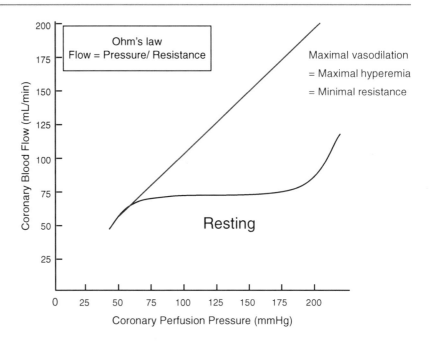

Fig. 20.10 The concept of maximal perfusion and hyperemia. In maximal vasodilation induced by pharmacologic agents, distal coronary pressure is directly proportional to maximum vasodilated perfusion and coronary flow. Therefore, in the hyperemic condition, coronary pressure can be interpreted as surrogate marker of coronary flow

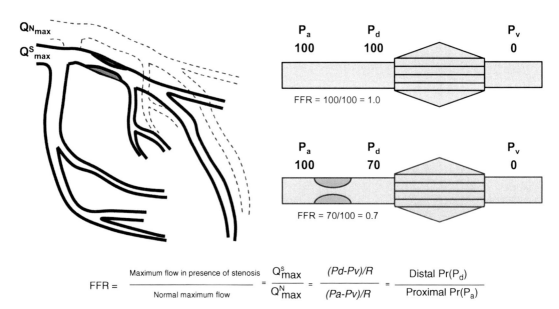

Fig. 20.11 The concept of fractional flow reserve. Fractional flow reserve is an indirect index determined by measuring driving pressure for microcirculatory flow distal to the stenosis (distal coronary pressure—coronary venous pressure), which is linearly proportional to stenotic coronary flow (Q^S_{max}), relative to the coronary driving pressure assumed in the absence of a stenosis (mean aortic pressure—coronary venous pressure), which is linearly proportional to assumed normal coronary flow (Q^N_{max}). As venous pressures are generally negligible compared to arterial pressure, this results in the simplified clinical index of mean distal coronary pressure/mean aortic pressure (P_d/P_a). Abbreviations: *FFR* fractional flow reserve, P_a aortic pressure, P_d distal coronary pressure, P_v venous pressure, Q coronary flow, R microvascular resistance

coronary blood flow and resistance are regulated by autoregulation, according to the myocardial demand, within the physiological range [20], induction of maximal hyperemia or minimizing microvascular resistance is mandatory for FFR measurement [13, 14]. To achieve maximal hyperemia for FFR measurement, continuous intravenous infusion of adenosine (140 μg/kg/min) is considered as a gold standard method [21]. However, the various hyperemic agents with comparable hyperemic efficacy along with the enhanced safety and feasibility have been recently introduced (the detailed descriptions of various hyperemic agents are described in the next chapter). A significant advantage of FFR is that there is now considerable prognostic information including large prospective randomized study and registry data, which will be further discussed in the later chapters [22–30].

References

1. Camici PG, Crea F. Coronary microvascular dysfunction. N Engl J Med. 2007;356:830–40.
2. Crea F, Camici PG, Bairey Merz CN. Coronary microvascular dysfunction: an update. Eur Heart J. 2014;35:1101–11.
3. Heller LI, Silver KH, Villegas BJ, Balcom SJ, Weiner BH. Blood flow velocity in the right coronary artery: assessment before and after angioplasty. J Am Coll Cardiol. 1994;24:1012–7.
4. Olsson RA, Gregg DE. Myocardial reactive hyperemia in the unanesthetized dog. Am J Phys. 1965;208:224–30.
5. Lilly LS, Braunwald E. Braunwald's heart disease: a textbook of cardiovascular medicine. 9th ed. Philadelphia: Elsevier; 2012.
6. Canty JM Jr. Coronary pressure-function and steady-state pressure-flow relations during autoregulation in the unanesthetized dog. Circ Res. 1988;63:821–36.
7. Klocke FJ. Coronary blood flow in man. Prog Cardiovasc Dis. 1976;19:117–66.
8. Gould KL, Kirkeeide RL, Buchi M. Coronary flow reserve as a physiologic measure of stenosis severity. J Am Coll Cardiol. 1990;15:459–74.
9. Kern MJ, Lerman A, Bech JW, et al. Physiological assessment of coronary artery disease in the cardiac catheterization laboratory: a scientific statement from the American Heart Association committee on diagnostic and interventional cardiac catheterization, council on clinical cardiology. Circulation. 2006;114:1321–41.
10. Klocke FJ. Measurements of coronary blood flow and degree of stenosis: current clinical implications and continuing uncertainties. J Am Coll Cardiol. 1983;1:31–41.
11. Pijls NH, Van Gelder B, Van der Voort P, et al. Fractional flow reserve. A useful index to evaluate the influence of an epicardial coronary stenosis on myocardial blood flow. Circulation. 1995;92:3183–93.
12. Spaan JA, Piek JJ, Hoffman JI, Siebes M. Physiological basis of clinically used coronary hemodynamic indices. Circulation. 2006;113:446–55.
13. Pijls NH, van Son JA, Kirkeeide RL, De Bruyne B, Gould KL. Experimental basis of determining maximum coronary, myocardial, and collateral blood flow by pressure measurements for assessing functional stenosis severity before and after percutaneous transluminal coronary angioplasty. Circulation. 1993;87:1354–67.
14. Pijls NH, De Bruyne B, Peels K, et al. Measurement of fractional flow reserve to assess the functional severity of coronary-artery stenoses. N Engl J Med. 1996;334:1703–8.
15. Young DF, Cholvin NR, Kirkeeide RL, Roth AC. Hemodynamics of arterial stenoses at elevated flow rates. Circ Res. 1977;41:99–107.
16. de Bruyne B, Bartunek J, Sys SU, Pijls NH, Heyndrickx GR, Wijns W. Simultaneous coronary pressure and flow velocity measurements in humans. Feasibility, reproducibility, and hemodynamic dependence of coronary flow velocity reserve, hyperemic flow versus pressure slope index, and fractional flow reserve. Circulation. 1996;94:1842–9.
17. Hwang D, Lee JM, Koo BK. Physiologic assessment of coronary artery disease: focus on fractional flow reserve. Korean J Radiol. 2016;17:307–20.
18. Kern MJ, Samady H. Current concepts of integrated coronary physiology in the catheterization laboratory. J Am Coll Cardiol. 2010;55:173–85.
19. Koo B-K. The present and future of fractional flow reserve. Circ J. 2014;78:1048–54.
20. Mosher P, Ross J Jr, McFate PA, Shaw RF. Control of coronary blood flow by an autoregulatory mechanism. Circ Res. 1964;14:250–9.
21. Pijls NH, van Nunen LX. Fractional flow reserve, maximum hyperemia, adenosine, and regadenoson. Cardiovasc Revasc Med. 2015;16:263–5.
22. Bech GJ, De Bruyne B, Pijls NH, et al. Fractional flow reserve to determine the appropriateness of angioplasty in moderate coronary stenosis: a randomized trial. Circulation. 2001;103:2928–34.
23. DeBruyne B, Pijls N, Kalesan B, et al. Fractional flow reserve-guided PCI versus medical therapy in stable coronary disease. N Engl J Med. 2012;367:991.
24. Di Serafino L, De Bruyne B, Mangiacapra F, et al. Long-term clinical outcome after fractional flow reserve- versus angio-guided percutaneous coronary intervention in patients with intermediate stenosis of coronary artery bypass grafts. Am Heart J. 2013;166:110–8.

25. Frohlich GM, Redwood S, Rakhit R, et al. Long-term survival in patients undergoing percutaneous interventions with or without intracoronary pressure wire guidance or intracoronary ultrasonographic imaging: a large cohort study. JAMA Intern Med. 2014;174:1360–6.
26. Li J, Elrashidi MY, Flammer AJ, et al. Long-term outcomes of fractional flow reserve-guided vs. angiography-guided percutaneous coronary intervention in contemporary practice. Eur Heart J. 2013;34:1375–83.
27. Park SJ, Ahn JM, Park GM, et al. Trends in the outcomes of percutaneous coronary intervention with the routine incorporation of fractional flow reserve in real practice. Eur Heart J. 2013;34:3353–61.
28. Pijls NH, Fearon WF, Tonino PA, et al. Fractional flow reserve versus angiography for guiding percutaneous coronary intervention in patients with multivessel coronary artery disease: 2-year follow-up of the FAME (fractional flow reserve versus angiography for multivessel evaluation) study. J Am Coll Cardiol. 2010;56:177–84.
29. van Nunen LX, Zimmermann FM, Tonino PA, et al. Fractional flow reserve versus angiography for guidance of PCI in patients with multivessel coronary artery disease (FAME): 5-year follow-up of a randomised controlled trial. Lancet. 2015;386:1853.
30. Zimmermann FM, Ferrara A, Johnson NP, et al. Deferral vs. performance of percutaneous coronary intervention of functionally non-significant coronary stenosis: 15-year follow-up of the DEFER trial. Eur Heart J. 2015;36:3182–8.

Setup for Fractional Flow Reserve and Hyperemia

Ho-Jun Jang and Sung Gyun Ahn

Fractional flow reserve (FFR) is defined as the ratio of flow in the stenotic artery to the flow in the same artery in the hypothetical absence of the stenosis during maximal hyperemia. Pressure-derived FFR is calculated as the ratio of distal coronary pressure (Pd) to aortic pressure (Pa) obtained during maximal hyperemia. To measure FFR, four things are mainly needed as the following: (1) Pa measured by the conventional catheter-based blood pressure system, (2) Pd measured by the pressure sensor, (3) induction of maximal hyperemia, and (4) the systematic step-by-step procedure algorithm (Fig. 21.1). Currently, five coronary pressure-measuring systems are commercially available to assess intracoronary pressure: (1) *PressureWire* (St. Jude Medical, St. Paul, Minnesota), (2) *WaveWire* (Philips, Eindhoven, the Netherlands), (3) *OptoWire* (Opsens Medical, Quebec, Canada), (4) *Comet Pressure Guidewire* (Boston Scientific, Marlborough, Massachusetts) (5) *Navvus* (Acist Medical Systems, Eden Prairie, Minnesota) [1]. These except the fifth are 0.014-in pressure-measuring guidewires, equipped with an electric (first two) or a fiber optic (latter two) pressure sensor 3–3.5 cm from the tip, at the junction between radiopaque and non-radiopaque portions of the wire. These can be used as a regular guidewire when percutaneous coronary intervention becomes indicated. *Navvus* is a 0.020-inch monorail pressure-measuring microcatheter, equipped with a fiber optic pressure sensor. It can be used over any regular guidewire. Adenosine, nicorandil (Sigmart®, Chugai Pharmaceutical, Tokyo, Japan), regadenoson (Rapidscan®, Pharma Solutions, London, UK), sodium nitroprusside, and papaverine can be used to induce hyperemia for FFR measurements (Table 21.1). Intravenous (IV) infusion of adenosine is the gold standard for steady state hyperemic induction. Alternative methods, including intracoronary (IC) bolus injection of adenosine and of nicorandil, are also available. In measurement of FFR for daily practice or clinical research, technical accuracy, avoidance of measurement artifacts, and maintaining reproducibility of FFR values are critical [2]. Therefore, standardization of FFR measurements with systematic step-by-step procedure is highly recommended [1].

H.-J. Jang
Department of Cardiology, Cardiovascular Center, Sejong General Hospital, Bucheon, South Korea

S.G. Ahn (✉)
Division of Cardiology, Department of Internal Medicine, Yonsei University Wonju College of Medicine, Wonju, South Korea
e-mail: sgahn@yonsei.ac.kr

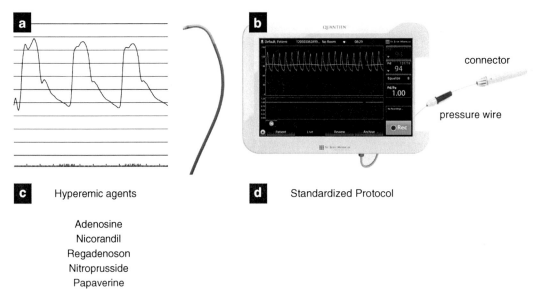

Fig. 21.1 Four basic requirements for FFR measurement. (1) Aortic pressure (Pa) measured by the conventional catheter-based blood pressure system. (2) Distal coronary pressure (Pd) measured by the pressure sensor. (3) Induction of maximal hyperemia. (4) Systematic step-by-step procedure algorithm

Table 21.1 Different hyperemic stimuli

Agent	Route	Dose	Time to peak hyperemia	Duration of plateau
Adenosine	IC bolus	100 μg in the RCA, 200 μg in the LCA	~10 s	12 s (RCA), 21 s (LCA)
	IV infusion (central or antecubital vein)	140 μg/kg/min	40 s	Steady state
Nicorandil	IC bolus	2 mg	15–20s	17–33 s
Regadenoson	IV bolus	400 μg	30–90 s	10–600 s
Nitroprusside	IC bolus	0.6 μg/kg	[a]	[a]
Papaverine	IC bolus	8 mg in the RCA, 12 mg in the LCA	[a]	[a]

[a]Currently, nitroprusside and papaverine are seldom used in real clinical practice because of their lower safety profiles compared to top three hyperemic stimuli. *IC* intracoronary, *IV* intravenous, *LCA* left coronary artery, *RCA* right coronary artery

21.1 Calibrating Pressure System

Like the catheter-based blood pressure system, the pressure transducer should be carefully calibrated before the pressure wire is inserted. Before connecting the pressure wire to the interface connector cable, the proximal tip of the wire where three flat electrodes are located should be gently wiped out with dry gauge to remove blood, contrast medium, and damp on the wire. Then, the proximal end of the wire is plugged into the interface connector cable (Fig. 21.1b). When it is connected successfully, the signal of the pressure wire will appear on the monitor. The wire and the sensor should be flushed with heparinized saline inside the plastic housing. Finally, located on the flat angiography table at the patient's heart level, the pressure transducer is zeroed and calibrated manometrically.

21.2 Equalizing Two Pressures

After calibration of the microchip transducer of the pressure guidewire for distal coronary pressure, the pressure wire is steered into the proximal part of the artery to be interrogated. It is essential to ensure the pressures by the guiding catheter and the pressure sensor equalized before advancing the wire over the coronary lesions. The radiopaque tip of the pressure wire should be visible outside the catheter, with the sensor positioned 1–2 mm distal to the guiding catheter. Catheter-induced pressure damping or ventricularization (Fig. 21.2) should be monitored which is caused by engagement of the catheter in significant ostial lesions, misalignment of the guides, during coronary spasm, or when a larger Fr guiding catheter is engaged in a smaller coronary artery (Fig. 21.3). When ventricularization or pressure damping appears on the arterial pressure tracings, the guiding catheter should be properly disengaged from the ostium of the coronary artery, and then the pressure sensor should be repositioned in the aorta outside the coronary artery. The guiding catheters with side holes are generally not recommended for FFR measurements because the pressure signal attained through these catheters reflects a mix between the coronary pressure (through the distal end) and the Pa (through the side holes). Indeed, FFR values are underestimated (i.e. lesion severity is overestimated) in side-hole catheters when FFR measurements are performed with engaged guide catheters [3]. If side-hole catheters are clinically indicated due to significant ostial stenosis, IV continuous infusion of adenosine (not IC injection) is mandatory for adequate hyperemia induction, and the catheter should be disengaged from the coronary artery. After positioning the pressure wire at the tip of the catheter, the guiding catheter is properly flushed with saline to remove any residual contrast. The introducer is removed from the Y connector to prevent the pressure leak. Finally, the two measured pressures are equalized electronically using that function of the console.

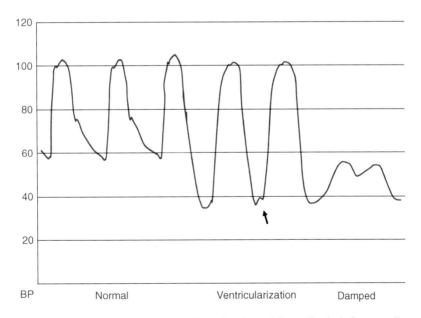

Fig. 21.2 Arterial pressure tracings from a guiding catheter. The arterial pressure waveform consists of systolic upstroke, systolic peak pressure, systolic decline, dicrotic notch, diastolic runoff, and end-diastolic pressure. The ventricularized pressure has a subtle decrease in systolic pressure, rapid diastolic decline, and small positive deflection (*arrow*) immediately before systolic upstroke. The damped pressure has sluggish oscillations following the downstroke and reduction in the pulse pressure (decreased systolic pressure and/or increased diastolic pressure) and loss of the dicrotic notch

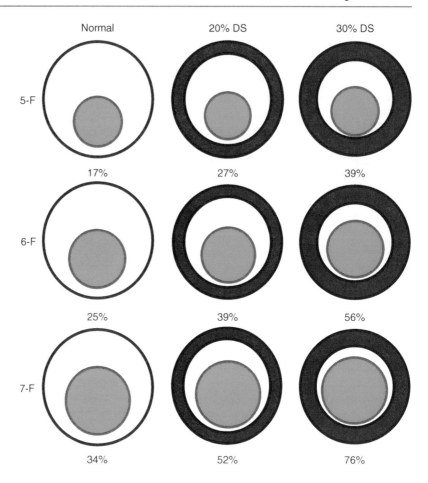

Fig. 21.3 Effect of guiding catheters in different coronary ostia. The presence of a guiding catheter in the coronary ostium can significantly impede coronary blood flow. The figure illustrates the potential area obstruction (values in *red*) of different sizes of guiding catheters (5 to 7 F from the first to the last row) in a 4-mm coronary ostium when it is intact (*left*), when it is 20% stenosed (*middle*), and when it is 30% stenosed (*right*). *DS* diameter stenosis. J Am Coll Cardiol 2016;68:742–53; with permission

21.3 Positioning the Pressure Sensor

Before advancing the pressure sensor, it is advisable to wait for 10–30 s to avoid spontaneous pressure signal drift. If it does occur, two pressures should be re-equalized. The pressure sensor is positioned as distally as possible in the coronary artery [4], at least 2–3 cm distal to the stenosis to be assessed. The exact position of the sensor should then be documented using side branches, stenoses, or coronary stent deployed as fiducial and recorded by fluoroscopy.

21.4 Inducing Maximal Hyperemia and Recording

For maximal vasodilation of the epicardial artery, IC bolus injection of nitroglycerine of 200 μg should precede hyperemic stimulus administration. In addition to vasodilators, anticoagulant should be given intravenously or subcutaneously as per local protocol. Before administering a hyperemic agent, the stable baseline Pd/Pa should be recorded for several heartbeats to measure resting coronary flow.

21.4.1 Intravenous Adenosine

Adenosine is a purine nucleoside composed of a molecule of adenine attached to a ribose sugar molecule and plays an important role in energy transfer process. The coronary vasodilating effect of adenosine occurs through *A2A* receptor, although adenosine is a nonselective stimulus to all types of adenosine receptors. As the plasma concentration is tightly controlled by several clearance mechanisms including uptake by red blood cells, its biological half-life is less than 10 s. Continuous infusion of adenosine via central vein is the current gold standard to induce maximal hyperemia. With the conventional dosage of 140 μg/kg/min, maximal hyperemia is usually achieved in approximately 40 s (Fig. 21.4). A transient rise in Pa often occurs before Pd reduces and adopts an ischemic waveform with diastolic blunting. It is advisable to fill the IV line with adenosine to avoid a delay between the start of the pump and the effect of adenosine. The hyperemic efficacy of IV adenosine via the forearm vein is comparable to that via the central vein [5]. During transradial coronary catheterization, peripheral IV infusion of adenosine is preferred to central IV infusion due to its convenience and similar hyperemic efficacy for FFR measurement. If FFR value is not stabilized by IV adenosine, increasing infusion dose of IV adenosine to 160–180 μg/kg/min can be helpful. Recording of Pd/Pa starts with adenosine infusion and ends when reaching stable hyperemic plateau. During the infusion of adenosine, patients will often complain of chest tightness and dyspnea. Blood pressure drop or atrioventricular (AV) block should be closely monitored [6].

21.4.2 Intracoronary Adenosine

IC injection of adenosine can also be given to induce maximal hyperemia (Fig. 21.5). IC adenosine of 40–200 μg for the left coronary artery and 20–100 μg for the right coronary artery are briskly injected (in 1–2 heart beats) which are diluted in 3–5 ml saline. The guiding catheter is properly engaged with the coronary ostium during the bolus injection and then quickly disengaged immediately after the injection to avoid any wedging phenomena. Recording of Pd/Pa starts before the bolus injection so that the baseline Pd/Pa can be seen and is traced till returning to the baseline value. Transient AV block can occur around 15% after injection of 200 μg in the left coronary artery (LCA) and 40% after injection of 100 μg in the right coronary artery (RCA) [7]. The hyperemic efficacy of IC adenosine injection is non-inferior to that of IV infusion [8]. However, the plateau of maximal hyperemia lasts only 12 ± 13 s after injection of 100 μg in the RCA and 21 ± 6 s after the injection of

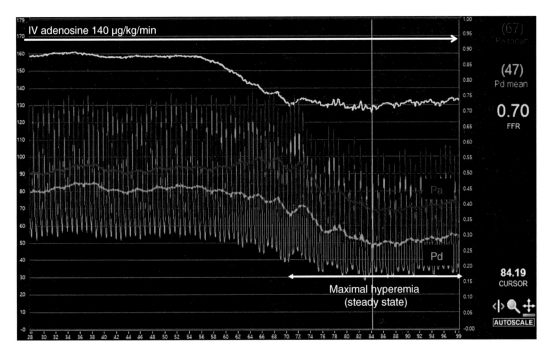

Fig. 21.4 FFR tracing after intravenous infusion of adenosine. Before reaching maximal hyperemia, a short-lasting increase in Pa is followed by a reduction in Pd. During maximal hyperemia, the fluctuations of Pa are paralleled by proportional fluctuations of Pd resulting in a stable Pd/Pa ratio

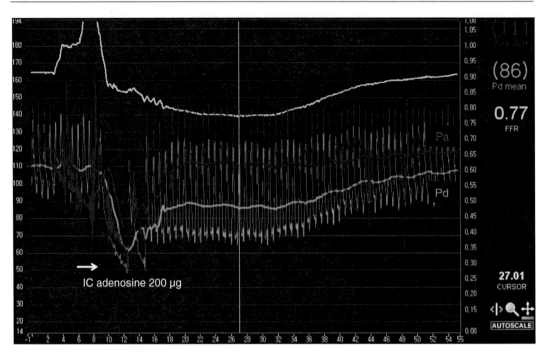

Fig. 21.5 FFR tracing after intracoronary bolus injection of adenosine. The injection of the bolus is brisk so that the aortic signal (*red*) is interrupted during few sec

200 μg in the LCA [7]. Accordingly, pressure pullback tracing and measurement of thermodilution-derived coronary flow reserve and index of microcirculatory resistance cannot be performed by using IC bolus administration of adenosine [9].

21.4.3 Intracoronary Nicorandil

Nicorandil is a nicotinamide ester with dual mechanisms of action on both macro- and microvascular circulation. Biological half-life of the drug is around 1 h, far longer than that of adenosine. IC injection of nicorandil of 2 mg has similar hyperemic efficacy compared to IV infusion of adenosine [8, 10]. Maximal hyperemia is usually achieved in 15–20 s and lasts for 17–33 s (Fig. 21.6). The detailed method of IC administration is identical to that of IC adenosine, except that IC nitroglycerin is not mandatory.

21.4.4 Intravenous Regadenoson

Regadenoson is a direct *A2A* receptor-selective hyperemic stimulus that can be administered as a single bolus IV injection either peripherally or centrally. Hyperemic efficacy of IV bolus injection of regadenoson 400 μg was comparable to that of IV infusion of adenosine. Maximal hyperemia is achieved in 30–90 s and the duration of hyperemic plateau is variable (10–600 s). Regadenoson produces less frequently tachycardia, blood pressure decrease, or AV block compared to adenosine [11].

Hyperemic efficacy is not significantly different according to the different route of adenosine administration (IC vs. IV) or the type of hyperemic stimuli (adenosine vs. regadenoson vs. nicorandil) [8, 10, 11]. If the submaximal hyperemic induction is suspected, remeasurement of FFR using another hyperemic induction modality is highly recommended for the clarification.

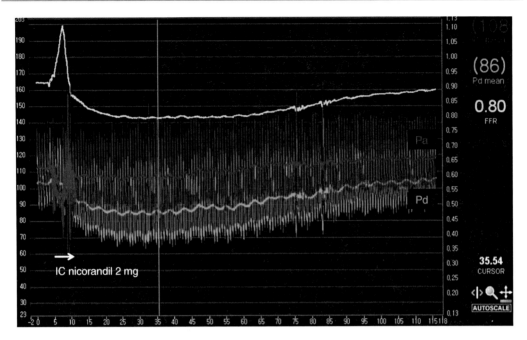

Fig. 21.6 FFR tracing after intracoronary bolus injection of nicorandil. The detailed method of intracoronary administration of nicorandil is quite similar to that of intracoronary adenosine. The plateau of maximal hyperemia with intracoronary nicorandil is somewhat longer compared to intracoronary injection of adenosine

Fig. 21.7 Patterns of Pd/Pa during hyperemia. Three major Pd/Pa response patterns of varying frequency occurred during steady intravenous adenosine infusion: "classic" (sigmoid), "humped" (sigmoid with superimposed bumps of varying height), and "unusual" (no pattern). For each observed example, the red dot marks the smart minimum FFR. The blue scale for Pd/Pa and time applies to the example tracings. JACC Cardiovasc Interv 2015;8:1018–27; with permission

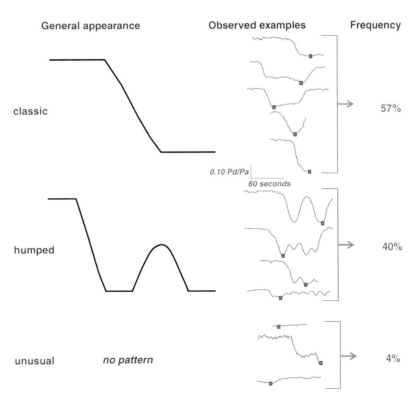

21.5 Determination of the FFR Value

Various hyperemic response patterns can be manifested after IV infusion of adenosine: (1) "classic" stable pattern (sigmoid shape, 57% of responses), (2) "humped" pattern (sigmoid with superimposed bumps, 39%), and (3) "unusual" pattern (no particular shape, 4%) (Fig. 21.7) [12]. Regardless of different hyperemia patterns, FFR is determined as the lowest value of the measured Pd/Pa under the maximal hyperemia. Although FFR value is automatically indicated on the pressure-measuring system, confirmation of its appropriateness should be made in every case. The operators should also make sure to avoid false calculations derived from artifacts in the coronary (drift, whipping, accordion effect) or the arterial pressure tracings (damping, ventricularization) [2].

21.6 Pullback Maneuver

A slow pullback of the pressure sensor provides important information on the distribution of the abnormal epicardial resistance in case of diffuse atherosclerotic disease or serial tandem stenoses (Fig. 21.8). It can be performed under steady state of hyperemia induced by IV infusion of adenosine.

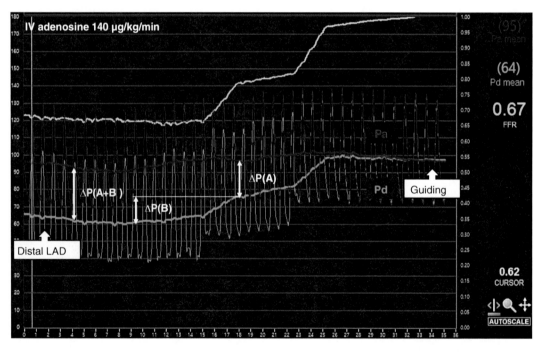

Fig. 21.8 Pressure wire pullback tracing. The pressure wire was pulled back from the distal left anterior descending artery (LAD) to the guiding catheter. FFR was 0.67 reflecting the ratio of coronary pressure generated by summation [[$\Delta P (A + B)$]] of proximal [$\Delta P (A)$] and distal [$\Delta P (B)$] stenoses to aortic pressure. Coronary stenting was performed in the proximal LAD (A) where the biggest pressure drop [$\Delta P (A)$] occurs

21.7 Final Check for Signal Drift

After finishing the FFR measurement, the pressure sensor is pulled back until positioned at the tip of the guiding catheter. For reliable FFR measurement, the pressure signal drift should be minimal. Although there is no consensus concerning the drift threshold for repeat FFR measurements, a small difference of less than 2–3 mmHg between the two pressures is usually acceptable. Pressure signal drift more than 4 mmHg can be corrected by calculation. However, repeated measurement is generally recommended.

In conclusion, FFR has a large body of evidence of its usefulness for detecting potentially flow-limiting coronary stenosis and anticipated benefit of revascularization according to FFR value. Attaining reliable and consistent FFR values are prerequisite for both clinician and patients to apply evidence-based medicine. Therefore, not only understanding basic physiology and clinical significance of FFR but also proper set up for FFR and hyperemia are crucial to maximize benefit of FFR-based decision-making.

References

1. Toth GG, Johnson NP, Jeremias A, Pellicano M, Vranckx P, Fearon WF, et al. Standardization of fractional flow reserve measurements. J Am Coll Cardiol. 2016;68(7):742–53.
2. Sharif F, Trana C, Muller O, De Bruyne B. Practical tips and tricks for the measurement of fractional flow reserve. Catheter Cardiovasc Interv. 2010;76(7):978–85.
3. Patel KS, Christakopoulos GE, Karatasakis A, Danek BA, Nguyen-Trong PK, Amsavelu S, et al. Prospective evaluation of the impact of side-holes and guide-catheter disengagement from the coronary ostium on fractional flow reserve measurements. J Invasive Cardiol. 2016;28(8):306–10.
4. Rodes-Cabau J, Gutierrez M, Courtis J, Larose E, Dery JP, Cote M, et al. Importance of diffuse atherosclerosis in the functional evaluation of coronary stenosis in the proximal-mid segment of a coronary artery by myocardial fractional flow reserve measurements. Am J Cardiol. 2011;108(4):483–90.
5. Seo MK, Koo BK, Kim JH, Shin DH, Yang HM, Park KW, et al. Comparison of hyperemic efficacy between central and peripheral venous adenosine infusion for fractional flow reserve measurement. Circ Cardiovasc Interv. 2012;5(3):401–5.
6. Pijls NH, De Bruyne B, Peels K, Van Der Voort PH, Bonnier HJ, Bartunek JKJJ, et al. Measurement of fractional flow reserve to assess the functional severity of coronary-artery stenoses. N Engl J Med. 1996;334(26):1703–8.
7. Adjedj J, Toth GG, Johnson NP, Pellicano M, Ferrara A, Flore V, et al. Intracoronary adenosine: dose-response relationship with hyperemia. JACC Cardiovasc Interv. 2015;8(11):1422–30.
8. Lim WH, Koo BK, Nam CW, Doh JH, Park JJ, Yang HM, et al. Variability of fractional flow reserve according to the methods of hyperemia induction. Catheter Cardiovasc Interv. 2015;85(6):970–6.
9. Ahn SG, Hung OY, Lee JW, Lee JH, Youn YJ, Ahn MS, et al. Combination of the thermodilution-derived index of microcirculatory resistance and coronary flow reserve is highly predictive of microvascular obstruction on cardiac magnetic resonance imaging after ST-segment elevation myocardial infarction. JACC Cardiovasc Interv. 2016;9(8):793–801.
10. Jang HJ, Koo BK, Lee HS, Park JB, Kim JH, Seo MK, et al. Safety and efficacy of a novel hyperaemic agent, intracoronary nicorandil, for invasive physiological assessments in the cardiac catheterization laboratory. Eur Heart J. 2013;34(27):2055–62.
11. van Nunen LX, Lenders GD, Schampaert S, van 't Veer M, Wijnbergen I, Brueren GR, et al. Single bolus intravenous regadenoson injection versus central venous infusion of adenosine for maximum coronary hyperaemia in fractional flow reserve measurement. EuroIntervention. 2015;11(8):905–13.
12. Johnson NP, Johnson DT, Kirkeeide RL, Berry C, De Bruyne B, Fearon WF, et al. Repeatability of fractional flow reserve despite variations in systemic and coronary hemodynamics. JACC Cardiovasc Interv. 2015;8(8):1018–27.

Validation of Fractional Flow Reserve

22

Sung Eun Kim and Jung-Won Suh

22.1 Introduction

Fractional flow reserve (FFR) is now considered a gold standard for the invasive assessment of myocardial ischemia. Since the concept of FFR was developed and introduced by Pijls and De Bruyne in the early 1990s, not only the benefit of FFR-guided revascularization strategy but also the issues of procedural feasibility have been thoroughly validated. In this chapter, evidences will be reviewed starting from the first validation study in animal and human to recent studies supporting the clinical benefit of FFR in daily practices.

22.2 First Animal and Human Validation

Pijls introduced the term "fractional flow reserve" (FFR) in 1993. In his foundational article for FFR [1], three reasons why pressure measurements have not been useful until then were clarified:

S.E. Kim
Division of Cardiology, Kangdong Sacred Heart Hospital, Hallym University College of Medicine, Seoul, South Korea

J.-W. Suh (✉)
Division of Cardiology, Seoul National University Bundang Hospital, Seoul National University College of Medicine, Seoul, South Korea
e-mail: suhjw1@gmail.com

unsuitable device to measure pressure, uncontrolled resistances in coronary circulation (i.e., lack of maximal vasodilation), and failure to account for collateral flow. They used thinner pressure monitoring wire (0.015 in.) and controlled myocardial resistance with maximal hyperemia using intracoronary administration of papaverine. The key component of their model was that it could distinguish between contributions from the epicardial conduit (fractional coronary artery flow reserve [FFRcor]) and collateral channel to myocardial blood flow (fractional myocardial flow reserve [FFRmyo]). In their experiment using five dogs, relative maximum blood flow through the stenotic artery (Qs) measured directly by Doppler flowmeter showed an excellent correlation with pressure-derived values (Qs) of the maximal myocardial blood flow and collateral blood flow (Fig. 22.1).

The first validation study in human was performed by De Bruyne in 1994 [2]. In 22 patients, myocardial and coronary fractional flow reserve was calculated from mean aortic, distal coronary, and right atrial pressures which was recorded during maximal vasodilation. Additionally, relative myocardial flow reserve, defined as the ratio of absolute myocardial perfusion during maximal vasodilation in the stenotic area to the absolute myocardial perfusion during maximal vasodilation (adenosine 140 μg/kg/min intravenously during 4 min) in the contralateral normally perfused area, was assessed by ^{15}O–labeled water and positron emission tomography (PET). Fractional flow

© Springer Nature Singapore Pte Ltd. 2018
M.-K. Hong (ed.), *Coronary Imaging and Physiology*,
https://doi.org/10.1007/978-981-10-2787-1_22

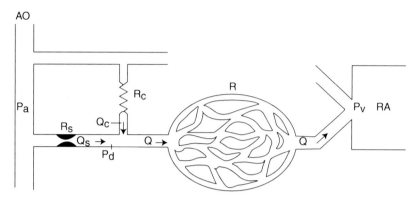

Fig. 22.1 Schematic model representing the coronary circulation. *AO* aorta, *Pa* arterial pressure, *Pd* distal coronary pressure, *Pv* venous pressure, *Q* blood flow through the myocardial vascular bed, *Qc* collateral blood flow, *Qs* blood flow through the supplying epicardial coronary artery, *R* resistance of the myocardial vascular bed, *Rc* resistance of the collateral circulation, *Rs* resistance of the stenosis in the supplying epicardial coronary artery, *RA* right atrium. Pijls NH, van Son JA, Kirkeeide RL, De Bruyne B, Gould KL. Experimental basis of determining maximum coronary, myocardial, and collateral blood flow by pressure measurements for assessing functional stenosis severity before and after percutaneous transluminal coronary angioplasty. Circulation. 1993; 86:1354–67

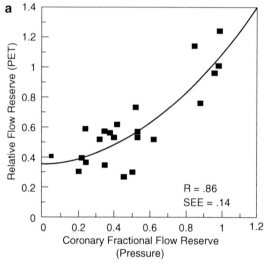
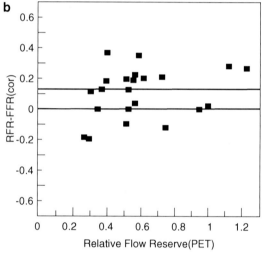

Fig. 22.2 (**a**) Plot shows relation between the relative myocardial flow reserve of the anterior region as determined by positron emission tomography (PET) and the coronary fractional flow reserve of the stenosis in the proximal left anterior descending coronary artery. (**b**) Plot of the difference between the relative flow reserve and the coronary fractional flow reserve [RFR-FFR(cor)] values. Solid line represents mean difference; dashed lines represent 2 SD from this mean. De Bruyne B, Baudhuin T, Melin JA, Pijls NH, Sys SU, Bol A, et al. Coronary flow reserve calculated from pressure measurements in humans. Validation with positron emission tomography. Circulation. 1994; 89:1013–22

reserve derived from pressure measurements correlated closely to the relative flow reserve derived from PET (Fig. 22.2). Furthermore, they also showed that the correlation between relative flow reserve obtained by PET and percentile stenosis measured from quantitative coronary angiography were markedly weaker.

22.3 Cutoff Value of 0.75 or 0.80

One of the specific features that make FFR particularly useful is that it has a normal value of 1 for every patient and every artery. Furthermore, any decrease in FFR has a direct clinical implication: For example, FFR of 0.60 means maximum blood flow to the myocardial distribution of the respective artery only reaches 60% of what it would be if that artery were completely normal [3]. However, for clinical decision-making such as whether to perform revascularization or not, we need a specific cutoff value of this continuous variable. The ischemic threshold of 0.75 was first proposed by Pijls et al. in 1995 [4]. In the report, they confirmed that normal FFR equals to 1.0 in five patients with normal coronary arteries. By utilizing FFR data of patients with stable angina, single-vessel disease, normal left ventricular function, and a positive exercise test before PTCA which normalized after angiographically successful PTCA, they showed that with the cutoff point of 0.74, there was only a minimal overlap between normal and pathological values (Fig. 22.3).

Subsequent clinical studies validated the diagnostic accuracy of FFR compared with other methods to evaluate myocardial ischemia. The specificity was reported as 82 ~ 100% and sensitivity as 68 ~ 88% with cutoff value of 0.66–0.78 (Table 22.1). In the FAME study, the investigators decided to choose 0.80 as a cutoff value based on the fact that many interventional cardiologists elect to perform PCI when the FFR value is between 0.75 and 0.80 if the clinical scenario is suggestive of myocardial ischemia [10]. Recently, many clinicians use FFR ≤ 0.80 as a cutoff value to guide revascularization, and current guidelines also recommend its clinical use based on FFR ≤ 0.80 [11, 12]. Because there is a gray zone in FFR value, which is between 0.76 and 0.80, sometimes clinicians feel confused. Repeating the measurement may not be helpful because there was a report that in this gray zone, even agreement among

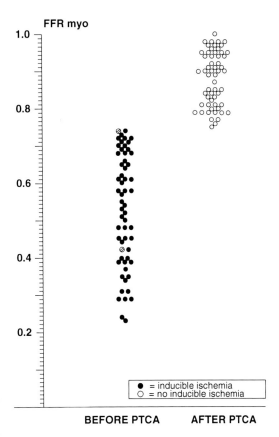

Fig. 22.3 Scatterplot showing values of FFRmyo before and after PTCA. Those values associated with proven ischemia are indicated by solid circles, and those values definitely not associated with ischemia are indicated by open circles.

Table 22.1 Summary of cutoff fractional flow reserve values which suggest myocardial ischemia

Authors	Year	Patients	Number	Comparator	Value	Specificity	Sensitivity	Reference
De Bruyne	1995	Single vessel	60	Bicycle ET	0.66	87%	87%	De Bruyne et al. [5]
Pijls	1996	Single vessel	45	ET + SPECT + DSE	0.75	100%	88%	Pijls et al. [6]
Abe	2000	Single vessel	46	SPECT	0.75	100%	83%	Abe et al. [7]
Chamuleau	2001	Multivessel	152	SPECT	0.74	82%	68%	Chamuleau et al. [8]
De Bruyne	2001	Infarct > 5 days	50	SPECT	0.78	88%	88%	De Bruyne et al. [9]

ET exercise test, *SPECT* single-photon emission computed tomography, *DSE* dobutamine stress echocardiography

each measurement falls, reaching nadir of approximately 50% around FFR value of 0.80 [13] (Fig. 22.4). Therefore, many experts recommend that decision-making should be based on sound clinical judgment, typicality of symptoms, presence of other test results, and technical issues related to the measurement of FFR [3].

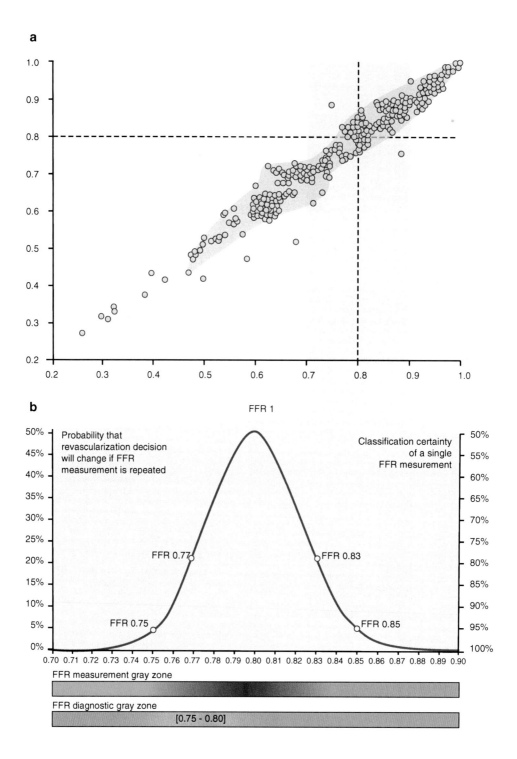

Fig. 22.4 Biological variability of FFR. Test-retest reproducibility of two repeated measurements of fractional flow reserve (FFR) taken 10 min apart is shown as a scatterplot (**a**, *gray dashed* envelope demarcates 99% of the data points from 0.5 to 1 and dotted lines show the 0.8 cutoffs). The classification certainty of a single FFR measurement is presented for FFR values from 0.70 to 0.90 (**b**, *right vertical axis*). Outside the 0.75 to 0.85 range, measurement certainty is higher than 95%. However, closer to its cut point, this certainty falls, reaching a nadir of approximately 50% around 0.8. In clinical practice, that means each time a single FFR value falls between 0.75 and 0.85, there is a chance that the dichotomous classification of a stenosis (and therefore the FFR-guided revascularization decision) will change if the test is repeated 10 min later. Within 0.77 to 0.83, this measurement certainty falls to <80%. The FFR diagnostic gray zone (0.75 to 0.85) is also displayed in (**b**) for comparison. FFR reproducibility data are from the landmark study DEFER, and data were obtained and digitized, from Kern et al. Classification certainty (**b**, *right vertical axis*) was calculated using the standard formula: $1 - \left(\frac{1}{2} e^{-\left(\frac{x - 0.80}{0.032} \right)^2} \right)$ with x representing each FFR value. Constant e is the base of the natural logarithm and equals 2.718. 0.8 is the currently established cutoff for FFR, and 0.032 is the standard deviation of the difference (SDD) between repeated FFR measurements, obtained from the digitized DEFER (Deferral Versus Performance of PTCA in Patients Without Documented Ischemia) reproducibility data. As this analysis was performed using the SDD of the overall population, it could be applied to any FFR cutoff. The chosen FFR cutoff of 0.8 follows current recommendations from clinical guidelines and is in line with the FAME (Fractional Flow Reserve Versus Angiography for Multivessel Evaluation) and FAME II (Fractional Flow Reserve Versus Angiography for Multivessel Evaluation 2) trials. Petraco R, Sen S, Nijjer S, Echavarria-Pinto M, Escaned J, Francis DP, Davies JE. Fractional flow reserve-guided revascularization: practical implications of a diagnostic gray zone and measurement variability on clinical decisions. JACC Cardiovasc Interv. 2013; 6:222–5

22.4 Validation Study of Outcomes

Over the last two decades, the favorable outcomes of FFR-guided PCI have been reported in many subsets of patients including intermediate stenosis, complex multivessel disease, stable coronary artery disease, left main disease, and bifurcation lesion.

22.4.1 DEFER Study

Initially, the FFR was used to decide upon the need for revascularization in patients with intermediate coronary artery stenosis. In the DEFER study, 325 patients for whom PCI was planned (>50% diameter stenosis by visual assessment) and who did not have documented ischemia, FFR of the stenotic lesion was measured. If FFR was >0.75, patients were randomly assigned to deferral (deferral group; $n = 91$) or performance (performance group; $n = 90$) of PCI. If FFR was <0.75, PCI was performed as planned (reference group; $n = 144$). Five-year outcome after deferral of PCI of an intermediate coronary stenosis was excellent. The incidence rate of death and acute myocardial infarction in the deferral group was only 3.3%. For angina-free symptom, there was no difference between deferral and performance group [14] (Fig. 22.5).

22.4.2 FAME 1 Study

In recent years, angiography of the majority of patients shows multivessel disease, confusing which lesion is responsible for symptom. In these patients, FFR can discriminate functionally significant lesion from nonsignificant lesion to be indicated for revascularization. In the FAME study, 1005 patients with multivessel coronary artery disease were randomly assigned to angiography-guided PCI group or FFR-guided PCI group. For FFR-guided PCI group, stenting was undergone if FFR was ≤0.80, whereas in angiography-guided PCI group, the investigator underwent stenting as planned before the randomization. At 1-year follow-up, FFR-guided group had a lower rate of primary

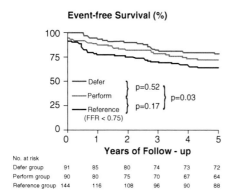
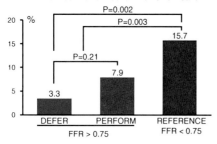
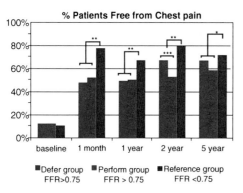

Fig. 22.5 Survival and adverse events. (*Top*) Kaplan-Meier survival curves for freedom from adverse cardiac events during 5-year follow-up for the three groups. (*Middle*) Cardiac death and acute myocardial infarction rate in the three groups after a follow-up of 5 years. (*Bottom*) Percentage of patients free from chest pain in the three groups at baseline and during follow-up. *$p = 0.028$; **$p = <0.001$; ***$p = 0.021$. *MI* myocardial infarction, *DEFER* deferral of percutaneous coronary intervention, *FFR* fractional flow reserve, *PCI* percutaneous coronary intervention, *PERFORM* performance of percutaneous coronary intervention, *REFERENCE* percutaneous coronary intervention anyway because of ischemic fractional flow reserve. Pijls NH, van Schaardenburgh P, Manoharan G, Boersma E, Bech JW, van't Veer M, Bär F, Hoorntje J, Koolen J, Wijns W, de Bruyne B. Percutaneous coronary intervention of functionally nonsignificant stenosis: 5-year follow-up of the DEFER Study. J Am Coll Cardiol. 2007; 49:2105–11

outcome end points which was a composite of death, nonfatal MI, and repeat revascularization (13.2% vs. 18.4%, $P = 0.02$) as compared with angiography-guided group even with fewer stents per patients (1.9 ± 1.3 vs. 2.7 ± 1.2, $P < 0.001$) [15]. After measurement of FFR, strategy of revascularization has been changed in ~35% of all stenotic lesions in the FAME study [16]. Similarly, there is a report that 32% of the coronary artery lesions and 48% of patients would have received a different treatment if the decision had been based on angiography only [17]. In multivessel disease, using FFR is cost-saving, saves contrast, and does not prolong the interventional procedure [18].

22.4.3 FAME 2 Study

The benefit of PCI as an initial treatment strategy in patients with stable coronary artery disease (CAD) is still controversial. In the COURAGE trial with 2287 stable CAD patients, PCI did not reduce the risk of death, myocardial infarction, or other major cardiovascular events when added to optimal medical therapy (OMT) (19.0% vs 18.5%, $P = 0.62$) [19]. However, previous clinical trial comparing PCI with OMT in patients with stable CAD, investigator did not use FFR guidance or drug-eluting stents. The FAME2 study showed that FFR-guided PCI improves the composite outcome of death from any cause, nonfatal MI, or urgent revascularization within 2 years (8.1% vs. 19.5%, $P < 0.001$) compared with OMT alone. This reduction was mainly driven by a lower rate of urgent revascularization in the PCI group (4.0% vs. 16.3%, $P < 0.001$) with no significant differences in the rates of death and MI [20, 21] (Table 22.2).

22.4.4 Left Main Study

With more liberal use of angiography, the incidental finding of intermediate left main (LM) coronary artery disease has increased. Several studies showed that angiography is not reliable for establishing functional significance of the LM

Table 22.2 Clinical events and triggers of urgent revascularization of FAME2 study[a]

Variable	PCI (N = 447)	Medical therapy (N = 441)	Hazard ratio (95% CI)[b]	P value[c]
	no. (%)			
Primary end point	36 (8.1)	86 (19.5)	0.39 (0.26–0.57)	<0.001
Death from any cause	6 (1.3)	8 (1.8)	0.74 (0.26–2.14)	0.58
Myocardial infarction	26 (5.8)	30 (6.8)	0.85 (0.50–1.45)	0.56
Urgent revascularization	18 (4.0)	72 (16.3)	0.23 (0.14–0.38)	<0.001
Death or myocardial infarction	29 (6.5)	36 (8.2)	0.79 (0.49–1.29)	0.35
Other end points				
Death from cardiac causes	3 (0.7)	3 (0.7)	0.99 (0.20–4.90)	0.99
Revascularization				
Any	36 (8.1)	179 (40.6)	0.16 (0.11–0.22)	<0.001
Nonurgent	18 (4.0)	117 (26.5)	0.13 (0.08–0.22)	<0.001
Stroke	7 (1.6)	4 (0.9)	1.74 (0.51–5.94)	0.37
Definite or probable stent thrombosis	7 (1.6)	2 (0.5)	3.48 (0.72–16.8)	0.10
Triggers of urgent revascularization according to Canadian Cardiovascular Society class[d]				
Any trigger				
All classes	18 (4.0)	72 (16.3)	0.23 (0.14–0.38)	<0.001
0, I, or II	4 (0.9)	7 (1.6)	0.56 (0.16–1.93)	0.35
III	3 (0.7)	20 (4.5)	0.14 (0.04–0.49)	<0.001
IV	11 (2.5)	47 (10.7)	0.22 (0.11–0.42)	<0.001
Myocardial infarction or changes on ECG				
All classes	15 (3.4)	31 (7.0)	0.47 (0.25–0.86)	0.01
0,1, or II	3 (0.7)	4 (0.9)	0.74 (0.17–3.31)	0.69
III	2 (0.4)	7 (1.6)	0.28 (0.06–1.35)	0.09
IV	10 (2.2)	21 (4.8)	0.46 (0.22–0.98)	0.04
Clinical features only				
All classes	3 (0.7)	43 (9.8)	0.07 (0.02–0.21)	<0.001
0,1, or II	1 (0.2)	3 (0.7)	0.33 (0.03–3.17)	0.31
III	1 (0.2)	14 (3.2)	0.07 (0.01–0.53)	0.001
IV	1 (0.2)	27 (6.1)	0.03 (0.00–0.26)	<0.001

[a]ECG denotes electrocardiography, and PCI denotes percutaneous coronary intervention
[b]Hazard ratios are for the PCI group as compared with the medical therapy group
[c]P values were calculated with the use of the log-rank test
[d]Patients could have more than one event. The Canadian Cardiovascular Society grades the severity of angina as follows: class I, angina only during strenuous or prolonged physical activity; class II, slight limitation, with angina only during vigorous physical activity; class III, symptoms with activities of everyday living (moderate limitation); and class IV, inability to perform any activity without angina or angina at rest (severe limitation)
De Bruyne B, Pijls NH, Kalesan B, Barbato E, Tonino PA, Piroth Z, et al. Fractional flow reserve-guided PCI versus medical therapy in stable coronary disease. New Engl J Med. 2012;367(11):991–1001

disease. With FFR measurement, patients with functionally nonsignificant stenosis can be safely deferred with similar outcome as compared with patients undergoing surgical revascularization for their functionally significant stenosis [22]. Hamilos et al. reported that the 5-year survival of deferred patients (n = 136) based on FFR ≥ 0.80 was 89.8%. The 5-year survival estimates of surgical group (n = 73) based on FFR < 0.80 was 85.4%, and there was no significant difference

Table 22.3 Fractional flow reserve and angiographic percent stenosis in jailed side branches

	% stenosis	
	≥50, <75	≥75
All lesions (n = 94)		
FFR < 0.75	0	20 (27%)
FFR ≥ 0.75	20	53
Vessel size ≥ 2.5 mm (n = 28)		
FFR < 0.75	0	8 (38%)
FFR ≥ 0.75	7	13

FFR fractional flow reserve
Koo BK, Kang HJ, Youn TJ, Chae IH, Choi DJ, Kim HS, et al. Physiologic assessment of jailed side branch lesions using fractional flow reserve. J Am Coll Cardiol. 2005;46(4):633–7

between the two groups. However, revascularization was more common in deferred group due to progression of the initial disease [23].

22.4.5 Bifurcation Study

One of the complex lesions treated with PCI is bifurcation lesion. The provisional side-branch intervention strategy is usually preferred. After stent implantation of main branch, the operator should decide upon whether to dilate the jailed side branch. Koo et al. showed FFR is safe and feasible tool to evaluate functional significance of the jailed side branch, and coronary angiography alone is unreliable in the assessment of the functional severity of jailed ostium, since no lesion with <75% stenosis in angiography had FFR < 0.75 [24] (Table 22.3). After performing kissing balloon dilation in patients with FFR < 0.75, follow-up FFR at 6 months was >0.75 in 95% of cases, and the outcome was excellent without further intervention if FFR was >0.75 [25].

22.5 Validation in Real-World Registry

With a stack of experience with FFR measurement, clinical benefit of FFR in daily practice was confirmed from three large registry data. With the use of FFR measurement, the treatment strategy of patients with CAD is changed, and the clinical outcome is improved. Park et al. reported that the first occurrence of death from any causes, MI, or any repeat revascularization was significantly lower in the routine FFR-guided PCI group as compared with non-FFR-guided PCI group (patient who underwent PCI before applying routine FFR strategy) in a propensity score-matched analysis ($n = 5097$, hazard ratio 0.55; 95% confidence interval 0.43–0.70; $P < 0.001$) [26]. The median number of stents implanted per patient decreased with the routine use of FFR since January 2008. Retrospective data of Mayo Clinic showed MACE at 7 years was significantly lower in FFR-guided group compared with PCI-only group ($n = 7358$, 50.0% vs. 57.0%, $P = 0.016$) [27]. From multicenter registry of France, nearly half of the patients (43%, $n = 1075$) received different strategy of treatment from angiography-based a priori decision [28].

References

1. Pijls NH, van Son JA, Kirkeeide RL, De Bruyne B, Gould KL. Experimental basis of determining maximum coronary, myocardial, and collateral blood flow by pressure measurements for assessing functional stenosis severity before and after percutaneous transluminal coronary angioplasty. Circulation. 1993;87(4):1354–67.
2. De Bruyne B, Baudhuin T, Melin JA, Pijls NH, Sys SU, Bol A, et al. Coronary flow reserve calculated from pressure measurements in humans. Validation with positron emission tomography. Circulation. 1994;89(3):1013–22.
3. Pijls NH, Tanaka N, Fearon WF. Functional assessment of coronary stenoses: can we live without it? Eur Heart J. 2013;34(18):1335–44.
4. Pijls NH, Van Gelder B, Van der Voort P, Peels K, Bracke FA, Bonnier HJ, et al. Fractional flow reserve. A useful index to evaluate the influence of an epicardial coronary stenosis on myocardial blood flow. Circulation. 1995;92(11):3183–93.
5. De Bruyne B, Bartunek J, Sys SU, Heyndrickx GR. Relation between myocardial fractional flow reserve calculated from coronary pressure measurements and exercise-induced myocardial ischemia. Circulation. 1995;92(1):39–46.
6. Pijls NH, De Bruyne B, Peels K, Van Der Voort PH, Bonnier HJ, Bartunek J, Koolen JJ, et al. Measurement of fractional flow reserve to assess the functional severity of coronary-artery stenoses. N Engl J Med. 1996;334(26):1703–8.

7. Abe M, Tomiyama H, Yoshida H, Doba N. Diastolic fractional flow reserve to assess the functional severity of moderate coronary artery stenoses: comparison with fractional flow reserve and coronary flow velocity reserve. Circulation. 2000;102(19):2365–70.
8. Chamuleau SA, Meuwissen M, van Eck-Smit BL, Koch KT, de Jong A, de Winter RJ, et al. Fractional flow reserve, absolute and relative coronary blood flow velocity reserve in relation to the results of technetium-99m sestamibi single-photon emission computed tomography in patients with two-vessel coronary artery disease. J Am Coll Cardiol. 2001;37(5):1316–22.
9. De Bruyne B, Pijls NH, Bartunek J, Kulecki K, Bech JW, De Winter H, et al. Fractional flow reserve in patients with prior myocardial infarction. Circulation. 2001;104(2):157–62.
10. Fearon WF, Tonino PA, De Bruyne B, Siebert U, Pijls NH, Investigators FS. Rationale and design of the fractional flow reserve versus angiography for multivessel evaluation (FAME) study. Am Heart J. 2007;154(4):632–6.
11. Fihn SD, Gardin JM, Abrams J, Berra K, Blankenship JC, Dallas AP, et al. 2012 ACCF/AHA/ACP/AATS/PCNA/SCAI/STS guideline for the diagnosis and management of patients with stable ischemic heart disease: a report of the American College of Cardiology Foundation/American Heart Association task force on practice guidelines, and the American College of Physicians, American Association for Thoracic Surgery, Preventive Cardiovascular Nurses Association, Society for Cardiovascular Angiography and Interventions, and Society of Thoracic Surgeons. Circulation. 2012;126(25):e354–471.
12. Windecker S, Kolh P, Alfonso F, Collet JP, Cremer J, Falk V, et al. 2014 ESC/EACTS guidelines on myocardial revascularization: the task force on myocardial revascularization of the European Society of Cardiology (ESC) and the European Association for Cardio-Thoracic Surgery (EACTS)developed with the special contribution of the European Association of Percutaneous Cardiovascular Interventions (EAPCI). Eur Heart J. 2014;35(37):2541–619.
13. Petraco R, Sen S, Nijjer S, Echavarria-Pinto M, Escaned J, Francis DP, et al. Fractional flow reserve-guided revascularization: practical implications of a diagnostic gray zone and measurement variability on clinical decisions. JACC Cardiovasc Interv. 2013;6(3):222–5.
14. Pijls NH, van Schaardenburgh P, Manoharan G, Boersma E, Bech JW, van't Veer M, et al. Percutaneous coronary intervention of functionally nonsignificant stenosis: 5-year follow-up of the DEFER study. J Am Coll Cardiol. 2007;49(21):2105–11.
15. Tonino PA, De Bruyne B, Pijls NH, Siebert U, Ikeno F, van't Veer M, et al. Fractional flow reserve versus angiography for guiding percutaneous coronary intervention. N Engl J Med. 2009;360(3):213–24.
16. Tonino PA, Fearon WF, De Bruyne B, Oldroyd KG, Leesar MA, Ver Lee PN, et al. Angiographic versus functional severity of coronary artery stenoses in the FAME study fractional flow reserve versus angiography in multivessel evaluation. J Am Coll Cardiol. 2010;55(25):2816–21.
17. Sant'Anna FM, Silva EE, Batista LA, Ventura FM, Barrozo CA, Pijls NH. Influence of routine assessment of fractional flow reserve on decision making during coronary interventions. Am J Cardiol. 2007;99(4):504–8.
18. Fearon WF, Bornschein B, Tonino PA, Gothe RM, Bruyne BD, Pijls NH, et al. Economic evaluation of fractional flow reserve-guided percutaneous coronary intervention in patients with multivessel disease. Circulation. 2010;122(24):2545–50.
19. Boden WE, O'Rourke RA, Teo KK, Hartigan PM, Maron DJ, Kostuk WJ, et al. Optimal medical therapy with or without PCI for stable coronary disease. N Engl J Med. 2007;356(15):1503–16.
20. De Bruyne B, Pijls NH, Kalesan B, Barbato E, Tonino PA, Piroth Z, et al. Fractional flow reserve-guided PCI versus medical therapy in stable coronary disease. N Engl J Med. 2012;367(11):991–1001.
21. De Bruyne B, Fearon WF, Pijls NH, Barbato E, Tonino P, Piroth Z, et al. Fractional flow reserve-guided PCI for stable coronary artery disease. N Engl J Med. 2014;371(13):1208–17.
22. Bech GJ, Droste H, Pijls NH, De Bruyne B, Bonnier JJ, Michels HR, et al. Value of fractional flow reserve in making decisions about bypass surgery for equivocal left main coronary artery disease. Heart. 2001;86(5):547–52.
23. Hamilos M, Muller O, Cuisset T, Ntalianis A, Chlouverakis G, Sarno G, et al. Long-term clinical outcome after fractional flow reserve-guided treatment in patients with angiographically equivocal left main coronary artery stenosis. Circulation. 2009;120(15):1505–12.
24. Koo BK, Kang HJ, Youn TJ, Chae IH, Choi DJ, Kim HS, et al. Physiologic assessment of jailed side branch lesions using fractional flow reserve. J Am Coll Cardiol. 2005;46(4):633–7.
25. Koo BK, Park KW, Kang HJ, Cho YS, Chung WY, Youn TJ, et al. Physiological evaluation of the provisional side-branch intervention strategy for bifurcation lesions using fractional flow reserve. Eur Heart J. 2008;29(6):726–32.
26. Park SJ, Ahn JM, Park GM, Cho YR, Lee JY, Kim WJ, et al. Trends in the outcomes of percutaneous coronary intervention with the routine incorporation of fractional flow reserve in real practice. Eur Heart J. 2013;34(43):3353–61.
27. Li J, Elrashidi MY, Flammer AJ, Lennon RJ, Bell MR, Holmes DR, et al. Long-term outcomes of fractional flow reserve-guided vs. angiography-guided percutaneous coronary intervention in contemporary practice. Eur Heart J. 2013;34(18):1375–83.
28. Van Belle E, Rioufol G, Pouillot C, Cuisset T, Bougrini K, Teiger E, et al. Outcome impact of coronary revascularization strategy reclassification with fractional flow reserve at time of diagnostic angiography: insights from a large French multicenter fractional flow reserve registry. Circulation. 2014;129(2):173–85.

Practical Learning in Coronary Pressure Measurement

Jin-Sin Koh and Chang-Wook Nam

Despite the usefulness of FFR in daily practice of coronary intervention, FFR measurement is greatly influenced by various factors of technical, hemodynamic, operational, and interpretational issues. Since the FFR measurement is performed through a wire operation with those issues, the potential pitfalls with the measurement operation might bring an erroneous result which can change a clinical or procedural decision. Thus, FFR measurement in daily practice should be as standardized as possible to avoid these kinds of pitfalls [1]. To get the reproducible stability in the procedure and reliable FFR data, the practitioner should know each issue and mechanism, which can be easily avoided.

In this chapter, we will overview the practical issues during FFR measurement which the operator should know and suggest methods on how to correct them.

J.-S. Koh
Division of Cardiology, Department of Internal Medicine, Gyeongsang National University Hospital, Jinju, South Korea

C.-W. Nam (✉)
Division of Cardiology, Department of Internal Medicine, Keimyung University Dongsan Hospital, Daegu, South Korea
e-mail: ncwcv@dsmc.or.kr

23.1 Basic Setting for FFR Measurement

Before specific setting for FFR measurement, there are several basic settings such as monitoring for vital sign, well-functional intravenous line for medication, etc. These are not just for FFR, but FFR measurement is an invasive diagnostic procedure. Therefore, all preparations for avoiding invasive procedure-related complications should be ready for immediate application.

In most catheterization laboratory, pressure transducer is usually ready as default setting. However, the height of pressure transducer can affect the value of FFR in a certain situation. Before wire insertion through the guide catheter, atmospheric pressure must be input to both transducers, so it can calculate the corrected pressure considering atmospheric pressure as reference value. For this matching atmosphere pressure to transducer (zeroing), all saline-filled pressure lines are located in the same level of the transducers with no pressure and calibrated manometrically.

Because the pressure difference of the lesion is measured based on the coincidence of coronary pressure and aortic pressure, the matching of the two pressures must be performed accurately before FFR measurement. The FFR value is calculated based on the aortic pressure measured through guide catheter. The pressure transducer is fixed on the table with height of the patient's heart (about 5 cm below the sternum) to prevent misreading of the aortic pressure due to the height

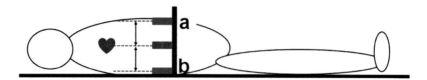

Fig. 23.1 Correction of transducer height. If the transducer is higher than aorta height level, the measured pressure can be lower than the real pressure (**a**) and vice versa (**b**)

change of transducer. If the height of transducer is improper to the aortic level, the wire pressure in front of coronary ostium will be incorrect by some mmHg. If this happens, the fixed transducer height can be adjusted higher or lower to the real value of the aortic pressure (Fig. 23.1).

The pressure sensor is positioned to the distal of the stenosis for FFR measurement, usually as far distal of the coronary artery as possible to determine the total degree of ischemia of the targeted coronary artery. It is also recommended to place the sensor at least 2–3 cm distal to the stenotic lesion to avoid the turbulence flow influence caused by the front stenosis lesion.

23.2 Reversed Gradient

When the wire sensor is located at the distal end of the normal coronary artery, the Pd value may exceed the Pa value by a few mmHg which is related to the difference of atmospheric pressure. In this case, the FFR value is displayed in excess of 1.00. In fact, this is not a true error but rather a phenomenon in which the pressure wire sensor is located at the distal point of the coronary artery, especially distal part of right coronary artery or circumflex artery, which is a lower height than the aorta level (Fig. 23.2). Usually this difference is so small that it does not confuse the interpretation of the FFR value and neither affects the clinical interpretation [2].

23.3 Issues for Guiding Catheter

Any size catheter can be attempted to measure FFR. However, because of the high level of friction in the inner coat of diagnostic catheter or smaller guiding catheter than 5 Fr, those catheters are not usually recommended in FFR measurement, or at least inspective measurement should be warranted. As the size of catheter increases, the coronary artery opening is more wedged which impede coronary blood flow (Figs. 21.2 and 21.3 in Chapter 21) [3]. This impeded flow can be noticed by ventricularization of Pa waveform especially in hyperemia. In this condition, the FFR value is incorrectly high, which underestimates the degree of coronary artery stenosis. Thus, the operator should pay attention to the change of wave morphology such as ventricularization or damping. Therefore, disengagement of guiding catheter from the ostium before hyperemia can be a good way to prevent this issue.

Contrast material in the catheter can subtly subside the aortic waveform especially in smaller catheters and can be easily corrected by flushing the guiding catheter with saline. Ideally, the dicrotic notch in pressure wave should be discernable on the aortic waveform to verify an adequate pressure tracing.

Other important issue related to catheter is usage of side-hole-guiding catheter. Because the coronary pressure measured through the side-hole catheter can be affected by the pressure interference through the side-hole-like pseudo-stenosis, it is not recommended for FFR measurement usually. However, in some situation such as concomitant proximal left main disease, the operator has to use the side-hole catheter. In that case, removing catheter from the ostium of coronary artery before FFR measurement and intravenous continuous hyperemia should be performed to get an adequate value.

23.4 Removing Wire Introducer and Equalization Before Measurement

To advance the FFR wire through the guide catheter, the introducer is inserted into the Y-connector, and the wire is manipulated together. Due to the volume occupied by the introducer, the

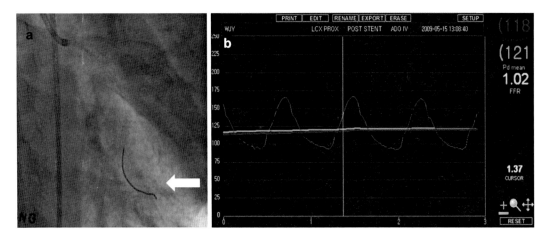

Fig. 23.2 Reverse gradient. The pressure wire sensor is located at the distal of the left circumflex coronary artery (**a**). Measured Pd value exceeds the Pa value and the FFR is over 1(**b**). This reverse gradient could happen when pressure wire sensor height (*green line*) is lower than the aorta level

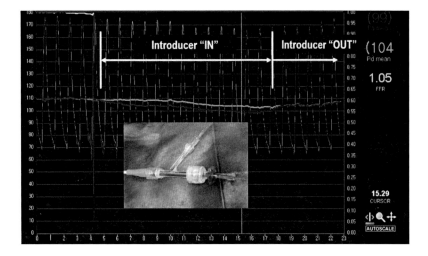

Fig. 23.3 The introducer in the Y-connector can influence pressure measurement. When the introducer is in the Y-connector, the pressure graph is displayed low. However, the pressure graph was elevated when the introducer was outside the Y-connector

Y-connector is partially opened, and consequently the aortic pressure can be measured low. In this state, the FFR measurement result will be different from the values obtained without the introducer (Fig. 23.3). This change can be amplified by hyperemia. Even if the difference is negligible, the meaning may change if the FFR value corresponds to a borderline zone. Therefore, the operator makes sure that the introducer is removed before measuring the coronary pressure.

Another forgetful step is the equalization of pressure wire. This pressure sensor is located at the 3 cm proximal end of opaque wire tip, at the junction between radiopaque and non-radiopaque portion. Pressure equalization between arterial pressure of coronary ostium and pressure sensor is performed by placing pressure sensor in front of catheter tip. When the guiding catheter is unstably engaged, some stenosis is observed in ostium of coronary artery, or significant catheter moving is observed, placing of additional guide wire can help the stabilization of pressure wire.

23.5 Pressure Damping During Pullback

The sensor is positioned to the distal of the stenosis for FFR measurement, usually as far distal of the coronary artery as possible to determine the total degree of ischemia of the targeted coronary artery. It is also recommended to place the sensor

Fig. 23.4 The catheter tip can be introduced into the coronary ostium during pulling back of the pressure wire. Loss of diastolic notch and ventricularization of Pa pressure can be observed (*white arrow*)

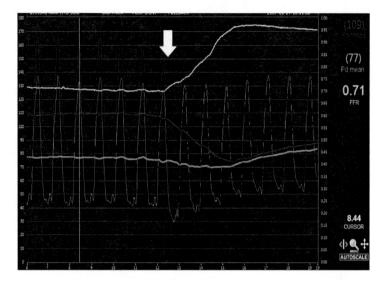

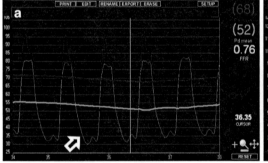 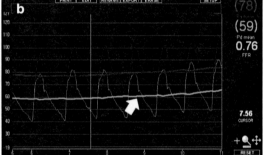

Fig. 23.5 Pressure drift (signal drift) should be suspected in the case of parallel pressure signals (similar morphology for Pa and Pd signal) throughout diastole and systole. Adequate signal shows ventricularizedPd with loss of dicrotic notch (**a**, *white empty arrow*). Unlike true gradient, the aortic dicrotic notch is preserved (*white arrow* at Pd pressure signal) despite a large pressure difference (**b**, *white arrow*)

at least 2–3 cm distal to the stenotic lesion to avoid the turbulence flow influence caused by the front stenosis lesion.

After FFR measurement under maximal hyperemia, pullback analysis is mostly performed to assess hemodynamic significant lesion and to exclude the possibility of drift. During pulling back of the pressure wire, the tip of catheter can be deeply engaged which can make a damping pressure curve (Fig. 23.4). In usual pressure pullback curve, FFR increases by Pd pressure increase. However, in damping situation, FFR increases incorrectly by Pa pressure decrease with loss of dicrotic notch or ventricularization of Pa pressure.

23.6 Pressure Drift

Pressure sensors in the FFR system are susceptible to drift from the initial calibrated state. This resetting of baseline pressure signal (pressure drift or signal drift) can be minimized with adequate device preparation such as the calibration and equalization. However, the initial setting can be distracted by mechanical and electric disturbance after long duration of pressure wire in vivo. During hyperemia, Pd pressure is ventricularized and loses the dicrotic notch (Fig. 23.5a, white empty arrow). However, if drift is happening, curve pattern of Pd pressure is identical as Pa

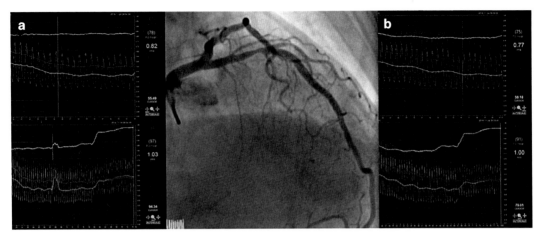

Fig. 23.6 FFR of intermediate lesion in proximal left anterior descending coronary artery was measured repeatedly (**a**, **b**). Initial FFR value is 0.82 and second 0.77. If cutoff value is taken as 0.80, decision for revascularization can be different. However, pullback curve showed a pressure drift in FFR measurement (**a**, *lower panel*). Therefore, the value of 0.77 is a correct FFR for this intermediate lesion

pressure (Fig. 23.5b, white arrow). This can be induced both in pressure measuring wire and pressure transducer. Drift of pressure wire can be easily detected by pullback measurement, which would be 1.00 at the tip of catheter. Although it is not frequent, drift of pressure transducer can be recognized by steady decrease of Pa pressure with FFR increase at the same location of pressure sensor. Therefore, checking the possible significant drift during measuring FFR in both systems is necessary [4]. Because of mechanical and electrical issues of pressure sensor, 1–2 mmHg drift can happen during measurement. This drift is acceptable to get exact FFR value. However, if drift is over 5 mmHg, it is usually recommended to re-equalize the pressure wire at the tip of the catheter and remeasure. Although it is just minor pressure change, the difference can change the decision of revascularization in some situation like a case in Fig. 23.6.

23.7 The Effect of Coronary Spasm and Accordion

Wiring of intracoronary procedure can easily make various degree of coronary spasm, which affects coronary pressure and FFR. To avoid these pseudo-stenotic effects, intracoronary bolus injection of 200 μg isosorbide dinitrate is needed after pressure wire advancement into the target location. After any kind of wire manipulation, repeated injection of nitroglycerine before FFR measurement is usually recommended.

Another inevitable limitation of FFR measurement is accordion effect which is induced by anatomical difference of coronary artery (Fig. 23.7). It can be expected when the baseline angiogram shows significantly twisted coronary artery especially with calcification and easily detected by single shot of angiogram when the pressure drop is over than expectation after wire advancement to distal part. Because accordion effect is a fundamental limitation of FFR measurement, it would be better to change to another method to evaluate the coronary lesion [5].

23.8 Whipping Artifact

In the long or tortuous coronary artery, the sensor of the pressure wire can be hit by the coronary wall, the so-called whipping artifact. The electrical spike of pressure signal (Fig. 23.8, white arrow) can be easily removed by moving the wire backward or forward (Fig. 23.8, white empty arrow).

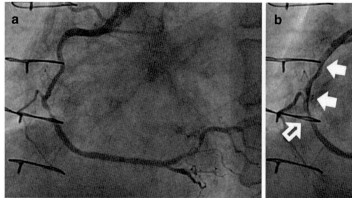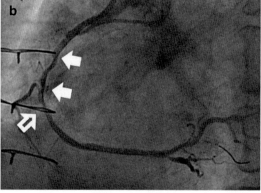

Fig. 23.7 (a) Angiography shows mid-right coronary lesion with disease-free severely angled proximal part. (b) After wiring, two pseudo (artificial)-stenosis were observed (*white arrow*) at the proximal part of true lesion (*white empty arrow*)

Fig. 23.8 Whipping artifact is the artificial spike which is caused by hitting of the pressure sensor by coronary vessel wall (*white arrow*). It can be corrected by moving the pressure wire a few millimeters backward or forward (*white empty arrow*)

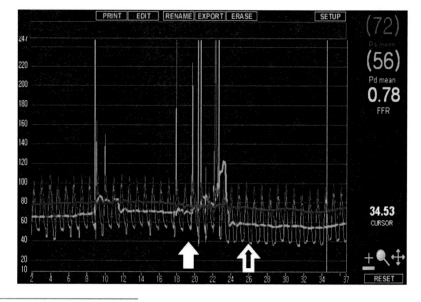

23.9 Issues for Hyperemia

Because this issue is one of the most important issues in FFR measurement, this issue will be handled in detail in another chapter. In here, we will discuss the basics of hyperemic method shortly. The commonly used method for hyperemia is continuous intravenous adenosine injection, 140 mg/kg of body weight/min. It usually takes 1 or 2 min of intravenous adenosine administration to reach a full hyperemic state. The advantages of continuous intravenous administration are stable maintenance of maximal hyperemia and feasibility of increase dose of adenosine. Since there is no evidence that administration of high dosage of intravenous adenosine over 180 is clinically more useful, increasing adenosine dose over 180 μg/min should only be considered when hyperemia is fluctuated [6]. The induction of hyperemia can be detected by observing the decrease of pressure and increase of heart rate on the monitor, and the stable plateau indicates to reach the maximal hyperemia. Adenosine should be used with caution in patients with low pressure or slow rate patient, especially older one because it can provoke hypotension and heart rate block. Other hyperemic methods are intracoronary bolus injection of adenosine or nicorandil and intravenous bolus injection of regadenoson. Although nitroprusside and dobutamine can be used for inducing hyperemia, those agents have issues for maximal microvascular dilation [7]. Because one method is not fit for all patients, it would be better to prepare and be ready to use another hyperemic method. This formatting practice simplifies FFR measurement and minimizes errors by increasing familiarity with FFR.

References

1. Vranckx P, Cutlip DE, McFadden EP, et al. Coronary pressure-derived fractional flow reserve measurements: recommendations for standardization, recording, and reporting as a core laboratory technique. Proposals for integration in clinical trials. Circ Cardiovasc Interv. 2012;5:312–7.
2. Pijls NHJ, De Bruyne B. Coronary pressure. Dordrecht: Kluwer Academic; 2000. p. 119–20.
3. Toth GG, Johnson NP, Jeremias A, Pellicano M, Vranckx P, Fearon WF, Barbato E, Kern MJ, Pijls NH, De Bruyne B. Standardization of fractional flow reserve measurements. J Am Coll Cardiol. 2016;68(7):742–53.
4. Spaan J, Piek J, Hoffman J, Siebes M. Physiological basis of clinically used coronary hemodynamic indices. Circulation. 2006;113:446–55.
5. Kern MJ, Lerman A, Bech JW, De Bruyne B, Eeckhout E, Fearon WF, Higano ST, Lim MJ, Meuwissen M, Piek JJ, Pijls NH, Siebes M, Spaan JA, American Heart Association Committee on Diagnostic and Interventional Cardiac Catheterization, Council on Clinical Cardiology. Physiological assessment of coronary artery disease in the cardiac catheterization laboratory: a scientific statement from the American Heart Association Committee on Diagnostic and Interventional Cardiac Catheterization, Council on Clinical Cardiology. Circulation. 2006;114:1321–41.
6. Adjedj J, Toth GG, Johnson NP, et al. Intracoronary adenosine: dose-response relationship with hyperemia. JACC Cardiovasc Interv. 2015;8:1422–30.
7. Jang HJ, Koo BK, Lee HS, et al. Safety and efficacy of a novel hyperaemic agent, intracoronary nicorandil, for invasive physiological assessments in the cardiac catheterization laboratory. Eur Heart J. 2013;34:2055–62.

24

Other Physiologic Indices for Epicardial Stenosis

Hong-Seok Lim and Hyoung-Mo Yang

While invasive coronary angiography (CAG) has been considered the diagnostic standard for evaluating patients with suspected or known coronary artery disease (CAD), adjunctive evaluations are proposed because of its inability to determine functional significance of coronary stenosis despite the importance of objective evidence of ischemia to improve patients' symptoms and outcomes [1–3]. Advances in intracoronary physiologic measurements allowed interventional cardiologists to have useful information to determine treatment strategies for patients with CAD. The cost-effectiveness is also improved when coronary physiology is used to guide coronary revascularization compared with that guided by CAG alone [4–7]. In particular, fractional flow reserve (FFR) by pressure-wire technology has been confirmed to provide useful guidance for determining treatment strategy in various clinical subsets of patients and coronary lesions and is recommended by current guidelines to detect ischemia-producing lesions after diagnostic CAG when objective evidence of inducible ischemia is not available [8–11]. Recently, as more refined methods for invasively determining the functional significance of CAD have been developed and are being extensively tested, interest in coronary physiology has been renewed and increasing. In this chapter, we review and summarize the main characteristics of other invasive functional indices of epicardial segment of coronary circulatory system, besides FFR.

24.1 Coronary Flow Reserve

As can be learned from its name, coronary flow reserve (CFR) is a physiologic index for evaluating the reserve—blood flow capacity—of coronary circulation according to myocardial demand. It is the value where the amount of coronary blood flow upon maximal hyperemia is divided by the amount of baseline blood flow (Fig. 24.1). Intracoronary Doppler wire can be used to measure coronary flow velocity, a method devised by applying Poiseuille's law in laminar flow field, which is a theoretical basis that well reflects actual coronary blood flow.

While epicardial segment of coronary system normally does not have any resistance on blood flow, myocardial blood flow becomes lacking as degree of stenosis increases. To offset this, coronary autoregulation works to reduce microvascular resistance to maintain blood supply to the subtended myocardium. However, as epicardial stenosis worsens, increase of coronary blood flow even upon maximal hyperemia becomes insufficient, resulting in reduced CFR. Criterion of CFR for predicting functional significance of coronary stenosis is known to be 2.0 [12, 13].

H.-S. Lim (✉) • H.-M. Yang
Department of Cardiology, Ajou University School of Medicine, Suwon, South Korea
e-mail: camdhslim@ajou.ac.kr

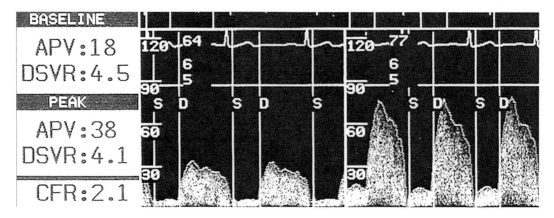

Fig. 24.1 Example of measuring coronary flow reserve. Coronary flow reserve is calculated as the ratio of hyperemic average peak flow velocity to baseline APV. (*APV* average peak velocity, *CFR* coronary flow reserve, *DSVR* diastolic systolic velocity ratio)

Meanwhile, since CFR is affected not only by epicardial stenosis but also by microvascular functions, these two factors must be considered together when interpreting an obtained result [14]. While the advantage of CFR is that it is an index which reflects epicardial artery and microvascular function as a whole, CFR is also limited in that it cannot specifically (or independently) evaluate epicardial artery or microvascular function [15]. Also, as CFR is affected by baseline blood flow, it might be lower in case of abnormal increase in baseline blood flow (i.e., high blood pressure, left ventricular hypertrophy, after interventional procedure, acute phase of myocardial infarction). In addition, considering that it is affected by factors such as left ventricular preload and heart rate and that the "normal" cutoff value is not clear, CFR has many limitations to be a specific index for assessing the functional significance of epicardial stenosis [14, 16]. Therefore, rather than functionally evaluating the degree of epicardial stenosis, it has been used for studies mainly focused on evaluating microvascular functions of the infarct-related artery or predicting myocardial viability and/or clinical prognosis after successful revascularization removing epicardial narrowing [17, 18]. According to a recent study that evaluated the prognostic value of CFR and index of microcirculatory resistance (IMR) in patients with FFR > 0.8, even for cases with negative FFR, both CFR and IMR independently showed improvements in risk stratification. Furthermore, prognosis was poorest in the case of low CFR with high IMR [19].

To overcome the aforementioned limitations of CFR, relative CFR (rCFR) was designed. The rCFR is a method which evaluates the degree of stenosis using the ratio of CFR of reference coronary artery without stenosis and the target vessel to be measured. While rCFR better reflects the functional significance of coronary stenosis than CFR, it is more complicated since two vessels are being measured and cannot be used in multivessel disease [20, 21]. Other limitations are that the reference vessel might not be truly normal and that the microvascular status of the two vessels may be different. Therefore, it may be better to use a more specific index for the purpose of evaluating functional significance of epicardial stenosis.

24.2 Hyperemic Stenosis Resistance

The hyperemic stenosis resistance (HSR) has been proposed by the groups of Spaan and Piek as an index indicating the resistance to coronary flow opposed by epicardial stenosis. It is calculated as the ratio of the pressure gradient across a stenosis divided by the coronary blood flow velocity at maximal hyperemia. As Meuwissen

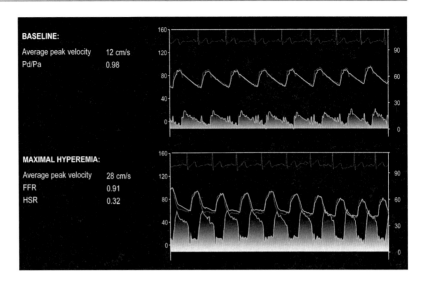

Fig. 24.2 Measurement of hyperemic stenosis resistance index. Hyperemic stenosis resistance is the ratio of transstenotic pressure gradient to average peak velocity during maximal hyperemia. (*APV* average peak velocity, *FFR* fractional flow reserve, *HSR* hyperemic stenosis resistance, *Pa* aortic pressure, *Pd* distal coronary pressure). Modified with copyright permission by Elsevier

et al. demonstrated that it would be a better predictor of reversible perfusion defects caused by a coronary stenosis than either CFR or FFR [22], it can be expected to provide the most accurate information of the functional significance of a given stenosis theoretically since it considers both the pressure gradient and the flow across epicardial narrowing. The measurements require both intracoronary pressure and flow velocity distal to the stenosis in the target vessel using a pressure-sensor guidewire and a Doppler-tipped guidewire. To overcome the limitation makes its practical application difficult and expensive, and a single wire with dual sensor (ComboWire®; Volcano Corp., San Diego, CA, USA) has been recently developed and became clinically available for combined measurements. Its accuracy has been proven to be higher than measuring with two separate single-sensor Doppler wire and pressure wire [23]. Figure 24.2 demonstrated an example of HSR measurement. High HSR values indicate poor clinical outcomes [24]; however, further study is needed to establish a simple cutoff value and its clinical utility.

24.3 Resting Indices

Although functional assessment of coronary narrowing at rest has been available since the early period of coronary intervention [25], it had not been used in clinical practice due to crucial limitation by bulky low-fidelity equipment. Meanwhile, maximum hyperemia ought to be induced to discriminate between stenoses by increasing flow across them using pharmacologic agents. However, there have been concerns regarding limitations of exogenous induction of hyperemia and needs and efforts for developing a reliable resting physiologic index [26–28]. By reducing procedural time and cost, avoiding adverse effects or patients' discomfort due to hyperemic agents, and allowing continuous online measurements, nonhyperemic resting indices are appealing to evaluate functional severity of epicardial stenosis.

24.3.1 Instantaneous Wave-Free Ratio

The instantaneous wave-free ratio (iFR) is the ratio of resting distal coronary pressure to aortic pressure, during diastole, when microcirculatory resistance is "naturally" constant and minimized compared with the rest of the cardiac cycle [29]. It is calculated as the mean distal coronary pressure (Pd) divided by the mean aortic pressure (Pa) during the diastolic wave-free period which extends from 25% of the way into diastole to 5 ms before the end of diastole. The concept is based on that resting blood flow was preserved across any given coronary stenosis, and this likely occurs by the vasodilatory compensation of microvasculature for the epicardial stenosis at the expense of Pd, which falls even at rest. During the wave-free period when the resistance waves are quiescent with constantly minimized (Fig. 24.3) [30], pressure and flow velocity linearly correlate and pressure ratios across a stenosis can reflect the flow limitation imposed by itself. No need for pharmacologic induction of hyperemia is the most significant practical advantage of iFR, facilitating the adoption of invasive coronary physiology. iFR has been rigorously validated to be closely correlated with FFR and proposed as a good surrogate for FFR. By comparing iFR and FFR in a routine clinical population, the ADVISE registry found a classification match of 80%, which is similar to the classification match between repeated measures of FFR in the DEFER trial (85% match) [31]. When iFR and FFR were compared to the HSR as third-party arbiter of ischemia in the CLARIFY study, both iFR and FFR had equal diagnostic efficiency to match an ischemic classification with HSR (both 92%, with no significant difference between the two tests) [32]. To verify its clinical utility more clearly, several clinical trials are now in progress to investigate the value of various clinical strategies based on iFR measurements (iFR SWEDEHEART NCT02166736, DEFINE-FLAIR NCT02053038, SYNTAX-2 NCT02015832, J-DEFINE NCT02002910). While awaiting clinical outcome data, the hybrid iFR-FFR strategy can be applied in daily clinical practice [33]. This strategy aims to achieve a high

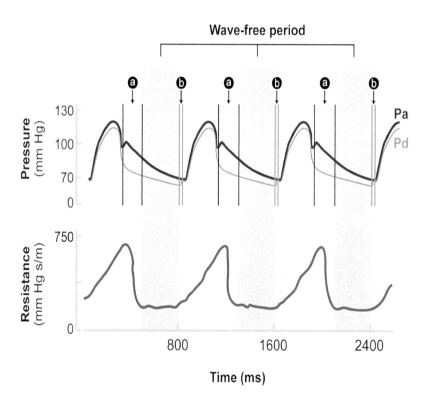

Fig. 24.3 Illustration of the distal pressure traces and instantaneous resistance for the instantaneous wave-free ratio measurement. Instantaneous wave-free ratio is the ratio of coronary pressure distal to the stenosis to aortic pressure during a specific period of diastole extending from 25% of the way into diastole (**a**) to 5 ms before the end of diastole (**b**). (*Pa* aortic pressure, *Pd* coronary pressure distal to the stenosis). Modified with copyright permission by Elsevier

diagnostic agreement with FFR that has validated outcome data, while reducing the need of inducing maximal hyperemia. After iFR is measured in all patients, if the value is within a narrow range, then a hyperemic agent such as adenosine is administered to measure FFR. It is possible to use various hybrid iFR-FFR approaches based on the classification match sought: a match of 95% requires adenosine if iFR values are between 0.86 and 0.93, sparing almost 60–70% of patients from adenosine administration [33, 34]. This hybrid approach is being used in SYNTAX-2. Furthermore, DEFINE-FLAIR and the iFR-SWEDEHEART are ongoing to assess the validity and safety of single threshold of significance, iFR 0.90, which was determined against an FFR threshold of 0.80 in the RESOLVE study [35].

24.3.2 Resting Pd/Pa

Interest has recently emerged as to whether nonhyperemic measure of pressure might be useful to evaluate the severity of coronary stenosis. Resting Pd/Pa is another nonhyperemic index, the ratio of distal coronary artery pressure to aortic pressure over the entire cardiac cycle at baseline. It is calculated in similar fashion to iFR except that the pressure measurements were time averaged over the entire cardiac cycle, thus including both systole and diastole. The most practical advantage of resting Pd/Pa is that it is always available before hyperemic measurements such as FFR. A cutoff value of Pd/Pa ≈ 0.90 has been shown to provide the best classification match, approximately 80%, with the clinically adopted FFR cutoff 0.8 [33, 35–37]. The combined application of resting Pd/Pa with FFR seems to provide a more comprehensive physiological assessment of coronary stenosis and a closer pressure-based appraisal of the flow reserve for the subtended myocardium [38]. iFR predicts the FFR value with the same accuracy as the mere baseline Pd/Pa ratio [39]; however, a recent study from Korea comparing resting indices demonstrated that iFR and the whole-cycle resting Pd/Pa had excellent diagnostic accuracy compared with FFR, with iFR demonstrating greater discriminatory power than resting Pd/Pa [40], which was similar to the result ADVISE registry [31].

24.4 Summary

Although FFR is the most verified index in evaluating inducible myocardial ischemia caused by epicardial narrowing and now regarded as the reference invasive method to functionally assess CAD, there is still room for further improvement in the diagnosis and treatment guidance of patients using other physiologic indices (Table 24.1). With rapidly advancing technology in catheterization laboratories, nonhyperemic indices are increasingly applied for functional assessment of coronary stenoses in place of hyperemic indices. In addition, using the combination of pressure and flow velocity is emphasized to understand the entire coronary physiology profoundly and improve accuracy of physiologic evaluations. Since the ischemic heart disease cannot be solely explained by epicardial stenosis, a more integrated physiologic approach considering pressure, flow, and resistance for the entire coronary system may lead to better treatment strategies in combination with epicardial indices such as FFR.

Table 24.1 Characteristics of FFR, iFR, and resting Pd/Pa

	FFR	iFR	Resting Pd/Pa
Intracoronary wire	Pressure	Pressure	Pressure
Hyperemia	Yes	No	No
Measurement time, min	5–10	2–3	1–2
Evidences against ischemia	+++	+	+/−
Evidences for clinical outcome	+++	Under evaluation	−

FFR fractional flow reserve, *iFR* instantaneous wave-free ratio, *Pd* coronary pressure distal to the stenosis, *Pa* aortic pressure

References

1. De Bruyne B, McFetridge K, Toth G. Angiography and fractional flow reserve in daily practice: why not (finally) use the right tools for decision-making? Eur Heart J. 2013;34(18):1321–2.
2. Meijboom WB, Van Mieghem CA, van Pelt N, Weustink A, Pugliese F, Mollet NR, et al. Comprehensive assessment of coronary artery stenoses: computed tomography coronary angiography versus conventional coronary angiography and correlation with fractional flow reserve in patients with stable angina. J Am Coll Cardiol. 2008;52(8):636–43.
3. Patel MR, Rao SV. Ischemia-driven revascularization: demonstrating and delivering a mature procedure in a mature way. Circ Cardiovasc Qual Outcomes. 2013;6(3):250–2.
4. Fearon WF, Bornschein B, Tonino PA, Gothe RM, Bruyne BD, Pijls NH, et al. Economic evaluation of fractional flow reserve-guided percutaneous coronary intervention in patients with multivessel disease. Circulation. 2010;122(24):2545–50.
5. Kim YH, Park SJ. Ischemia-guided percutaneous coronary intervention for patients with stable coronary artery disease. Circ J. 2013;77(8):1967–74.
6. Pijls NH, Fearon WF, Tonino PA, Siebert U, Ikeno F, Bornschein B, et al. Fractional flow reserve versus angiography for guiding percutaneous coronary intervention in patients with multivessel coronary artery disease: 2-year follow-up of the FAME (fractional flow reserve versus angiography for multivessel evaluation) study. J Am Coll Cardiol. 2010;56(3):177–84.
7. Tonino PA, De Bruyne B, Pijls NH, Siebert U, Ikeno F, van' t Veer M, et al. Fractional flow reserve versus angiography for guiding percutaneous coronary intervention. N Engl J Med. 2009;360(3):213–24.
8. Nallamothu BK, Tommaso CL, Anderson HV, Anderson JL, Cleveland JC Jr, Dudley RA, et al. ACC/AHA/SCAI/AMA-Convened PCPI/NCQA 2013 performance measures for adults undergoing percutaneous coronary intervention: a report of the American College of Cardiology/American Heart Association Task Force on Performance Measures, the Society for Cardiovascular Angiography and Interventions, the American Medical Association-Convened Physician Consortium for Performance Improvement, and the National Committee for Quality Assurance. Circulation. 2014;129(8):926–49.
9. Patel MR, Dehmer GJ, Hirshfeld JW, Smith PK, Spertus JA. ACCF/SCAI/STS/AATS/AHA/ASNC/HFSA/SCCT 2012 appropriate use criteria for coronary revascularization focused update: a report of the American College of Cardiology Foundation Appropriate Use Criteria Task Force, Society for Cardiovascular Angiography and Interventions, Society of Thoracic Surgeons, American Association for Thoracic Surgery, American Heart Association, American Society of Nuclear Cardiology, and the Society of Cardiovascular Computed Tomography. J Am Coll Cardiol. 2012;59(9):857–81.
10. Task Force on Myocardial Revascularization of the European Society of Cardiology, the European Association for Cardio-Thoracic Surgery, European Association for Percutaneous Cardiovascular Interventions, Wijns W, Kolh P, Danchin N, et al. Guidelines on myocardial revascularization. Eur Heart J. 2010;31(20):2501–55.
11. Task Force M, Montalescot G, Sechtem U, Achenbach S, Andreotti F, Arden C, et al. 2013 ESC guidelines on the management of stable coronary artery disease: the task force on the management of stable coronary artery disease of the European Society of Cardiology. Eur Heart J. 2013;34(38):2949–3003.
12. Doucette JW, Corl PD, Payne HM, Flynn AE, Goto M, Nassi M, et al. Validation of a Doppler guide wire for intravascular measurement of coronary artery flow velocity. Circulation. 1992;85(5):1899–911.
13. Joye JD, Schulman DS, Lasorda D, Farah T, Donohue BC, Reichek N. Intracoronary Doppler guide wire versus stress single-photon emission computed tomographic thallium-201 imaging in assessment of intermediate coronary stenoses. J Am Coll Cardiol. 1994;24(4):940–7.
14. de Bruyne B, Bartunek J, Sys SU, Pijls NH, Heyndrickx GR, Wijns W. Simultaneous coronary pressure and flow velocity measurements in humans. Feasibility, reproducibility, and hemodynamic dependence of coronary flow velocity reserve, hyperemic flow versus pressure slope index, and fractional flow reserve. Circulation. 1996;94(8):1842–9.
15. Kern MJ, Samady H. Current concepts of integrated coronary physiology in the catheterization laboratory. J Am Coll Cardiol. 2010;55(3):173–85.
16. Ng MK, Yeung AC, Fearon WF. Invasive assessment of the coronary microcirculation: superior reproducibility and less hemodynamic dependence of index of microcirculatory resistance compared with coronary flow reserve. Circulation. 2006;113(17):2054–61.
17. Takahashi T, Hiasa Y, Ohara Y, Miyazaki S, Ogura R, Miyajima H, et al. Usefulness of coronary flow reserve immediately after primary coronary angioplasty for acute myocardial infarction in predicting long-term adverse cardiac events. Am J Cardiol. 2007;100(5):806–11.
18. Yoon MH, Tahk SJ, Yang HM, Woo SI, Lim HS, Kang SJ, et al. Comparison of accuracy in the prediction of left ventricular wall motion changes between invasively assessed microvascular integrity indexes and fluorine-18 fluorodeoxyglucose positron emission tomography in patients with ST-elevation myocardial infarction. Am J Cardiol. 2008;102(2):129–34.
19. Lee JM, Jung JH, Hwang D, Park J, Fan Y, Na SH, et al. Coronary flow reserve and microcirculatory resistance in patients with intermediate coronary stenosis. J Am Coll Cardiol. 2016;67(10):1158–69.
20. Baumgart D, Haude M, Goerge G, Ge J, Vetter S, Dagres N, et al. Improved assessment of coronary ste-

20. nosis severity using the relative flow velocity reserve. Circulation. 1998;98(1):40–6.
21. Chamuleau SA, Meuwissen M, van Eck-Smit BL, Koch KT, de Jong A, de Winter RJ, et al. Fractional flow reserve, absolute and relative coronary blood flow velocity reserve in relation to the results of technetium-99m sestamibi single-photon emission computed tomography in patients with two-vessel coronary artery disease. J Am Coll Cardiol. 2001;37(5):1316–22.
22. Meuwissen M, Siebes M, Chamuleau SA, van Eck-Smit BL, Koch KT, de Winter RJ, et al. Hyperemic stenosis resistance index for evaluation of functional coronary lesion severity. Circulation. 2002;106(4):441–6.
23. Verberne HJ, Meuwissen M, Chamuleau SA, Verhoeff BJ, van Eck-Smit BL, Spaan JA, et al. Effect of simultaneous intracoronary guidewires on the predictive accuracy of functional parameters of coronary lesion severity. Am J Physiol Heart Circ Physiol. 2007;292(5):H2349–55.
24. Meuwissen M, Chamuleau SA, Siebes M, de Winter RJ, Koch KT, Dijksman LM, et al. The prognostic value of combined intracoronary pressure and blood flow velocity measurements after deferral of percutaneous coronary intervention. Catheter Cardiovasc Interv. 2008;71(3):291–7.
25. Gruntzig AR, Senning A, Siegenthaler WE. Nonoperative dilatation of coronary-artery stenosis: percutaneous transluminal coronary angioplasty. N Engl J Med. 1979;301(2):61–8.
26. Echavarria-Pinto M, Gonzalo N, Ibanez B, Petraco R, Jimenez-Quevedo P, Sen S, et al. Low coronary microcirculatory resistance associated with profound hypotension during intravenous adenosine infusion: implications for the functional assessment of coronary stenoses. Circ Cardiovasc Interv. 2014;7(1):35–42.
27. Jeremias A, Filardo SD, Whitbourn RJ, Kernoff RS, Yeung AC, Fitzgerald PJ, et al. Effects of intravenous and intracoronary adenosine 5′-triphosphate as compared with adenosine on coronary flow and pressure dynamics. Circulation. 2000;101(3):318–23.
28. Jeremias A, Whitbourn RJ, Filardo SD, Fitzgerald PJ, Cohen DJ, Tuzcu EM, et al. Adequacy of intracoronary versus intravenous adenosine-induced maximal coronary hyperemia for fractional flow reserve measurements. Am Heart J. 2000;140(4):651–7.
29. Sen S, Escaned J, Malik IS, Mikhail GW, Foale RA, Mila R, et al. Development and validation of a new adenosine-independent index of stenosis severity from coronary wave-intensity analysis: results of the ADVISE (adenosine vasodilator independent stenosis evaluation) study. J Am Coll Cardiol. 2012;59(15):1392–402.
30. Wilson RF, White CW. Measurement of maximal coronary flow reserve: a technique for assessing the physiologic significance of coronary arterial lesions in humans. Herz. 1987;12(3):163–76.
31. Petraco R, Escaned J, Sen S, Nijjer S, Asrress KN, Echavarria-Pinto M, et al. Classification performance of instantaneous wave-free ratio (iFR) and fractional flow reserve in a clinical population of intermediate coronary stenoses: results of the ADVISE registry. EuroIntervention. 2013;9(1):91–101.
32. Sen S, Asrress KN, Nijjer S, Petraco R, Malik IS, Foale RA, et al. Diagnostic classification of the instantaneous wave-free ratio is equivalent to fractional flow reserve and is not improved with adenosine administration. Results of CLARIFY (classification accuracy of pressure-only ratios against indices using flow study). J Am Coll Cardiol. 2013;61(13):1409–20.
33. Petraco R, Park JJ, Sen S, Nijjer SS, Malik IS, Echavarria-Pinto M, et al. Hybrid iFR-FFR decision-making strategy: implications for enhancing universal adoption of physiology-guided coronary revascularisation. EuroIntervention. 2013;8(10):1157–65.
34. Escaned J, Echavarria-Pinto M, Garcia-Garcia HM, van de Hoef TP, de Vries T, Kaul P, et al. Prospective assessment of the diagnostic accuracy of instantaneous wave-free ratio to assess coronary stenosis relevance: results of ADVISE II international, multicenter study (adenosine vasodilator independent stenosis evaluation II). JACC Cardiovasc Interv. 2015;8(6):824–33.
35. Jeremias A, Maehara A, Genereux P, Asrress KN, Berry C, De Bruyne B, et al. Multicenter core laboratory comparison of the instantaneous wave-free ratio and resting Pd/Pa with fractional flow reserve: the RESOLVE study. J Am Coll Cardiol. 2014;63(13):1253–61.
36. Echavarria-Pinto M, van de Hoef TP, Garcia-Garcia HM, de Vries T, Serruys PW, Samady H, et al. Diagnostic accuracy of baseline distal-to-aortic pressure ratio to assess coronary stenosis severity: a post-hoc analysis of the ADVISE II study. JACC Cardiovasc Interv. 2015;8(6):834–6.
37. Kim JS, Lee HD, Suh YK, Kim JH, Chun KJ, Park YH, et al. Prediction of fractional flow reserve without hyperemic induction based on resting baseline Pd/Pa. Korean Circ J. 2013;43(5):309–15.
38. Echavarria-Pinto M, van de Hoef TP, van Lavieren MA, Nijjer S, Ibanez B, Pocock S, et al. Combining baseline distal-to-aortic pressure ratio and fractional flow reserve in the assessment of coronary stenosis severity. JACC Cardiovasc Interv. 2015;8(13):1681–91.
39. Plein S, Motwani M. Fractional flow reserve as the reference standard for myocardial perfusion studies: fool's gold? Eur Heart J Cardiovasc Imaging. 2013;14(12):1211–3.
40. Park JJ, Petraco R, Nam CW, Doh JH, Davies J, Escaned J, et al. Clinical validation of the resting pressure parameters in the assessment of functionally significant coronary stenosis; results of an independent, blinded comparison with fractional flow reserve. Int J Cardiol. 2013;168(4):4070–5.

Comparison Between Anatomic and Physiologic Indices

Eun-Seok Shin

In patients with coronary artery disease (CAD), clinical outcomes depend on the extent of reversible myocardial ischemia, and alleviating this ischemia decreases symptom and improves outcome (Fig. 25.1) [1, 2]. Angiographic severity of coronary artery stenoses has historically been used as the primary guide to decide between treating CAD with revascularization or medical therapy, but its inability to identify those lesions responsible for myocardial ischemia, particularly those of intermediate diameter stenosis, remains a major limitation [3]. To address this issue, there has been a long-standing interest in coronary wire-based methods for assessing coronary artery physiology. The goal of this review is to compare the anatomic and physiologic indices used to diagnose lesions responsible for inducing myocardial ischemia.

The Clinical Outcomes Utilizing Revascularization and Aggressive Drug Evaluation (COURAGE) trial demonstrated that in patients with stable CAD, up-front percutaneous coronary intervention (PCI) on the basis of angiographic stenosis severity only does not reduce coronary events more than optimal medical therapy (OMT) [4]. In contrast, randomized trials of PCI report better outcomes when revascularization is guided by fractional flow reserve (FFR), when compared to using angiographic severity only [5] or initial medical treatment [6]. Physiologic-based revascularization may even reduce adverse events compared with anatomic-based revascularization (Fig. 25.2) [7]. However, the validity of these conclusions has been questioned that the significant reductions seen in the primary composite end points were driven by reductions in urgent revascularizations which are subject to interventional bias, while no differences were observed in myocardial infarction or death.

In addition, it has become increasingly clear that it is not the lesion stenosis alone but the plaque morphology and its composition that is the basis of adverse events in atherosclerotic disease. High-risk plaques are positively remodeled and contain a large lipid-rich necrotic core covered by a thin and inflamed fibrous cap.

E.-S. Shin
Department of Internal Medicine, Division of Cardiology, Ulsan University Hospital, University of Ulsan College of Medicine, Ulsan, South Korea
e-mail: sesim1989@gmail.com

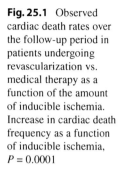
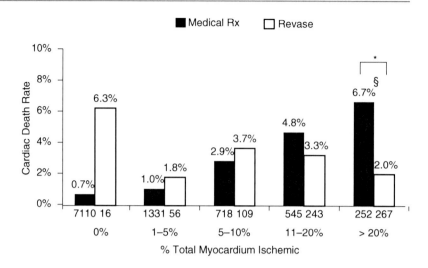

Fig. 25.1 Observed cardiac death rates over the follow-up period in patients undergoing revascularization vs. medical therapy as a function of the amount of inducible ischemia. Increase in cardiac death frequency as a function of inducible ischemia, $P = 0.0001$

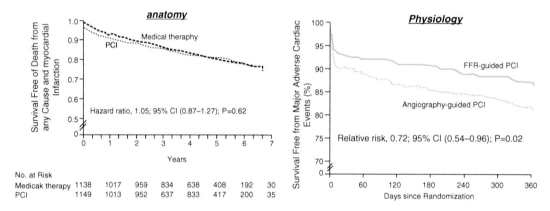

Fig. 25.2 Anatomy-based trials of revascularization such as COURAGE (*left*) have failed to improve survival. In contrast, physiology-based trials of revascularization such as FAME (*right*) demonstrate a survival advantage

25.1 Anatomic Versus Physiologic Severity

Anatomic and physiologic measures of stenosis severity have evolved in parallel over the past 40 years. Stenosis severity and their pressure or flow effects have been integrated into fluid dynamic equations and validated in experimental models. Based on animal stenosis models, the concept that a 70% diameter narrowing identifies a "critical stenosis" which reduces coronary flow capacity [8] persists as an anatomic threshold for revascularization. However, the limitations of percent stenosis are well established, particularly with documentation of diffuse disease, multiple stenoses, heterogeneous remodeling, and endothelial dysfunction which all have complex cumulative effects on coronary flow and pressure not accounted for by a single percent diameter narrowing [9]. Evidence over the intervening years has proven that percent diameter stenosis is an inadequate measure of severity for guiding management.

25.2 Coronary Anatomy and Prognosis

Anatomic measures such as diameter stenosis and location, coronary plaque volume, and the overall extent of disease substantially contribute to individual cardiovascular risk. The number of

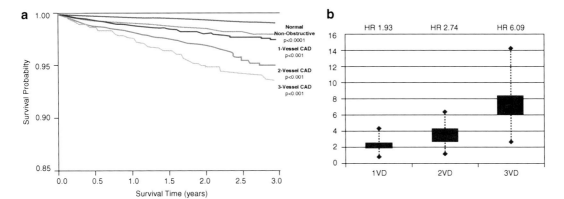

Fig. 25.3 (a) Unadjusted all-cause 3-year Kaplan–Meier survival by the presence, extent, and severity of coronary artery disease by coronary computerized tomographic angiography (CCTA). (b) Hazard ratios (HRs) for all-cause mortality stratified by number of vessels with any nonobstructive coronary atherosclerosis on CCTA. In this study, 2583 patients who underwent CCTA and had <50% stenosis were followed up for 3.1 years. There HR increases significantly when going from nonobstructive one-vessel disease (1VD) to nonobstructive 2VD to nonobstructive 3VD

vessels with a stenosis of ≥50% was the most robust predictor of outcomes, beyond that provided by traditional risk factors and left ventricular ejection fraction [10] (Fig. 25.3a). Because atherosclerosis is the substrate of most myocardial infarctions, sudden deaths, and strokes, even commonly identified nonobstructive lesions (<50% diameter stenosis) portend additional risk when compared with the excellent prognosis known to be associated with the absence of coronary atherosclerosis on coronary computerized tomographic angiography (CCTA) [11] (Fig. 25.3b).

25.3 Coronary Anatomy and the Decision to Revascularize

Given the robust prognostic power of coronary anatomy in determining future events, a simplistic and straightforward approach would be to treat all patients with stable CAD with either elective PCI or coronary artery bypass graft (CABG), as appropriate. However, the COURAGE trial revealed that an initial approach of OMT was equally as effective as PCI plus OMT in preventing death or myocardial infarction (MI) and that revascularization could be safely deferred in approximately two-thirds of patients with stable CAD. No current randomized trial data support the concept that coronary anatomy alone should dictate therapeutic strategy in stable CAD. An exception may be patients with ischemic cardiomyopathy in whom improved survival with CABG compared with OMT alone was recently reported from the 10-year extension of the NHLBI-sponsored Surgical Treatment for Ischemic Heart Failure (STICH) trial [12].

Overall, coronary anatomy provides prognostic utility beyond traditional risk factors and risk estimate scores. However, in most cases anatomy alone does not help guide revascularization decisions when improving survival and freedom from MI are the major goals.

25.4 Coronary Physiology and Prognosis

Numerous stress imaging studies have demonstrated a gradient between the extent and the severity of ischemia and subsequent risk of cardiac events [13]. Thus, enrollment of patients with lower levels of ischemia in published trials of stable CAD may explain why revascularization did not improve prognosis. In the COURAGE nuclear sub-study, the average amount of left ventricle ischemia was only

8.2% (with 10% usually accepted as representing moderate ischemia). In several reports from the Cedars-Sinai registry, patients with ≈10% or more (i.e., moderate to severe) ischemic myocardium had a nearly doubling of mortality when treated medically as compared with a demonstrable reduction in death among patients undergoing coronary revascularization [2, 14].

25.5 Coronary Physiology and the Decision to Revascularize

To date, definitive data that revascularization improves the prognosis of patients with stable CAD and noninvasively detected ischemia are absent. Although progressive narrowing of a canine coronary artery produced a predictable decline in coronary flow reserve, in clinical studies, the relationship between anatomy (including intravascular ultrasound (IVUS) and optical coherence tomography (OCT)) and physiology has been far from perfect (Fig. 25.4) [15].

To overcome the fundamental limitations of anatomical imaging, sensor guide wires have been developed to enable intracoronary measurements of pressure and flow. The physiological impact of a stenosis may be characterized by its effect on post-stenotic pressure (and flow) transmission. The post-stenotic pressure is a function of stenosis flow and resistance specific to unique morphological features that include minimal lumen area (MLA), lesion length, the stenosis

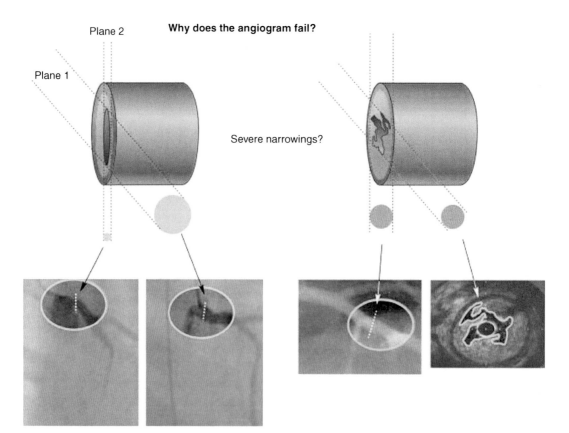

Fig. 25.4 Why does the angiogram fail to predict physiology? The angiogram is a two-dimensional image of three-dimensional structures. Most intermediate lesions are oval shaped with 2 diameters, one narrow and one wide dimension. The angiogram of an eccentric lesion cannot reliably indicate flow adequacy. Other lesions (*lower right*) may appear hazy but widely patent, only to be responsible for angina due to plaque rupture, as demonstrated by intravascular ultrasound cross section (*far right corner*)

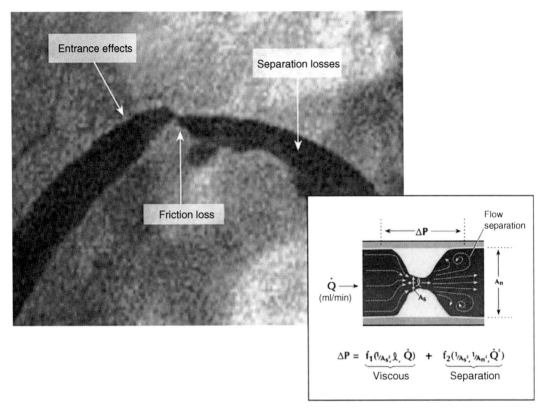

Fig. 25.5 Factors producing resistance to coronary blood flow. The angiographic two-dimensional images cannot account for the multiple factors that produce resistance to coronary blood flow and loss of pressure across a stenosis. The eccentric and irregular stenosis (*upper panel*) shows arrows designating entrance effects, friction, and zones of turbulence accounting for separation energy loss. The calculation of pressure loss (ΔP) across a stenosis (*lower right panel*) incorporates length (l), areas stenosis (As), reference area (An), flow (Q), and coefficients of viscous friction and laminar separation (f_1 and f_2) as contributors to resistance and hence pressure loss

entrance and exit orifice configurations, and the shape and size of the normal reference vessel segment (Fig. 25.5) [16].

Early studies suggested that intracoronary Doppler flow velocity measurements could determine the significance of a coronary lesion [17]. However, these approaches were never adopted into clinical practice because of difficulty in obtaining a valid flow velocity signal and the unknown status of the microcirculation in interpreting an abnormal coronary flow reserve. However, FFR, the ratio of post-stenotic pressure/aortic pressure obtained at maximal pharmacologically induced hyperemia, closely correlates to indices of ischemia to a greater degree than the resting trans-stenotic pressure gradient. Because its derivation was based on pressure at maximal flow and excluded the microcirculatory resistance, FFR was largely independent of changes in basal flow, systemic hemodynamics, or contractility [18], although an intact microcirculation is required for the hemodynamic effects of full hyperemia to be established. FFR thresholds for hemodynamic significance were established by comparisons with ischemic stress testing modalities and subsequently validated in numerous clinical outcome studies. Compared with traditional angiographic PCI guidance, FFR-guided decisions have demonstrated clinical and economic superiority in numerous single- and multi-center interventional trials. In the FAME trial, an FFR-guided PCI strategy was superior to an angiography-guided PCI therapy in reducing both stent use and the rates of future urgent revascularization because of unstable angina and MI [1]. In the FAME 2 trial, FFR-guided revascularization resulted in lower rates of progressive ischemic symptoms and the need for urgent or elective revascularization within 2 years, compared with OMT alone. Notably, these trials were not blinded, and rates of death or MI were not significantly reduced with revascularization. Nonetheless, the totality of the evidence supports the strong guideline-based recommendations for the use of FFR to guide PCI revascularization decisions.

Of note, although intravascular imaging techniques provide an extra dimension beyond angiography in assessing plaque geometry and extent, such techniques are still not accurate correlates of ischemia. For example, MLA assessed by IVUS and OCT correlates better with FFR than with simple diameter stenosis measured by angiography [7, 10, 19]. However, among 25 studies that compared IVUS or OCT imaging to FFR, the best cutoff value for MLA ranged from 1.8 to 4.0 mm^2 (excluding the left main for which the best cutoff values were 4.8–5.9 mm^2), with areas under the curve ranging from 0.63 to 0.90 (Fig. 25.6) [3]. However, although an MLA of >4 mm^2 in non-left main lesions predicted an FFR of >0.80 in 91% of cases, an MLA of <4 mm^2 correlated poorly to FFR, with most studies reporting FFR of <0.8 in ≈50% of cases [20–22]. IVUS thresholds are also dependent on lesion location, whereas the FFR threshold is not (Fig. 25.7). The major reason why location strongly influences the IVUS/FFR relationship is

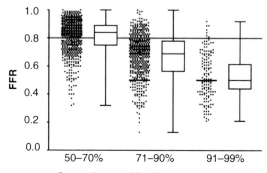

Fig. 25.6 Angiographic severity versus functional severity of coronary artery stenoses. Box-and-whisker plot showing the fractional flow reserve (FFR) values of all lesions in the categories of 50–70%, 71–90%, and 91–99% diameter stenosis. The *red horizontal line* corresponds to the FFR cutoff value for myocardial ischemia (FFR ≤ 0.80 corresponds with myocardial ischemia)

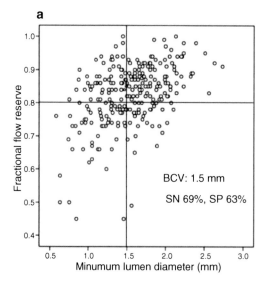
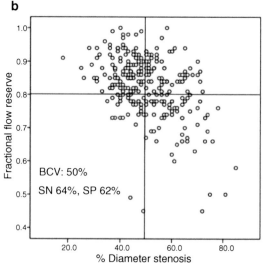

Fig. 25.7 Plots of relationship between FFR and angiographic and IVUS parameters. There was a positive correlation between fractional flow reserve (FFR) and minimum lumen diameter (**a**) and a negative correlation between FFR and percent diameter stenosis (**b**)/minimum lumen area (**c**)/percent plaque burden (**d**). The best cutoff value (BCV) to predict the functional significance was 50% for angiographic percent diameter stenosis and 2.75 mm^2 for minimum lumen area by intravascular ultrasound

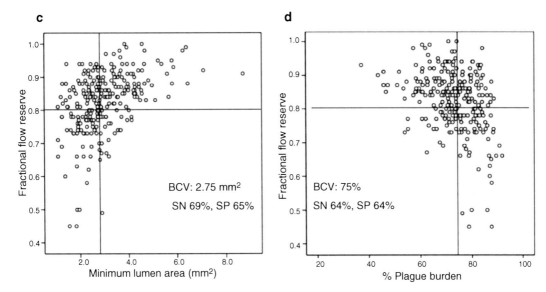

Fig. 25.7 (continued)

that both the size of the reference vessel and the flow volume of the myocardial bed subtended by the stenotic vessel are important variables needed to compute the trans-stenotic pressure loss. Thus, IVUS is not a replacement for FFR as a valid measure of ischemia.

Despite the FFR which has become the standard for physiological lesion assessment in the catheterization laboratory, its adoption by the clinical community has not been widespread. One issue is the requirement of pharmacological hyperemia that produces additional time, cost, and patient discomfort. Clinical studies comparing instantaneous wave-free ratio to FFR demonstrate concordance in ≈80% of cases [23, 24]. Instantaneous wave-free ratio by a brief wave-free period of diastole wherein coronary resistance is low and constant, wherein flow and pressure are linearly related as was assumed for the FFR derivation, has the potential to reduce the use of pharmacological hyperemia for lesion assessment [25]. Several large-scale outcome trials of instantaneous wave-free ratio versus FFR for PCI guidance are in progress. Finally, there are numerous causes of ischemia beyond a fixed stenosis, including diffuse small vessel and microvascular disease, primary endothelial dysfunction, and coronary spasm. In this regard, either invasive or noninvasive assessment of coronary flow may be complementary to translational pressure measurement [26, 27].

25.6 Plaque Morphology

Noninvasive and invasive imaging modalities can distinguish the morphological structure, physical characteristics, and chemical components of high-risk plaques. Invasive IVUS, OCT, and near-infrared spectroscopy can detect distinct features of high-risk plaques. In the PROSPECT study [28], the highest risk plaques were an IVUS-classified thin-capped fibroatheroma with a large plaque burden (≥70%) and MLA of ≤4 mm². Plaques with these three characteristics had an 18.2% likelihood of causing an event within the 3.4-year follow-up period (hazard ratio, 11.1; 95% confidence interval, 4.4–27.8; $P < 0.001$). The relationship between plaque burden and subsequent events arising from untreated lesions in the PROSPECT study was particularly striking [29]; the event rate rose exponentially with increasing plaque burden and was 9.5% in lesions with ≥70% plaque burden. No such events arose from several thousand plaques with burdens <40%, and the 3-year event rate arising from lesions with plaque burden ≥40% to <60% was <1%. The mean angiographic diameter stenosis of lesions responsible for future events was only 32% at baseline but progressed rapidly during follow-up to a mean 65% diameter stenosis, usually with thrombus. The PROSPECT study thus demonstrated that although angiographically mild, vulnerable plaques are actually severe stenoses with large plaque burden and sizable necrotic cores [30].

25.7 Summary

Anatomy and physiology all variably contribute to worsening prognostic outcomes among patients with stable CAD. The physiology-based revascularization approach has been the most successful strategy to date. However, all strategies relying solely on either anatomic, physiological, or morphological lesion characterization have been associated with a low positive predictive value for predicting future events, which can lead to unnecessary revascularization in a high percentage of patients. Ideally all three features (anatomy, physiology, and plaque morphology) need to be combined for the most accurate prognostication and therapeutic decision-making. In this regard, it is important to recognize that each of obstructive, ischemic, and pathologically high-risk groups is a mix of both benign and malignant lesions, amenable to further risk stratification. Any given lesion could have combinations of high-risk features that pertain to anatomy, physiology, and morphology, and therefore the attempt to predict the lesion's prognosis based on only one of these features leads to an incomplete assessment.

References

1. De Bruyne B, Fearon WF, Pijls NH, et al. Fractional flow reserve-guided PCI for stable coronary artery disease. N Engl J Med. 2014;371:1208–17.
2. Hachamovitch R, Hayes SW, Friedman JD, Cohen I, Berman DS. Comparison of the short-term survival benefit associated with revascularization compared with medical therapy in patients with no prior coronary artery disease undergoing stress myocardial perfusion single photon emission computed tomography. Circulation. 2003;107:2900–7.
3. Tonino PA, Fearon WF, De Bruyne B, et al. Angiographic versus functional severity of coronary artery stenoses in the FAME study fractional flow reserve versus angiography in multivessel evaluation. J Am Coll Cardiol. 2010;55:2816–21.
4. Boden WE, O'Rourke RA, Teo KK, et al. Optimal medical therapy with or without PCI for stable coronary disease. N Engl J Med. 2007;356:1503–16.
5. Pijls NH, Fearon WF, Tonino PA, et al. Fractional flow reserve versus angiography for guiding percutaneous coronary intervention in patients with multivessel coronary artery disease: 2-year follow-up of the FAME (fractional flow reserve versus angiography for multivessel evaluation) study. J Am Coll Cardiol. 2010;56:177–84.
6. De Bruyne B, Pijls NH, Kalesan B, et al. Fractional flow reserve-guided PCI versus medical therapy in stable coronary disease. N Engl J Med. 2012;367:991–1001.
7. Johnson NP, Kirkeeide RL, Gould KL. Coronary anatomy to predict physiology: fundamental limits. Circ Cardiovasc Imaging. 2013;6:817–32.
8. Gould KL, Lipscomb K, Hamilton GW. Physiologic basis for assessing critical coronary stenosis. Instantaneous flow response and regional distribution during coronary hyperemia as measures of coronary flow reserve. Am J Cardiol. 1974;33:87–94.
9. White CW, Wright CB, Doty DB, et al. Does visual interpretation of the coronary arteriogram predict the physiologic importance of a coronary stenosis? N Engl J Med. 1984;310:819–24.
10. Kang SJ, Lee JY, Ahn JM, et al. Validation of intravascular ultrasound-derived parameters with fractional flow reserve for assessment of coronary stenosis severity. Circ Cardiovasc Interv. 2011;4:65–71.
11. Lin FY, Shaw LJ, Dunning AM, et al. Mortality risk in symptomatic patients with nonobstructive coronary artery disease: a prospective 2-center study of 2583 patients undergoing 64-detector row coronary computed tomographic angiography. J Am Coll Cardiol. 2011;58:510–9.
12. Velazquez EJ, Lee KL, Jones RH, et al. Coronary-artery bypass surgery in patients with ischemic cardiomyopathy. N Engl J Med. 2016;374:1511–20.
13. Shaw LJ, Berman DS, Picard MH, et al. Comparative definitions for moderate-severe ischemia in stress nuclear, echocardiography, and magnetic resonance imaging. JACC Cardiovasc Imaging. 2014;7:593–604.
14. Hachamovitch R, Rozanski A, Shaw LJ, et al. Impact of ischaemia and scar on the therapeutic benefit derived from myocardial revascularization vs medical therapy among patients undergoing stress-rest myocardial perfusion scintigraphy. Eur Heart J. 2011;32:1012–24.
15. Park SJ, Kang SJ, Ahn JM, et al. Visual-functional mismatch between coronary angiography and fractional flow reserve. JACC Cardiovasc Interv. 2012;5:1029–36.
16. Kern MJ, Samady H. Current concepts of integrated coronary physiology in the catheterization laboratory. J Am Coll Cardiol. 2010;55:173–85.

17. Kern MJ. Coronary physiology revisited : practical insights from the cardiac catheterization laboratory. Circulation. 2000;101:1344–51.
18. de Bruyne B, Bartunek J, Sys SU, Pijls NH, Heyndrickx GR, Wijns W. Simultaneous coronary pressure and flow velocity measurements in humans. Feasibility, reproducibility, and hemodynamic dependence of coronary flow velocity reserve, hyperemic flow versus pressure slope index, and fractional flow reserve. Circulation. 1996;94:1842–9.
19. Gonzalo N, Escaned J, Alfonso F, et al. Morphometric assessment of coronary stenosis relevance with optical coherence tomography: a comparison with fractional flow reserve and intravascular ultrasound. J Am Coll Cardiol. 2012;59:1080–9.
20. Ahmadi A, Stone GW, Leipsic J, et al. Prognostic determinants of coronary atherosclerosis in stable ischemic heart disease: anatomy, physiology, or morphology? Circ Res. 2016;119:317–29.
21. Koo BK, Yang HM, Doh JH, et al. Optimal intravascular ultrasound criteria and their accuracy for defining the functional significance of intermediate coronary stenoses of different locations. JACC Cardiovasc Interv. 2011;4:803–11.
22. Waksman R, Legutko J, Singh J, et al. FIRST: fractional flow reserve and intravascular ultrasound relationship study. J Am Coll Cardiol. 2013;61:917–23.
23. Jeremias A, Maehara A, Genereux P, et al. Multicenter core laboratory comparison of the instantaneous wave-free ratio and resting Pd/Pa with fractional flow reserve: the RESOLVE study. J Am Coll Cardiol. 2014;63:1253–61.
24. Sen S, Escaned J, Malik IS, et al. Development and validation of a new adenosine-independent index of stenosis severity from coronary wave-intensity analysis: results of the ADVISE (adenosine vasodilator independent stenosis evaluation) study. J Am Coll Cardiol. 2012;59:1392–402.
25. Petraco R, Park JJ, Sen S, et al. Hybrid iFR-FFR decision-making strategy: implications for enhancing universal adoption of physiology-guided coronary revascularisation. EuroIntervention. 2013;8:1157–65.
26. Johnson NP, Gould KL, Di Carli MF, Taqueti VR. Invasive FFR and noninvasive CFR in the evaluation of ischemia: what is the future? J Am Coll Cardiol. 2016;67:2772–88.
27. van de Hoef TP, Siebes M, Spaan JA, Piek JJ. Fundamentals in clinical coronary physiology: why coronary flow is more important than coronary pressure. Eur Heart J. 2015;36:3312–9a.
28. Calvert PA, Obaid DR, O'Sullivan M, et al. Association between IVUS findings and adverse outcomes in patients with coronary artery disease: the VIVA (VH-IVUS in vulnerable atherosclerosis) study. JACC Cardiovasc Imaging. 2011;4:894–901.
29. Motoyama S, Kondo T, Sarai M, et al. Multislice computed tomographic characteristics of coronary lesions in acute coronary syndromes. J Am Coll Cardiol. 2007;50:319–26.
30. Cheng JM, Garcia-Garcia HM, de Boer SP, et al. In vivo detection of high-risk coronary plaques by radiofrequency intravascular ultrasound and cardiovascular outcome: results of the ATHEROREMO-IVUS study. Eur Heart J. 2014;35:639–47.

Fractional Flow Reserve in Intermediate or Ambiguous Lesion

Bong-Ki Lee

26.1 Definition of Intermediate Lesion

An intermediate coronary lesion is defined as a luminal narrowing between diameter stenosis of 40% and 70% on angiography [1–3]. In an animal study with 12 dogs, Gould et al. demonstrated that resting coronary flow is not altered until a constriction of at least 85% diameter stenosis is present, whereas maximal coronary flow is affected by constriction as small as 30–45% (Fig. 26.1) [2]. In a human study with 35 patients, Uren et al. also showed that basal flow was unchanged regardless of the stenosis severity (range, 17–87%), and hyperemic flow correlated inversely and significantly with the degree of stenosis. The "coronary vasodilator reserve" (defined as the ratio of flow during hyperemia to flow at base line; currently called as coronary flow reserve, CFR) began to decline from the diameter stenosis of 40% and approached unity when stenosis ≥80% (Fig. 26.2) [3]. In a meta-analysis of 31 studies comparing the results of fractional flow reserve (FFR) against quantitative coronary angiography (QCA) in human, overall concordances were 61% for lesions with diameter stenosis 30–70%, 67% for stenoses > 70%, and 95% for stenoses < 30% [1]. In this zone of intermediate narrowing, anatomical stenosis and physiologic flow correlates poorly, as coronary angiography alone cannot assess such lesions [4]. So, determining the functional significance of an intermediate coronary lesion is often challenging.

26.2 Limitations of Angiography in Intermediate Lesion

Angiographic evidence of arterial stenosis is usually not detected until the cross-sectional area of plaque approaches 40–50% of the total vascular cross-sectional area as Glagov et al. reported in a histopathological autopsy study with 136 left main coronary arteries from the human heart [5]. The outer wall of the artery, encompassed by the external elastic membrane (EEM), dilates to accommodate the growing plaque. This compensatory enlargement process seems to be limited, and as the plaque area exceeds 40–50% of the EEM area, the plaque begins to encroach on the lumen. At this point, an angiogram might reveal minimal luminal narrowing [6].

B.-K. Lee
Division of Cardiology, Department of Internal Medicine, Kangwon National University Hospital, Kangwon National University School of Medicine, Chuncheon, South Korea
e-mail: mdbklee@kangwon.ac.kr

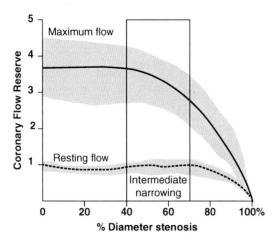
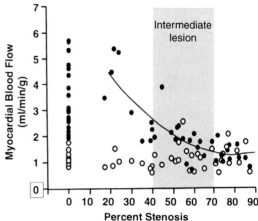

Fig. 26.1 The relation of percent diameter stenosis of the left circumflex artery to resting mean flow (*dashed line*) and hyperemic response (*solid line*) after intracoronary injection of Hypaque in 12 consecutive dogs. Flows are expressed as ratios to control resting mean values at the beginning of each experiment. The shaded area indicates the limits of the relation plotted for individual dogs (Modified, with permission, from Gould KL, Lipscomb K, Hamilton GW: Physiologic basis for assessing critical coronary stenosis. Am J Cardiol. 1974;33:87–94. Copyright 1974 Elsevier)

Fig. 26.2 Myocardial blood flow in relation to stenosis expressed as a percentage of vessel diameter in human. There was no significant correlation between blood flow in the 35 patients at base line (*open circles*) and their degree of stenosis; flow during hyperemia (*solid circles*) decreased significantly as stenosis increased. Between percent stenosis of 40 and 70 (*shaded zone*), ratio of hyperemic to basal flow showed ambiguous distribution. The values in the 21 controls are shown at 0% stenosis (Modified, with permission, from Uren NG et al. Relation Between Myocardial Blood Flow and The Severity of Coronary Artery Stenosis. N Engl J Med. 1994;330:1782–8. Copyright 1994 Massachusetts Medical Society)

Similar to a flashlight projection of a tube in three-dimensional space, an angiogram is a two-dimensional X-ray shadow of the arterial lumen along the vessel length (Fig. 26.3). So, the eccentric lumen produces conflicting degrees of angiographic diameter stenosis from different viewing angles and introduces uncertainty related to lumen size and its relationship to coronary blood flow [7]. Arterial narrowing might be incorrectly assessed owing to angulation or tortuosity, artery overlap, a short "napkin-ring stenosis," contrast streaming or separation as it enters an ectatic area, or X-ray beam angulation that is not perpendicular to the stenosis. Moreover, a long, moderate narrowing can be as or more hemodynamically significant than a short, focal severe narrowing (Fig. 26.4). Additional artifacts including vessel foreshortening, branch overlap, ostial origins, and calcifications further contribute to the uncertainty of the angiographic interpretation.

The degree of stenosis is judged by comparison with a "normal" reference segment that is theoretically free of disease, while the reference segment often has significant disease as demonstrated by IVUS or histopathology [8]. Furthermore, significant intra- and interobserver variability exists in the assessment of coronary narrowing [9].

26.3 What Makes the FFR Discrepancies Between Different Intermediate Lesions?

As regards FFR, features such as lesion length, entrance angle, exit angle, plaque rupture, blood viscosity, and absolute flow relative to the perfusion territory are important in determining translesional hemodynamic responses to hyperemia (Fig. 26.5) [10–13]. These might explain the discrepancy between the epicardial visual luminal narrowing and FFR-based physiologic significance of the lesion in many cases (Figs. 26.6 and 26.7).

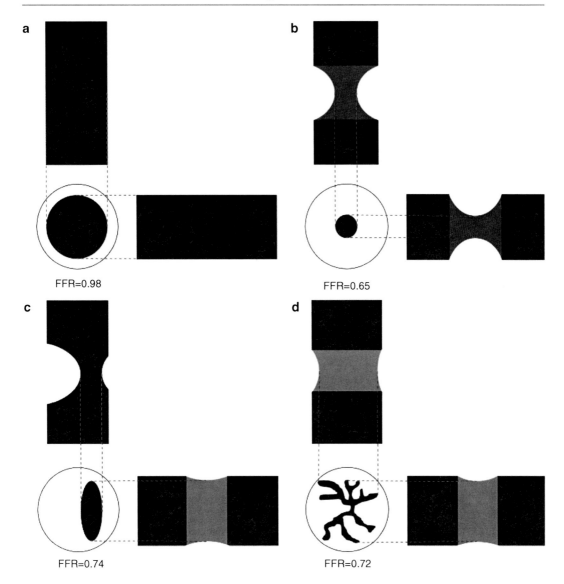

Fig. 26.3 A schematic demonstration of discrepancies between angiographic morphology and fractional flow reserve (FFR) by various lesion shapes. (**a**) A normal coronary artery shows similarly wide angiographic diameter at different angle and normal FFR; (**b**) a concentric luminal narrowing shows similarly narrow angiographic diameter at different angle and low FFR; (**c**) an asymmetrically narrowed lesion shows different diameter and gradation at different angle and low FFR; (**d**) an irregular-shaped lesion shows normal-looking diameter with diminished gradation at different angle and low FFR (Figure illustrated by Bong-Ki Lee)

26.4 FFR for Intermediate Lesion

To overcome the limitations of angiography, FFR technique is a useful modality to assess the functional significance of an intermediate or ambiguous coronary lesion. For example, in the FAME study, only 35% of intermediate coronary stenoses (between 50% and 70% diameter stenosis on angiography) had an FFR ≤ 0.8 [14]. Therefore, in angiographically intermediate lesions, it is important to determine the potential flow impediment before attempting a revascularization. A meta-analysis of 66 studies revealed that FFR-based strategy improved the prognosis of coronary artery disease (CAD) patients by decreasing 20% cardiovascular events and 10%

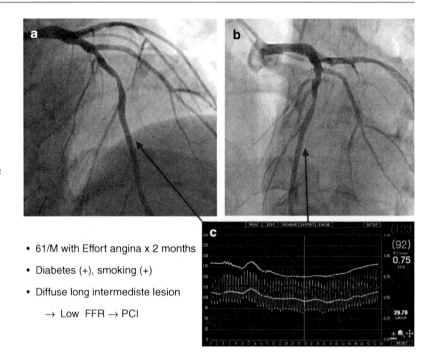

Fig. 26.4 A demonstrative case of a diffuse intermediate lesion underwent fractional flow reserve (FFR)-guided PCI. (**a**) Anterior-posterior cranial view and (**b**) left anterior oblique view images of coronary angiography showed diffuse long intermediate lesion, and (**c**) FFR was significantly low. Diffuse long atherosclerosis caused continuous pressure fall along arterial length

- 61/M with Effort angina x 2 months
- Diabetes (+), smoking (+)
- Diffuse long intermediste lesion
  → Low FFR → PCI

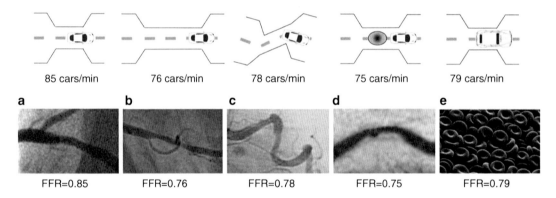

Fig. 26.5 A schematic demonstration of various features causing fractional flow reserve (FFR) discrepancies between different intermediate lesions. The blood flow is compared to the traffic of vehicles. (**a**) A simple lesion with intermediate narrowing without significant pressure drop; (**b**) a diffuse long intermediate lesion with low FFR; (**c**) an intermediate lesion with acute entrance angle and exit angle showing low FFR; (**d**) an intermediate lesion with plaque rupture with low FFR; (**e**) higher blood viscosity causes more pressure drop (Figure illustrated by Bong-Ki Lee)

better angina relief and avoiding unnecessary revascularizations in nearly 50% of cases [15].

Several studies have demonstrated that percutaneous coronary intervention (PCI) can be safely deferred in patients with an intermediate lesion and an FFR ≥ 0.75 (or ≥0.80) [16–21]. Cardiac event rates were extremely low in this cohort of patients and even lower than that predicted if a PCI had been performed in bare metal stent era owing to the avoidance of restenosis in the deferred treatment group [16, 22]. In comparison with noninvasive techniques such as exercise electrography, stress echocardiography, and myocardial perfusion scintigraphy, FFR is more accurate in predicting the hemodynamic significance of a lesion [19]. FFR application remains the most standard indication in intermediate stenosis with unclear hemodynamic significance [9]. Thus, FFR is considered as the gold standard for the evaluation of intermediate grade stenosis.

Fig. 26.6 A demonstrative case of anatomical-functional mismatch. (**a**) Right coronary artery (RCA) angiogram showed tandem intermediate lesions. (**b**) Intravascular ultrasound showed small minimal lumen area (2.22 mm²), but (**c**) fractional flow reserve (FFR) was 0.82, and these lesions are deferred by FFR guidance. Small perfusion territory of RCA in this patient might cause this anatomical-functional discrepancy

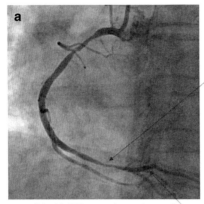
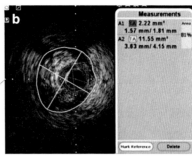
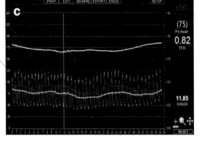

- 61/M with Effort angina x 1 months
- Hyppertension (+), Diabetes (+)
- Concurrently, proximal LAD showed 90% tubular stenosis → PCI

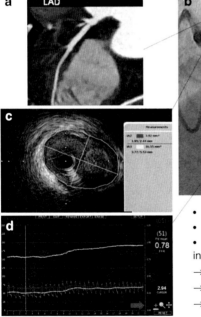
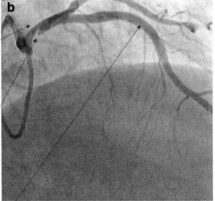

- 61/M with Effort angina x 6 months
- Hyppertension (+), Diabetes (+)
- Coronary CT angiogram showed insignificant disease → observation
- → Symptom continued
- → CAG with IVUS & FFR
- → PCI

Fig. 26.7 An example of ambiguous lesion. Ruptured plaque at ostial left anterior descending artery was not delineated in the (**a**) coronary computed tomography angiogram. The (**b**) coronary angiogram and the (**c**) intravascular ultrasound angiogram showed ruptured plaque with intermediate diameter stenosis (50%) and minimal lumen area of 3.92 mm². But the (**d**) fractional flow reserve (FFR) was 0.78, so this lesion was treated by percutaneous coronary intervention. Complex geometry at the lesion and large perfusion territory might cause flow disturbance and pressure drop

26.5 Prognostic Role of FFR for Intermediate Lesions

In patients with intermediate coronary lesion, even in multivessel disease, FFR has been proven to be an effective strategy with superior clinical outcomes compared to angiographically guided PCI [17, 20, 22–26].

26.5.1 DEFER Trial

The DEFER trial is a randomized clinical trial performed in the bare metal stent era, and 325 patients scheduled for PCI were randomly assigned into three groups and reported the 2-, 5-, and 15-year outcomes [21, 22, 25]. If FFR was >0.75, patients were assigned to the Defer group ($n = 91$, medical therapy for CAD) or the Perform group ($n = 90$, stenting 46%). If FFR was ≤ 0.75, PCI was performed as planned, and patients were entered into the reference group ($n = 144$, stenting 59%). Primary end point was absence of major adverse cardiac events (MACE) including death, MI, and revascularization during 24 months.

At 24 months, a complete follow-up was obtained in 98% of patients. Event-free survival was similar between the deferral and performance groups (92% vs. 89% at 12 months and 89% vs. 83% at 24 months; $p = 0.27$) but was significantly lower in the reference group than deferral group (80% at 12 months and 78% at 24 months; $p = 0.03$) [22].

At 5 years, follow-up was completed in 97% of patients, and the event-free survival was not different between the Defer and Perform groups (80% and 73%, $p = 0.52$) but was significantly worse in the Reference group (63%; $p = 0.03$). The composite rates of cardiac death and acute myocardial infarction (MI) in the Defer, Perform, and Reference groups were 3.3%, 7.9%, and 15.7%, respectively ($p = 0.21$ for Defer vs. Perform group; $p = 0.003$ for the reference vs both other groups). The 5-year risk of cardiac death or MI in patients with normal FFR is <1% per year and is not decreased by stenting [25]. Treating patients by FFR guidance is associated with a low event rate, comparable to event rates in patients with normal noninvasive testing.

At 15-year follow-up, complete follow-up was obtained in 92% of patients. After 15 years of follow-up, the mortality was not different between the three groups: 33.0% in the Defer group, 31.1% in the Perform group, and 36.1% in the Reference group (Defer vs. Perform, RR 1.06, 95% CI: 0.69–1.62, $P = 0.79$). The MI incidence was significantly lower in the Defer group (2.2%) compared with the Perform group (10.0%, $p = 0.03$). Among stable angina patients, functionally insignificant coronary stenosis as indicated by FFR $\geq$ 0.75 showed an excellent prognosis only with medical treatment, even after 15 years. Performing PCI of such hemodynamically nonsignificant stenosis has no benefit than medical treatment [21].

26.5.2 FAME Trial

The FAME trial randomized 1005 patients scheduled for PCI with drug-eluting stents into two groups as angiography-guided (angiography group) or FFR-guided group (FFR group). Patients assigned to angiography group underwent stenting of all indicated lesions, and those assigned to FFR group underwent stenting only for lesions with FFR $\leq$ 0.80. The primary end point was the MACE at 1 year and was reported after 1 and 5 years [20, 26].

At 1 year, the event rate was 18.3% in the angiography group and 13.2% in the FFR group ($p = 0.02$). Patients free from angina was 78% in the angiography group and 81% in the FFR group ($p = 0.20$). Routine measurement of FFR in patients with multivessel coronary artery disease who are undergoing PCI with drug-eluting stents significantly reduces the MACE [20].

At 5-year, MACE occurred in 31% of patients in the angiography group vs. 28% in the FFR group (relative risk 0.91, 95% CI 0.75–1.10; $p = 0.31$). The stents number implanted per patient was significantly higher in the angiography group than in the FFR group (mean 2.7 [SD 1.2] vs 1.9 [1.3], $p < 0.0001$). The results confirm the long-term safety of FFR-guided PCI in patients with multivessel disease. A strategy of FFR-guided PCI resulted in a significant reduction of MACE for up to 2 years after the index procedure. From 2 years to 5 years, the risks for both groups developed similarly. This clinical outcome in the FFR-guided group was achieved with a lower number of stented arteries and less resource use. These results indicate that FFR-guided PCI should be the standard of care in most patients [26].

26.5.3 DEFER-DES Trial

The DEFER-DES trial randomized 229 patients scheduled for PCI with drug-eluting stents into two groups as FFR-guided or Routine-DES group. For FFR-guided group ($n = 114$), treatment strategy was determined according to the target vessel FFR (FFR < 0.75: DES implantation [FFR-DES group]; FFR ≥ 0.75: deferral of stenting [FFR-Defer group]). Routine-DES group underwent DES implantation without FFR measurement ($n = 115$). The primary end point was the MACE incidence. Of lesions assigned to FFR guidance, only 25% had functional significance (FFR < 0.75).

At a 2-year follow-up, the cumulative incidence of MACE was 7.9% in the FFR-guided group and 8.8% in Routine-DES group ($P = 0.80$).

At a 5-year follow-up, the cumulative MACE incidence was 11.6% vs. 14.2% for the FFR group and the Routine-DES group ($P = 0.55$). There was no difference in MACE incidence between the two groups during a 5-year follow-up (hazard ratio, 1.25; 95% confidence interval, 0.60–2.60). In this study, FFR guidance provided a tailored approach for patients with intermediate coronary stenosis, which is comparable to angiography-guided Routine-DES implantation strategy, and avoided unnecessary DES stenting in a considerable part of the patients [24].

As shown in Table 26.1, FFR guidance provides useful prognostic information to decide how to treat patients with intermediate coronary lesions in daily practice in catheterization laboratory (Table 26.1).

Table 26.1 Randomized trials reported clinical outcomes after FFR-based treatment decision in intermediate coronary lesions

Study	n	FFR cutoff	MACE (%)					p value (FFR-guided groups)	p value (All groups)	Follow-up (Months)
			FFR-Medical	FFR-Defer	FFR-PCI	FFR-All	Angio-PCI			
Bech et al. [22] (DEFER)	325	>0.75	NR	11.1	29.2	20.3	17.8	0.27 (D vs P)	0.03	24
Pijls et al. [25] (DEFER 5 years)	325	>0.75	NR	21	39	29.2	27	0.52 (D vs P)	0.03	60
Courtis et al. [23]	107	0.75–0.80	23	NR	5	12.1	NR	0.005 (M vs P)	NR	13
Tonino et al. [20] (FAME)	1005	≤0.80	NR	NR	13.2	NR	18.3	NR	0.02	12
De Bruyne et al. [17] (FAME 2)	1220	≤0.80	12.7	3.0	4.3	6.6	NR	<0.001 (M vs P)	NR	12
Van Nunen et al. [26] (FAME 5 years)	1005	≤0.80	NR	NR	28	NR	31	NR	0.31	60
Park et al. [24] (DEFER-DES)	229	≥0.75	NR	7.1	24.1	11.6	14.2	0.69 (F vs A)	0.05	60

FFR indicates fractional flow reserve, *MACE* major adverse cardiac events, *FFR-Medical* group of medical treatment for lesions under FFR cutoff, *FFR-Defer* group of medical treatment for lesions above FFR cutoff, *FFR-PCI* group of percutaneous coronary intervention (PCI) for lesions under FFR cutoff, *FFR-All* all FFR-measured subjects, *Angio-PCI* group of coronary angiography-guided PCI, *NR* not reported, *D vs P* FFR-Defer vs FFR-PCI, *M vs P* FFR-Medical vs FFR-PCI, *F vs A* FFR guided vs angiography guided

26.6 Current Guidelines Recommendation (Table 26.2)

The American College of Cardiology Foundation/American Heart Association/Society for Cardiac Angiography and Interventions (ACCF/AHA/SCAI) guidelines recommend FFR as a reasonable option to assess angiographic intermediate coronary lesions and for guiding revascularization decisions as level IIaA recommendation [27].

The European Society of Cardiology (ESC) guidelines recommend FFR for functional assessment of coronary lesions without available evidence of ischemia or multivessel disease as level IA. Revascularization is recommended for stenosis with FFR < 0.80 (level IB). The guidelines discourage revascularization of an intermediate stenosis without related ischemia or without FFR < 0.80 as IIIB.

The European Society of Cardiology and European Association for Cardiothoracic Surgery (ESC/EACTS), in their 2014 guidelines, proposed utilization of FFR in hemodynamically relevant coronary lesions in stable patients when evidence of ischemia is not available as a class IA recommendation. A class IIA recommendation was made regarding the use of FFR-guided PCI in patients with multivessel disease [29].

Table 26.2 Recommendations for FFR in coronary artery disease with intermediate lesion

Provider (Year)	Recommendation	COR	LOE	Reference
ACCF/AHA/SCAI (2011)	FFR is reasonable to assess angiographic intermediate coronary lesions (50–70% diameter stenosis) and can be useful for guiding revascularization decisions in patients with SIHD	IIa	A	Levine et al. [27]
ESC (2013)	FFR is recommended to identify hemodynamically relevant coronary lesion(s) when evidence of ischemia is not available	I	A	Task Force Members et al. [28]
	FFR-guided PCI in patients with multivessel disease	IIa	B	
	Revascularization of stenoses with FFR < 0.80 is recommended in patients with angina symptoms or a positive stress test	I	B	
	Revascularization of an angiographically intermediate stenosis without related ischemia or without FFR < 0.80 is not recommended	III	B	
ESC/EACTS (2014)	FFR to identify hemodynamically relevant coronary lesion(s) in stable patients when evidence of ischemia is not available	I	A	Authors/Task Force Members et al. [29]
	FFR-guided PCI in patients with multivessel disease	IIa	B	

COR indicates class of recommendation, *LOE* level of evidence, *ACCF* American College of Cardiology Foundation, *AHA* American Heart Association, *SCAI* Society for Cardiovascular Angiography and Interventions, *FFR* fractional flow reserve, *SIHD* stable ischemic heart disease, *ESC* European Society of Cardiology, *EACTS* European Association for Cardiothoracic Surgery

References

1. Christou MA, Siontis GC, Katritsis DG, Ioannidis JP. Meta-analysis of fractional flow reserve versus quantitative coronary angiography and noninvasive imaging for evaluation of myocardial ischemia. Am J Cardiol. 2007;99(4):450–6.
2. Gould KL, Lipscomb K, Hamilton GW. Physiologic basis for assessing critical coronary stenosis. Instantaneous flow response and regional distribution during coronary hyperemia as measures of coronary flow reserve. Am J Cardiol. 1974;33(1):87–94.
3. Uren NG, Melin JA, De Bruyne B, Wijns W, Baudhuin T, Camici PG. Relation between myocardial blood flow and the severity of coronary-artery stenosis. N Engl J Med. 1994;330(25):1782–8.
4. van de Hoef TP, Siebes M, Spaan JA, Piek JJ. Fundamentals in clinical coronary physiology: why coronary flow is more important than coronary pressure. Eur Heart J. 2015;36(47):3312–9a.
5. Glagov S, Weisenberg E, Zarins CK, Stankunavicius R, Kolettis GJ. Compensatory enlargement of human atherosclerotic coronary arteries. N Engl J Med. 1987;316(22):1371–5.
6. Mintz GS, Popma JJ, Pichard AD, Kent KM, Satler LF, Chuang YC, et al. Limitations of angiography in the assessment of plaque distribution in coronary artery disease: a systematic study of target lesion eccentricity in 1446 lesions. Circulation. 1996;93(5):924–31.
7. Meijboom WB, Van Mieghem CA, van Pelt N, Weustink A, Pugliese F, Mollet NR, et al. Comprehensive assessment of coronary artery stenoses: computed tomography coronary angiography versus conventional coronary angiography and correlation with fractional flow reserve in patients with stable angina. J Am Coll Cardiol. 2008;52(8):636–43.
8. Mintz GS, Painter JA, Pichard AD, Kent KM, Satler LF, Popma JJ, et al. Atherosclerosis in angiographically "normal" coronary artery reference segments: an intravascular ultrasound study with clinical correlations. J Am Coll Cardiol. 1995;25(7):1479–85.
9. Fisher LD, Judkins MP, Lesperance J, Cameron A, Swaye P, Ryan T, et al. Reproducibility of coronary arteriographic reading in the coronary artery surgery study (CASS). Catheter Cardiovasc Diagn. 1982;8(6):565–75.
10. Johnson NP, Kirkeeide RL, Gould KL. Coronary anatomy to predict physiology: fundamental limits. Circ Cardiovasc Imaging. 2013;6(5):817–32.
11. Kern MJ, Samady H. Current concepts of integrated coronary physiology in the catheterization laboratory. J Am Coll Cardiol. 2010;55(3):173–85.
12. Kimball BP, Dafopoulos N, LiPreti V. Comparative evaluation of coronary stenoses using fluid dynamic equations and standard quantitative coronary arteriography. Am J Cardiol. 1989;64(1):6–10.
13. Park SJ, Kang SJ, Ahn JM, Shim EB, Kim YT, Yun SC, et al. Visual-functional mismatch between coronary angiography and fractional flow reserve. JACC Cardiovasc Interv. 2012;5(10):1029–36.
14. Tonino PA, Fearon WF, De Bruyne B, Oldroyd KG, Leesar MA, Ver Lee PN, et al. Angiographic versus functional severity of coronary artery stenoses in the FAME study fractional flow reserve versus angiography in multivessel evaluation. J Am Coll Cardiol. 2010;55(25):2816–21.
15. Johnson NP, Toth GG, Lai D, Zhu H, Acar G, Agostoni P, et al. Prognostic value of fractional flow reserve: linking physiologic severity to clinical outcomes. J Am Coll Cardiol. 2014;64(16):1641–54.
16. Bech GJ, De Bruyne B, Bonnier HJ, Bartunek J, Wijns W, Peels K, et al. Long-term follow-up after deferral of percutaneous transluminal coronary angioplasty of intermediate stenosis on the basis of coronary pressure measurement. J Am Coll Cardiol. 1998;31(4):841–7.
17. De Bruyne B, Pijls NH, Kalesan B, Barbato E, Tonino PA, Piroth Z, et al. Fractional flow reserve-guided PCI versus medical therapy in stable coronary disease. N Engl J Med. 2012;367(11):991–1001.
18. Kern MJ, Donohue TJ, Aguirre FV, Bach RG, Caracciolo EA, Wolford T, et al. Clinical outcome of deferring angioplasty in patients with normal translesional pressure-flow velocity measurements. J Am Coll Cardiol. 1995;25(1):178–87.
19. Pijls NH, De Bruyne B, Peels K, Van Der Voort PH, Bonnier HJ, Bartunek JKJJ, et al. Measurement of fractional flow reserve to assess the functional severity of coronary-artery stenoses. N Engl J Med. 1996;334(26):1703–8.
20. Tonino PA, De Bruyne B, Pijls NH, Siebert U, Ikeno F, van' t Veer M, et al. Fractional flow reserve versus angiography for guiding percutaneous coronary intervention. N Engl J Med. 2009;360(3):213–24.
21. Zimmermann FM, Ferrara A, Johnson NP, van Nunen LX, Escaned J, Albertsson P, et al. Deferral vs. performance of percutaneous coronary intervention of functionally non-significant coronary stenosis: 15-year follow-up of the DEFER trial. Eur Heart J. 2015;36(45):3182–8.
22. Bech GJ, De Bruyne B, Pijls NH, de Muinck ED, Hoorntje JC, Escaned J, et al. Fractional flow reserve to determine the appropriateness of angioplasty in moderate coronary stenosis: a randomized trial. Circulation. 2001;103(24):2928–34.
23. Courtis J, Rodes-Cabau J, Larose E, Dery JP, Nguyen CM, Proulx G, et al. Comparison of medical treatment and coronary revascularization in patients with moderate coronary lesions and borderline fractional flow

reserve measurements. Catheter Cardiovasc Interv. 2008;71(4):541–8.
24. Park SH, Jeon KH, Lee JM, Nam CW, Doh JH, Lee BK, et al. Long-term clinical outcomes of fractional flow reserve-guided versus routine drug-eluting stent implantation in patients with intermediate coronary stenosis: five-year clinical outcomes of DEFER-DES trial. Circ Cardiovasc Interv. 2015;8(12):e002442.
25. Pijls NH, van Schaardenburgh P, Manoharan G, Boersma E, Bech JW, van't Veer M, et al. Percutaneous coronary intervention of functionally nonsignificant stenosis: 5-year follow-up of the DEFER study. J Am Coll Cardiol. 2007;49(21):2105–11.
26. van Nunen LX, Zimmermann FM, Tonino PA, Barbato E, Baumbach A, Engstrom T, et al. Fractional flow reserve versus angiography for guidance of PCI in patients with multivessel coronary artery disease (FAME): 5-year follow-up of a randomised controlled trial. Lancet. 2015;386(10006):1853–60.
27. Levine GN, Bates ER, Blankenship JC, Bailey SR, Bittl JA, Cercek B, et al. 2011 ACCF/AHA/SCAI guideline for percutaneous coronary intervention. A report of the American College of Cardiology Foundation/American Heart Association Task Force on Practice Guidelines and the Society for Cardiovascular Angiography and Interventions. J Am Coll Cardiol. 2011;58(24):e44–122.
28. Task Force Members, Montalescot G, Sechtem U, Achenbach S, Andreotti F, Arden C, et al. 2013 ESC guidelines on the management of stable coronary artery disease: the Task Force on the management of stable coronary artery disease of the European Society of Cardiology. Eur Heart J. 2013;34(38):2949–3003.
29. Authors/Task Force Members, Windecker S, Kolh P, Alfonso F, Collet JP, Cremer J, et al. 2014 ESC/EACTS guidelines on myocardial revascularization: the Task Force on Myocardial Revascularization of the European Society of Cardiology (ESC) and the European Association for Cardio-Thoracic Surgery (EACTS) Developed with the special contribution of the European Association of Percutaneous Cardiovascular Interventions (EAPCI). Eur Heart J. 2014;35(37):2541–619.

FFR in Complex Lesions

27

Hyun-Jong Lee and Joon-Hyung Doh

27.1 Bifurcation Lesion

Side branch (SB) ostial stenosis has a high likelihood of interobserver variation [1] and visual-functional mismatch [2, 3] in prediction of myocardial ischemia. The interobserver variation in assessment for stenotic severity in SB ostium is mostly secondary to angled takeoff of SB and image foreshortening. The previous studies demonstrated anatomical imaging such as coronary angiogram, and intravascular ultrasound (IVUS) cannot predict functional significance of SB ostial stenosis unlike main vessel (MV) stenosis (Fig. 27.1) [4]. Usually, visual estimation tended to overestimate the severity of jailed SB lesions compared to fractional flow reserve (FFR). The prevalence of FFR < 0.75–0.8 was very low even though significant stenosis (Table 27.1). Pre-procedural FFR measurement of SB can assess the need of complex procedure such as two-stent technique in true bifurcation lesion. Also, FFR-guided decision-making for SB treatment after MV stenting reduces unnecessary SB intervention. The recently published DKCRUSH-VI study randomly compared FFR-guided versus angiography-guided SB treatment in 320 patients with true bifurcation lesions and SB diameter ≥ 2.5 mm [5]. The angiography-guided group received more side branch intervention (angioplasty: 63.1% vs 56.3%, p = 0.07; stenting: 38.1% vs. 25.9%, p = 0.01) without any benefit in major adverse cardiac events at 1 year (18.1% vs. 18.1%, p = 1.00) (Fig. 27.2). Rather, the restenosis rate at the distal main vessel was higher in the angiography-guided group (9.2% vs. 1.7%, p = 0.01). Low FFR in SB does not mean always clinically relevant ischemia in subtended myocardium corresponded with SB stenosis unlike MV stenosis. SB stenosis with myocardial ischemia is associated with less ischemic burden and high recruitability of collateral vessels compared to MV stenosis [6]. Therefore, FFR-guided strategy for SB was warranted in an intermediate to significant stenosis with SB ostial diameter more than 2.0 mm at least or subtended myocardium more than 10%. Up to date, there was no study to prove long-term clinical benefit of FFR-guided SB intervention compared to angiography-guided SB intervention.

27.2 Left Main Lesion

In patients with left main (LM) coronary artery stenosis which produce myocardial ischemia, revascularization therapy confers a survival benefit over medical therapy alone, both for symptomatic and asymptomatic [7–9]. Therefore, it is

H.-J. Lee
Sejong General Hospital, Bucheon, South Korea

J.-H. Doh (✉)
Inje University Ilsan Paik Hospital,
Goyang, South Korea
e-mail: joon.doh@gmail.com

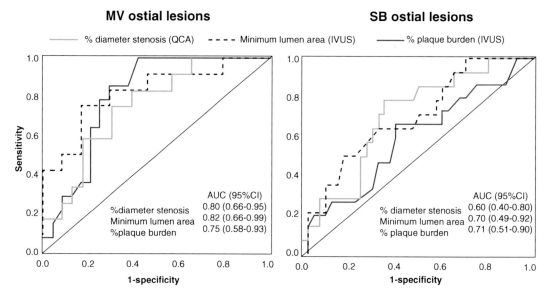

Fig. 27.1 The receiver-operator characteristic curve analysis to assess the diagnostic accuracy of angiographic/IVUS parameters for the prediction of functional significance in both MV and SB ostial lesions. In MV ostial lesions, best cutoff value (BCV) of angiographic percent diameter stenosis, IVUS MLA, percent plaque burden was 53%, 3.5 mm^2, and 70%, respectively. Their area under the curves (AUC) was 0.80, 0.82, and 0.75, respectively. However, there was no statistically significant BCV with good accuracy to predict functional significance in SB ostial lesions. Koh et al. JACC Cardiovasc Interv. 2012;5:409–15

Table 27.1 Incidence of SB FFR < 0.75–0.8 after MV stenting

References	Cutoff value	Prevalence of true bifurcation lesion	Incidence of low FFR
Koo et al. [3]	0.75	69% (n = 65)	27% (n = 20)
Ahn et al. [2]	0.8	27% (n = 61)	17.8% (n = 41)
Chen et al. [5]	0.8	100% (n = 145)	52% (n = 75)

very crucial to exactly evaluate functional significance in intermediate left main lesion compared to intermediate non-LM lesion. Coronary angiography has limited accuracy in assessing actual stenosis severity, and there is great interobserver variability in lesions of the left main coronary artery [10, 11]. Hamilos et al. compared FFR values and the angiographic stenosis by two reviewer's visual estimations in 213 equivocal left main coronary artery stenosis. In 55 (26%), the two reviewers disagreed whether the stenosis of

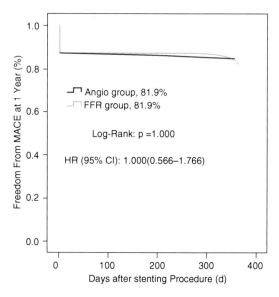

Fig. 27.2 DKCRUSH-VI study showed angiographic and FFR guidance of provisional SB stenting of true coronary bifurcation lesions provided similar 1-year clinical outcomes. Chen et al. JACC Cardiovasc Interv. 2015 20;8:536–46

LMCA was significant, insignificant, or unsure. In 158 (74%) whom the two reviewers agreed, 48 (23%) were misclassified on the basis of visual estimate of the angiogram; 23 patients had an estimated DS > 50% while the FFR was >0.80, and 25 patients had an estimated DS < 50% while the FFR was <0.80. In those, the sensitivity, specificity, and diagnostic accuracy of the visual estimate of DS > 50% to predict an FFR < 0.80 were 46%, 79%, and 69%, respectively. There was either disagreement or misclassification in 49% of all lesions. Therefore, in patients with equivocal stenosis of the left main coronary artery, angiography alone does not allow appropriate individual decision-making about the need for revascularization and often underestimates the functional significance of the stenosis (Fig. 27.3). The prevalence of visual-functional "mismatch" which means FFR > 0.80 even with luminal stenosis >50% in coronary angiogram is lower than non-LM lesion (35% vs. 57%, LM vs. non-LM). Also, the prevalence of visual-functional "reverse mismatch" which means FFR < 0.80 even with luminal stenosis <50% in coronary angiogram is higher than non-LM lesion (40% vs. 16%, LM vs. non-LM) (Fig. 27.4) [12]. Kang et al. reported the cutoff value of IVUS minimal lumen area (MLA) to predict FFR < 0.8 was 4.5 mm^2, and the diagnostic accuracy was good (77% sensitivity, 82% specificity, 84% positive predictive value, 75% negative predictive value, area under the curve: 0.83) in 112 patients with isolated, ostial, or mid-shaft LM intermediate lesion (Fig. 27.5) [13]. The current guideline defined that IVUS are also reasonable for the prediction of myocardial ischemia in intermediate left main disease (Class IIA, Level of Evidence B) [14]. It has been unknown which cutoff value (0.75 vs. 0.8) of FFR is optimal to decide to revascularize intermediate LM lesion. Considering the great concern about the safety of deferred LM stenosis, the use of higher cutoff value for LM FFR with high sensitivity is preferred. The deferral of revascularization in FFR-negative LM stenosis has been reported to be safe [10, 15]. Up to date, there was no prospective study to compare clinical outcomes between angiogram and FFR-guided PCI in this subset. Two thirds of patients with significant LM lesion had multiple stenotic lesions beyond LMCA. Therefore, we should consider the influence of downstream LAD or LCX disease on LM FFR in these cases. Recently, Fearon et al. created an intermediate LMCA stenosis using a deflated balloon catheter after PCI of the LAD, LCX, or both to validate the effect of downstream disease on LM FFR. They measured true FFR of the LMCA via non-diseased downstream vessel, while creating of downstream

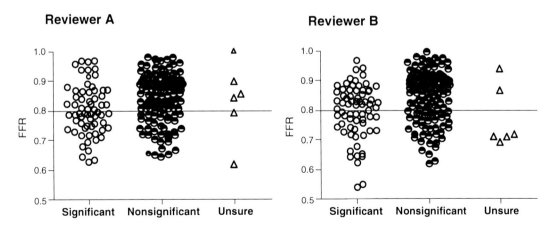

Fig. 27.3 Relation between FFR values and the two reviewers' visual estimations (lesions were classified as significant, nonsignificant, and unsure). Hamilos et al. Circulation 2009;120:1505–12

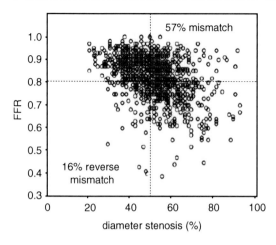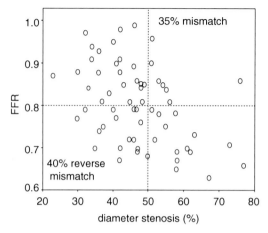

Fig. 27.4 Correlation between angiographic diameter stenosis and FFR in 1066 non-LMCA lesions and 63 LMCA lesions. There was a significant, but modest, correlation between angiographic DS and FFR in the non-LMCA ($r = -0.395$, $p < 0.001$) and LMCA ($r = -0.428$, $p < 0.001$) groups. In 57% of non-LMCA lesions with angiographic DS > 50%, FFR > 0.80 (*mismatch*). Conversely, in 15% of non-LMCA lesions with DS ≤ 50%, FFR < 0.80 (*reverse mismatch*). In the LMCA group, mismatch was observed in 35% of lesions, whereas reverse mismatch was seen in 40% lesions. The LMCA group showed significantly lower frequency of mismatch (35% vs. 57%, $p = 0.032$) and much higher frequency of reverse mismatch (40% vs. 16%, $p = 0.001$) compared with the non-LMCA group. In other words, visually insignificant but functionally significant stenosis was frequent in intermediate LM stenosis. This finding was associated with large myocardial territory supplied by LMCA. Park et al. JACC Cardiovasc Interv. 2012;5:1029–36

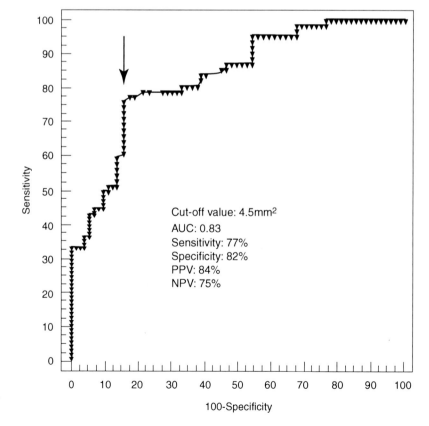

Fig. 27.5 Cutoff value and corresponding diagnostic accuracy of IVUS MLA (minimal luminal area) of an FFR of ≤0.80 in 112 patients with isolated ostial and shaft intermediate LMCA stenosis. The optimal cutoff value of IVUS MLA for an FFR of ≤0.80 was 4.5 mm^2 (77% sensitivity, 82% specificity, area under the curve = 0.83). IVUS-derived MLA had a relatively good accuracy to predict functional significance in intermediate LM stenosis, compared to intermediate non-LM stenosis. Park et al. JACC Cardiovasc Interv. 2014;7:868–74

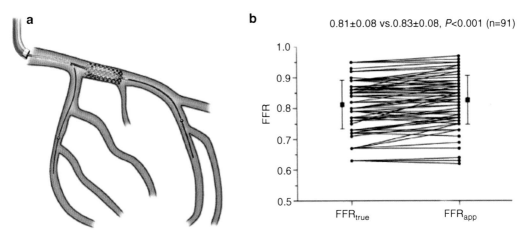

Fig. 27.6 (**a**) Intermediate LM stenosis was made by deflated balloon in the left main coronary artery. The balloon within the stented segment of LAD was then gradually inflated to create a variety of downstream LAD disease, while the apparent FFR (FFRapp) of the LMCA from the pressure wire in the non-diseased LCX was recorded simultaneously. (**b**) Effect of downstream disease on left main coronary artery FFR. FFR_{true} means FFR value for LM lesion itself without downstream disease. FFR_{app} means FFR value for LM and downstream disease. The FFR value of the LMCA after creation of downstream disease was significantly increased but numerically small, with an absolute mean difference of 0.015 (FFR_{true} vs. FFR_{app} was 0.81 ± 0.08 vs. 0.83 ± 0.08). In most case, the influence of downstream disease on LM FFR seems to be clinically irrelevant. Fearon et al. JACC Cardiovasc. Interv. 2015;8:398–403

stenosis by inflating an angioplasty balloon within the newly placed stent (Fig. 27.6) [16]. They demonstrate that the presence of significant downstream disease in LAD or LCX increases the true FFR value of LM lesion itself in their study. However, the difference between LM FFR_{true} and FFR_{app} was small. (0.81 ± 0.08 vs. 0.83 ± 0.08) [11]. This difference correlated with the severity of the downstream disease.

27.3 Serial Lesion

The coronary atherosclerosis is usually a diffuse process; serial stenosis in one epicardial coronary artery is very frequent. It is important to assess which lesion is functionally significant or which lesion is more ischemic when several stenoses exist within one coronary artery and the ischemia was proven. FFR measurements with pullback pressure tracing under maximal hyperemia is useful to identify the culprit ischemic lesion among serial intermediate lesions in a culprit vessel. The interaction between proximal and distal stenosis can change lesion-specific FFR in each stenosis [17]. The functional significance of each stenosis is usually underestimated due to hemodynamic interaction among the lesions. The lesion with the largest pressure step-up on pullback pressure tracing is recommended to treat first, and then repetitive FFR measurement for remnant intermediate lesions guides to treat. Kim et al. performed FFR with pullback pressure tracing in 141 vessels and 298 lesions in 131 patients with serial lesions in one coronary artery [18]. They deferred PCI in 182 lesions (61.1%) after stenting of primary target lesion, and only one deferred lesion was treated during mean follow-up of 501 days. True FFR (ratio of the pressures distal to proximal to each stenosis but after removal of the other one) was lower than apparent FFR (ratio of the pressure just distal to that just proximal to each stenoses) in both proximal and distal stenoses, and the pressure step-up of a non-primary target lesion was increased after stenting the primary target lesion. Interestingly, when the primary target lesion was a distal lesion, the increase of FFR after stenting of primary target lesion was larger than when was a proximal lesion (31.8% vs. 21.7% increment) (Fig. 27.7). It is very difficult to predict true FFR of lesion b before stenting of lesion even through the

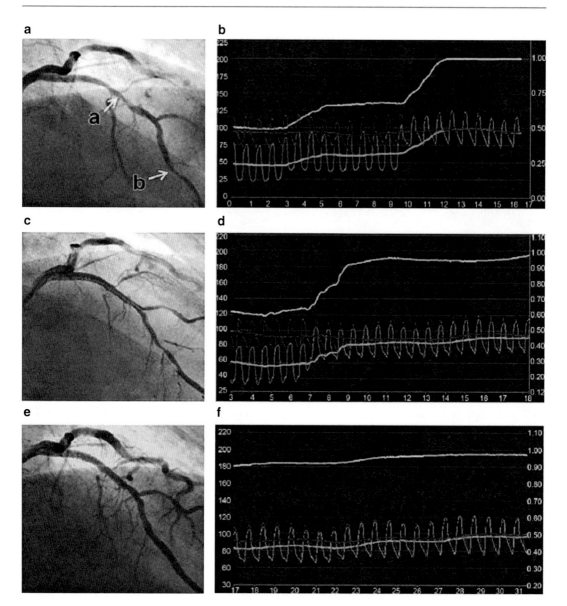

Fig. 27.7 Two consecutive intermediate stenosis (labeled **a** and **b** with *arrows*) were observed in the left anterior descending artery (**a**). As the FFR was 0.48, pullback pressure tracing was performed while simultaneously monitoring the intracoronary pressure (*green line*), aortic pressure (*red line*), and FFR (*yellow line*). Two step-ups of intracoronary pressure were observed during pullback pressure tracing under maximal hyperemia (**b**). Apparent FFR of lesions a and b were 0.67 (ratio of pressure across lesion a = 60/90) and 0.75 (ratio of pressures across lesion b = 45/60), respectively. As the larger pressure step-up was observed across lesion a (30 mmHg) than lesion b (16 mmHg), the proximal stenosis was regarded as the primary target lesion and stenting was performed (**c**). After stenting lesion a, pullback pressure tracings (**d**) were performed again. FFR was 0.59 and intracoronary pressure step-up across lesion b was 20 mmHg. Therefore, stenting to the distal stenosis followed (**e**). True FFR of lesion b was 0.73 (55/75 mmHg). In other words, true FFR of lesion b without lesion a got increased with the presence of lesion a; 0.73 (55/75 mmHg) → 0.75 (45/60 mmHg); pressure step-up across lesion b without lesion a got decreased with the presence of lesion a (20 mmHg → 15 mmHg). After stenting both proximal and distal lesions, FFR was 0.85, and no significant pressure step-up was found across lesion a or b (**f**)

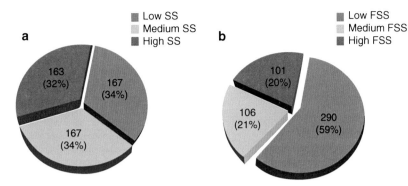

Fig. 27.8 Proportions of the FAME study according to the tertiles of the classic SYNTAX score (SS). By recalculating the SS after counting only ischemia-producing lesions with FFR < 0.80, termed "functional SYNTAX score" (FSS), 32% of studied patients moved from higher-risk groups by SS to lower-risk groups by FSS. In particular, 23% of patients in the highest SS tertile moved to the middle group, 15% of the highest tertile moved to the lowest group, and 59% of patients in the middle SS tertile moved to the lowest group. These changes were driven in large part by the conversion of angiographic three-vessel CAD to functional one- or two-vessel CAD. Nam et al. J Am Coll Cardiol. 2011;58(12):1211–8

development of few mathematical prediction models. Therefore, FFR measurement and repetitive pullback pressure tracing in serial lesion were mandatory to understand lesion-specific ischemia and treat properly.

27.4 Multivessel Disease

Previous studies demonstrated routine measurement of FFR in addition to angiographic guidance, as compared with PCI guided by angiography alone, results in a significant improvement of long-term clinical outcomes in patients with multivessel coronary artery disease [19]. According to FAME study, only 35% among stenotic lesions between 50 and 70% were functionally significant. On the other hand, 20% among stenotic lesions between 71 and 90% were functionally nonsignificant. Indeed, FFR-guided arm could avoid unnecessary stenting by 37%, compared to angiography-guided arm. Therefore, routine FFR measurement in multivessel disease can improve the outcomes by allowing more judicious use of stents and achievement of functionally complete revascularization. The SYNTAX score (SS) is a well-validated anatomic scoring system based on the coronary angiogram, which not only quantifies lesion complexity but also predicts outcome after PCI in patients with multivessel CAD [20]. Recalculating SS by only incorporating ischemia-producing lesions as assessed by FFR decreases the number of higher-risk patients and improve prognostic implication for clinical outcomes in patients with multivessel CAD [21] (Fig. 27.8). The interaction of blood flow between diseased vessels exist in patients with multivessel disease. The FFR value for intermediate lesion can be affected by a significant stenosis in other coronary artery. Multivessel disease with CTO lesion is an extreme subset, which can identify the interaction of stenotic severity between diseased vessels. Sachdeva et al. reported that the FFR value of lesion in donor artery got increased after CTO revascularization in 9 (64%) patients among 14 patients with CTO and intermediate lesion in donor artery [22]. However, it has not been evaluated whether the interaction of blood flow between diseased vessels in patients with multivessel disease with non-CTO lesion exist or not, and, if exist, it is clinically relevant or not. The FFR measurement of intermediate lesion was usually recommended after the PCI of severe stenotic lesion in other vessel in multivessel disease.

Conclusions
FFR also can guide the need of revascularization and reduce unnecessary complicated intervention in lesion subsets including

bifurcation, left main, serial lesions, and multivessel disease. The application of FFR to complex lesion subsets requires a comprehensive understanding of physiologic and anatomical findings and pitfalls in specific complex lesions.

References

1. Shin DH, Koo BK, Waseda K, Park KW, Kim HS, Corral M, et al. Discrepancy in the assessment of jailed side branch lesions by visual estimation and quantitative coronary angiographic analysis: comparison with fractional flow reserve. Catheter Cardiovasc Interv. 2011;78:720–6.
2. Ahn JM, Lee JY, Kang SJ, Kim YH, Song HG, Oh JH, et al. Functional assessment of jailed side branches in coronary bifurcation lesions using fractional flow reserve. JACC Cardiovasc Interv. 2012;5:155–61.
3. Koo BK, Kang HJ, Youn TJ, Chae IH, Choi DJ, Kim HS, et al. Physiologic assessment of jailed side branch lesions using fractional flow reserve. J Am Coll Cardiol. 2005;46:633–7.
4. Koh JS, Koo BK, Kim JH, Yang HM, Park KW, Kang HJ, et al. Relationship between fractional flow reserve and angiographic and intravascular ultrasound parameters in ostial lesions: major epicardial vessel versus side branch ostial lesions. JACC Cardiovasc Interv. 2012;5:409–15.
5. Chen SL, Ye F, Zhang JJ, Xu T, Tian NL, Liu ZZ, Lin S, et al. Randomized comparison of FFR-guided and angiography-guided provisional stenting of true coronary bifurcation lesions: the DKCRUSH-VI trial (double kissing crush versus provisional stenting technique for treatment of coronary bifurcation lesions VI). JACC Cardiovasc Interv. 2015;8:536–46.
6. Koo BK, Lee SP, Lee JH, Park KW, Suh JW, Cho YS, et al. Assessment of clinical, electrocardiographic, and physiological relevance of diagonal branch in left anterior descending coronary artery bifurcation lesions. JACC Cardiovasc Interv. 2012;5:1126–32.
7. Campeau L, Corbara F, Crochet D, Petitclerc R. Left main coronary artery stenosis: the influence of aortocoronary bypass surgery on survival. Circulation. 1978;57:1111–5.
8. Chaitman BR, Fisher LD, Bourassa MG, Davis K, Rogers WJ, Maynard C, et al. Effect of coronary bypass surgery on survival patterns in subsets of patients with left main coronary artery disease. Report of the collaborative study in coronary artery surgery (CASS). Am J Cardiol. 1981;48:765–77.
9. Taylor HA, Deumite NJ, Chaitman BR, Davis KB, Killip T, Rogers WJ. Asymptomatic left main coronary artery disease in the coronary artery surgery study (CASS) registry. Circulation. 1989;79:1171–9.
10. Hamilos M, Muller O, Cuisset T, Ntalianis A, Chlouverakis G, Sarno G, et al. Long-term clinical outcome after fractional flow reserve-guided treatment in patients with angiographically equivocal left main coronary artery stenosis. Circulation. 2009;120:1505–12.
11. Lindstaedt M, Spiecker M, Perings C, Lawo T, Yazar A, Holland-Letz T, et al. How good are experienced interventional cardiologists at predicting the functional significance of intermediate or equivocal left main coronary artery stenoses? Int J Cardiol. 2007;120:254–61.
12. Park SJ, Kang SJ, Ahn JM, Shim EB, Kim YT, Yun SC, et al. Visual-functional mismatch between coronary angiography and fractional flow reserve. JACC Cardiovasc Interv. 2012;5:1029–36.
13. Park SJ, Ahn JM, Kang SJ, Yoon SH, Koo BK, Lee JY, et al. Intravascular ultrasound-derived minimal lumen area criteria for functionally significant left main coronary artery stenosis. JACC Cardiovasc Interv. 2014;7:868–74.
14. Levine GN, Bates ER, Blankenship JC, Bailey SR, Bittl JA, Cercek B, et al. 2011 ACCF/AHA/SCAI guideline for percutaneous coronary intervention: executive summary: a report of the American College of Cardiology Foundation/American Heart Association Task Force on Practice Guidelines and the Society for Cardiovascular Angiography and Interventions. Circulation. 2011;124:2574–609.
15. Mallidi J, Atreya AR, Cook J, Garb J, Jeremias A, Klein LW, et al. Long-term outcomes following fractional flow reserve-guided treatment of angiographically ambiguous left main coronary artery disease: a meta-analysis of prospective cohort studies. Catheter Cardiovasc Interv. 2015;86:12–8.
16. Fearon WF, Yong AS, Lenders G, Toth GG, Dao C, Daniels DV, et al. The impact of downstream coronary stenosis on fractional flow reserve assessment of intermediate left main coronary artery disease: human validation. JACC Cardiovasc Interv. 2015;8:398–403.
17. Pijls NH, De Bruyne B, Bech GJ, Liistro F, Heyndrickx GR, Bonnier HJ, et al. Coronary pressure measurement to assess the hemodynamic significance of serial stenoses within one coronary artery: validation in humans. Circulation. 2000;102:2371–7.
18. Kim HL, Koo BK, Nam CW, Doh JH, Kim JH, Yang HM, et al. Clinical and physiological outcomes of fractional flow reserve-guided percutaneous coronary intervention in patients with serial stenoses within one coronary artery. JACC Cardiovasc Interv. 2012;5:1013–8.
19. Tonino PA, De Bruyne B, Pijls NH, Siebert U, Ikeno F, van' t Veer M, FAME Study Investigators, et al. Fractional flow reserve versus angiography for guiding percutaneous coronary intervention. N Engl J Med. 2009;360:213–24.
20. Serruys PW, Morice MC, Kappetein AP, Colombo A, Holmes DR, Mack MJ, SYNTAX Investigators, et al. Percutaneous coronary intervention versus

coronary-artery bypass grafting for severe coronary artery disease. N Engl J Med. 2009;360:961–72.
21. Nam CW, Mangiacapra F, Entjes R, Chung IS, Sels JW, Tonino PA, FAME Study Investigators, et al. Functional SYNTAX score for risk assessment in multivessel coronary artery disease. J Am Coll Cardiol. 2011;58:1211–8.
22. Sachdeva R, Agrawal M, Flynn SE, Werner GS, Uretsky BF. Reversal of ischemia of donor artery myocardium after recanalization of a chronic total occlusion. Catheter Cardiovasc Interv. 2013; 82:E453–8.

FFR in Acute Coronary Syndrome and Noncoronary Disease

28

Jang Hoon Lee and Dong-Hyun Choi

FFR in Acute Coronary Syndrome

Fractional flow reserve (FFR) is a well-validated method to guide coronary intervention decision in patients with stable coronary artery disease [1, 2]. However, the role of FFR to assess culprit and/or non-culprit lesions in the setting of acute coronary syndrome (ACS) remains unclear due to potential disruption of microcirculation. In this chapter, we explored all relevant publications to date, and summarized the current utility of FFR for ACS.

28.1 ST-Segment Elevation Myocardial Infarction

28.1.1 Culprit Vessel

In ST-segment elevation myocardial infarction (STEMI), there are multiple contributors affecting microcirculatory function including cell death, peri-infarct edema, inflammation, distal embolization, and local vasospasm [3]. Microcirculatory dysfunction results in diminished maximal hyperemia, which can underestimate the hemodynamic severity of a given stenosis and falsely elevate FFR (Fig. 28.1). In an observational study, FFR was evaluated in 33 patients with STEMI who underwent primary percutaneous coronary intervention (PCI) within 12 hours of onset and 15 patients with stable angina pectoris who underwent elective PCI [4]. Although there were no significant differences in intravascular ultrasound parameters, post-PCI FFR was higher in STEMI patients compared with stable angina pectoris (0.95 ± 0.04 vs. 0.90 ± 0.04; $p = 0.002$). In STEMI subgroups, post-PCI FFR was greater in the patients with Thrombolysis In Myocardial Infarction (TIMI) flow grade 2 than those with TIMI 3 (0.98 ± 0.02 vs. 0.93 ± 0.05; $p = 0.017$). Thus, in patients with STEMI, the assessment of FFR in the culprit vessel is not recommended. However, with time, the microvascular dysfunction may recover, maximum achievable flow may increase, and a larger gradient with a lower FFR may be measured across a given stenosis. De Bruyne et al. demonstrated FFR measurement made at 6 days or older after acute myocardial infarction (AMI) reflects a definitive reduction in perfused myocardium [5]. When they compared FFR and myocardial perfusion single-photon emission computed tomography (SPECT) obtained before and after PCI in 57 patients, the 0.75 cutoff value of FFR had 100% specificity against truly posi-

J.H. Lee (✉)
Department of Internal Medicine, Kyungpook National University Hospital, Daegu, South Korea
e-mail: ljhmh75@chol.com

D.-H. Choi
Department of Internal Medicine, Chosun University Hospital, Gwangju, South Korea

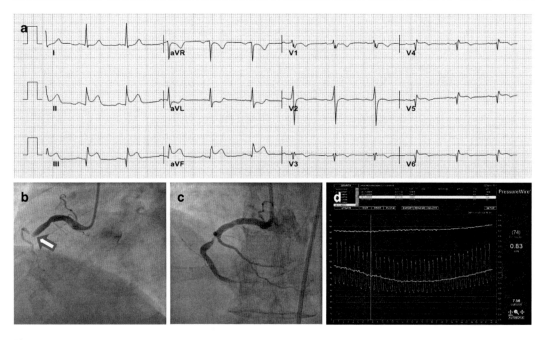

Fig. 28.1 Representative case of fractional flow reserve (FFR) measurement in culprit vessel of ST-segment elevation myocardial infarction. (**a**) Electrocardiogram showed ST-segment elevation in lead II, III, aVF, and V4–V6. (**b**) Coronary angiogram showed total occlusion of right coronary artery (*arrow*). (**c**) After thrombus aspiration, coronary angiogram showed significant stenosis with filling defect in right coronary artery. (**d**) The measured FFR of the culprit vessel was 0.83, indicating underestimation of the hemodynamic severity and falsely elevate FFR due to diminished maximal hyperemia related to microcirculatory dysfunction

tive and truly negative SPECT. Samady et al. also reported that FFR of an infarct-related artery (IRA) accurately identifies reversibility on noninvasive imaging early after AMI [6]. FFR and SPECT were performed in 48 patients 3.7 days after MI. Among them, 23 patients also had myocardial contrast echocardiography (MCE). Follow-up SPECT was performed 11 weeks later to identify true positive and negatives. The sensitivity, specificity, positive and negative predictive value, and concordance of FFR ≤ 0.75 for detecting reversibility on either SPECT or MCE were 88%, 93%, 91%, 91%, and 91% (chi-square <0.001), respectively. The optimal FFR value was 0.78 or less for discriminating inducible ischemia on noninvasive imaging. However, it is still uncertain how long do we have to wait for "microvasculature stunning" to resolve because the time to recovery of the microvasculature is variable. Sometimes, microcirculation remains persistently abnormal up to 6 months following MI [7]. In addition, the mass of viable myocardium being perfused by culprit artery can influence on the validity of FFR. If there is a large area of infarction with less viable myocardium, FFR can be expected to be higher for the same degree of stenosis (Fig. 28.2). Therefore, optimal timing to get a reproducible FFR is variable and seems to depend on the size of infarct.

28.1.2 Non-Culprit Vessels

At the time of primary PCI, coronary angiography reveals multivessel disease (MVD) with at least one angiographically significant lesion in a non-culprit artery in about half of the STEMI patients [8–11]. However, immediate revascularization of non-culprit lesion at the time of primary PCI is highly debatable with only limited evidence of advantage or disadvantage (Table 28.1). The 2014 European Society of Cardiology (ESC) guideline

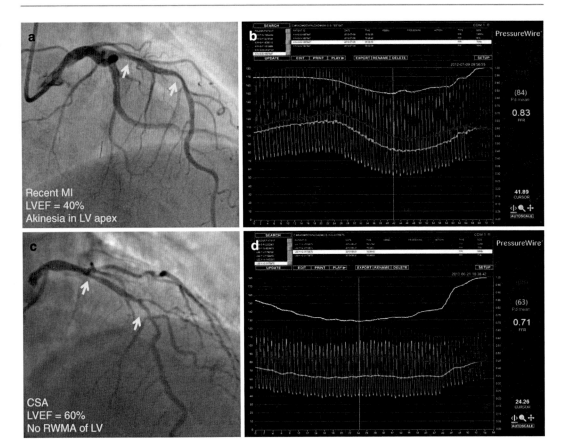

Fig. 28.2 Hypothetical case of fractional flow reserve (FFR) measurement of infarcted myocardium. The mass of viable myocardium being perfused by culprit artery can influence on the validity of FFR. If there is a large area of infarction with less viable myocardium, FFR can be expected to be higher for the same degree of stenosis (*arrows*) in patients with recent myocardial infarction (**a** and **b**) compared to those of chronic stable angina (**c** and **d**). *MI* myocardial infarction, *LVEF* left ventricular ejection fraction, *LV* left ventricular, *CSA* chronic stable angina, *RWMA* regional wall motion abnormality

recommends that primary PCI should be limited to the culprit vessel with the exception of cardiogenic shock and persistent ischemia after PCI of the supposed culprit lesion (Table 28.2). Immediate revascularization of a significant non-culprit lesion during the same procedure as primary PCI of the culprit vessel may be considered in selected [12]. The 2015 American College of Cardiology/American Heart Association guideline upgraded multivessel PCI of a non-IRA in selected hemodynamically stable patient with MVD at the time of primary PCI from class III to class IIb [13]. These current guideline recommendations were based on large observational studies. In the observational data, PCI of non-culprit artery at the time of primary PCI showed increased mortality at 90 days [14, 15]. In contrast, randomized controlled trials (RCTs) comparing culprit artery-only PCI with multivessel PCI have demonstrated conflicting results. In the PRAMI trial, there was absolute 14% reduction in the primary outcome in favor of preventive PCI in non-culprit arteries versus culprit-only PCI (Hazard Ratio = 0.35, 95% Confidence Interval 0.21–0.58; $p < 0.001$) [16]. As such, FFR guidance has the potential to identify lesions which may benefit from immediate multivessel PCI.

There are several theoretical concerns over the validity of FFR measurement of non-culprit vessel during STEMI (Fig. 28.3). First of all,

Table 28.1 Advantage and disadvantage of complete revascularization in patients with ST-segment elevation myocardial infarction and multivessel disease

Advantage of complete revascularization
Plaque instability is not limited to the culprit only
Reducing infarct size results in improving myocardial recovery and better long-term prognosis
Reducing ischemic burden results in less subsequent revascularization
More comfortable feelings after knowing that residual stenoses have been treated
Disadvantage of complete revascularization
Lesion severity of non-culprit artery can be overestimated due to diffuse vasoconstriction
Longer procedure time can increase the risk of contrast-induced nephropathy
Additional PCI of non-culprit artery can result in unnecessary complication
Multiple vessel PCI can increase the risk of no-reflow and stent thrombosis

PCI percutaneous coronary intervention

Table 28.2 Recent recommendations regarding percutaneous coronary intervention of non-culprit artery at the time of primary percutaneous coronary intervention

Recommendations	Class of recommendation	Level of evidence
European Society of Cardiology (12)		
Primary PCI should be limited to the culprit vessel with the exception of cardiogenic shock and persistent ischemia after PCI of the supposed culprit lesion	IIa	B
Staged revascularization of non-culprit lesions should be considered in STEMI patients with multivessel disease in case of symptoms or ischemia within days to weeks after primary PCI	IIa	B
Immediate revascularization of significant non-culprit lesions during the same procedure as primary PCI of the culprit vessel may be considered in selected patients	IIb	B
In patients with continuing ischemia and in whom PCI of the infarct-related artery cannot be performed, CABG should be considered	IIa	C
ACC/AHA/SCAI (13)		
PCI of a noninfarct artery may be considered in selected patients with STEMI and multivessel disease who are hemodynamically stable, either at the time of primary PCI or as a planned procedure	IIb	B-R

PCI percutaneous coronary intervention, *STEMI* ST-segment elevation myocardial infarction, *CABG* coronary artery bypass grafting, *ACC* American College of Cardiology Foundation, *AHA* American Heart Association, *SCAI* Society for Cardiovascular Angiography and Interventions

FFR is critically dependent on the ability to achieve maximal hyperemia. In STEMI, there are multiple factors to influence on microcirculatory function including neurohormonal activation of resistance vessel, increased left ventricular (LV) diastolic pressure, impaired LV systolic function and hypoxic vascular stunning, and thereby compromise the accuracy of FFR assessments in non-culprit vessels. Second, maximal myocardial flow maybe reduced in non-IRAs resulting from remote effect of significantly stenosed culprit artery [17]. However, this result was not consistent with other study measuring Doppler-derived coronary flow reserve (CFR). CFR was preserved in non-IRAs, even in the presence of previous remote MI [18]. Ntalianis et al. investigated the reliability of FFR of non-culprit coronary stenoses during PCI in 75 acute STEMI patients and 26 non-ST elevation myocardial infarction (NSTEMI) patients [19]. The FFR measurements in 112 non-culprit stenoses were obtained immediately after PCI of the culprit stenosis and were repeated 35 ± 4 days later. FFR remained unchanged between the acute and the follow-up

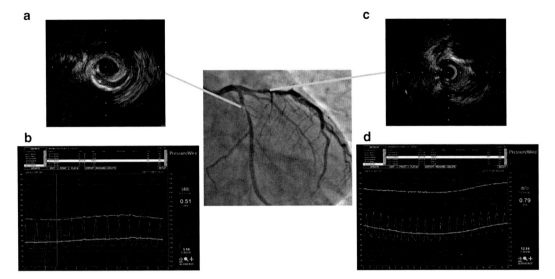

Fig. 28.3 A case showing validity of fractional flow reserve (FFR) measurement of culprit vessel and non-culprit vessel during non-ST segment elevation acute coronary syndrome. (**a**) Intravascular ultrasound (IVUS) showed ruptured plaque at proximal portion of left circumflex artery. This finding indicates culprit vessel is left circumflex artery. (**b**) The measured FFR of culprit vessel was 0.51, indicating significant myocardial ischemia. (**c**) The IVUS image was taken at the narrowest site of left anterior descending artery on coronary angiogram (*arrows*) and showed large amount of plaque burden without rupture or thrombus formation, indicating non-culprit vessel. (**d**) The measured FFR of non-culprit vessel was 0.79

phases despite a significant improvement in LV ejection fraction. In only two patients, the FFR value was higher than 0.8 at the acute phase and lower than 0.75 at follow-up. Index of microcirculatory resistance (IMR) also remained unchanged in a small subgroup. In the other study, reproducibility of FFR in non-culprit lesion has been tested in 47 STEMI patients who had 55 non-culprit stenoses with at least 50% diameter stenosis by visual estimation [20]. FFR measurement was obtained immediately after primary PCI and repeated at 42 ± 10 days. Although there was a small decrease in FFR over time (0.84 ± 0.08 vs. 0.82 ± 0.08, $p = 0.025$), there was a good correlation between paired FFR measurements ($R = 0.85$, $p < 0.001$). Recently, DANAMI-3-PRIMULTI study, which is a randomized clinical trial to compare complete FFR-guided revascularization versus treatment of the culprit lesion only in patients with STEMI and MVD, has demonstrated that FFR-guided complete revascularization significantly reduced the risk of future events compared with no further invasive intervention after primary PCI [21]. This effect is mainly driven by significantly fewer repeat revascularizations. Thus, to avoid repeat revascularization, patients can safely have all their lesions treated during the index admission.

28.2 Non-ST Segment Elevation Acute Coronary Syndrome

28.2.1 Culprit Vessel

There are theoretical concerns regarding the utility of FFR-guided decision making in patients with non-ST segment elevation (NSTE)–ACS because of global microcirculatory dysfunction. In a prospective study, resistance reserve ratio, which is a measure of the ability to achieve maximal hyperemia, of 50 patients with NSTEMI were compared to those of 50 patients with stable angina and 40 patients with STEMI [3]. There was no significant difference between non-culprit vessels instable angina (2.9 [2.3 − 3.9]) and either culprit vessels in stable angina (2.8 [1.7 − 4.8], $p = 0.75$) or culprit vessels in NSTEMI (2.46 [1.6 − 3.9], $p = 0.61$). As expected, IMR was greater in NSTEMI compared with the non-culprit vessels instable angina (22.7 ± 11.36 vs. 16.9 ± 9.06, $p = 0.015$). These data imply that, in selected patients with NSTEMI, microcirculation can dilate sufficiently to enable maximal hyperemia and measuring FFR in both culprit and the non-culprit vessels may be as reliable as in stable angina. The FAME study included 328 patients with unstable angina (UA) or NSTEMI, of whom 178 were randomized to angiographically guided PCI and 150 to FFR-guided PCI [22]. FFR to guide PCI resulted in similar risk reductions of major adverse cardiac events and its components in patients with UA or NSTEMI, compared with patients with chronic stable angina (absolute risk reduction of 5.1% vs. 3.7%, respectively, $p = 0.92$). The benefit of using FFR to guide PCI in MVD does not differ between patients with UA or NSTEMI, compared with patients with chronic stable angina. However, this was a secondary analysis of original FAME study [23], and these patients were generally stable prior to the study. Therefore, it is hard to apply this study to general NSTEMI population and should be confirmed in a prospective trial. Carrick et al. tested clinical utility of using FFR to guide decision making in NSTEMI in a retrospective analysis [24]. Five interventional cardiologists independently reviewed the clinical history and coronary angiogram of 100 patients and then made a treatment decision. Following FFR disclosure, the same cardiologists were asked to reevaluate their initial decisions. Cardiologists changed their initial treatment plans in 46% of patients ($p = 0.0016$). The use of FFR led to increase of medical therapy (24%, $p = 0.0016$). In a French FFR registry (1075 patients, 19% with recent ACS), FFR disclosure was associated with reclassification of their treatment decision in 43% of the patient [25]. In the RCT, Lessar et al. demonstrated FFR-guided decision making in 70 patients with recent UA or NSTEMI markedly reduces the duration and cost of hospitalization compared with stress perfusion scintigraphy [26]. However, patients in this study were medically stable for 48 h or more, and thereby this was not genuine NSTE-ACS population. The FAMOUS-NSTEMI trial is a prospective multicenter RCT which is designed to assess whether management decisions guided by routine FFR measurement in patients with NSTEMI would be feasible and safe, and would optimize clinical outcome compared with angiography-guided standard care [27]. All patients were to obtain FFR measurement to each vessel containing at least 30% stenosis, but only in the FFR-guided arm was the result revealed to the operators. This study demonstrated that the proportion of patients that were assigned to medical therapy was significantly higher in the FFR-guided group than in the angiography-guided group (22.7% vs. 13.2%, difference 95% (95% CI: 1.4–17.7%), $p = 0.022$). FFR disclosure resulted in a change in management decision in 38 (21.6%) patients. In terms of clinical outcome, revascularization remained lower in the FFR-guided group (79.0 vs. 86.8%, difference 7.8% (−0.2%, 15.8%), $p = 0.054$) at 12 months.

The other concern is that whether using contemporary threshold for FFR is safe for deferring PCI in the culprit lesions in patients with NSTE-ACS. In an all-comer ACS population including NSTEMI and STEMI, Potvin et al. demonstrated 201 unselected patients with non-flow-limiting lesions (FFR threshold ≤ 0.75) had 7.5% cardiac events related to the deferred coronary lesion [28]. Although it seems to be safe to allow deferral of PCI, the use of FFR was neither blinded nor randomized. In particular, plaque rupture can occur at the site of a moderate stenosis with less flow-limiting lesion after thrombus resolution in NSTE-ACS. In case, the stenosis might result in a less significant pressure gradient and higher FFR. Therefore, there is a concern that medical therapy for deferred biologically active plaque may not be effective to prevent ischemic events compared to stable plaque with a similar nonischemic FFR value in stable angina. Although FAME substudy and FAMOUS-NSTEMI trial support the safety and effectiveness of deferring PCI in lesion with

Table 28.3 The 2015 European Society of Cardiology guidelines for the management of acute coronary syndrome in patients presenting without ST-segment elevation

2015 European Society of Cardiology guideline
5.6.1.3 Fractional flow reserve (FFR)
The achievement of maximal hyperemia may be unpredictable in NSTEMI because of the dynamic nature of coronary lesions and the associated acute microvascular dysfunction. As a result, FFR may be overestimated and the hemodynamic relevance of a coronary stenosis underestimated. So far, the value of FFR-guided PCI in this setting has not been properly addressed
5.6.5.1 Technical aspects and challenges
While FFR is considered the invasive gold standard for the functional assessment of lesion severity in stable CAD, its role in NSTE-ACS still needs to be defined

NSTEMI non-ST segment elevation myocardial infarction, *PCI* percutaneous coronary intervention, *CAD* coronary artery disease, *NSTE-ACS* non-ST elevation–acute coronary syndrome

FFR >0.80 in non-culprit vessels of patients with NSTE-ACS, there are few data regarding medical treatment of similar culprit lesions. Hakeem et al. compared outcomes in NSTE-ACS patients who did not undergo PCI of any lesion on the basis of FFR to those in a similar group of patients with stable angina [29]. The long-term major adverse cardiovascular events were higher in ACS group than in the stable angina group (25% versus 12%, $p < 0.0001$). Best cutoffs to predict accuracy for major adverse cardiovascular events is less than 0.84 for ACS and less than 0.81 for stable angina. The 2015 ESC guidelines for the management of NSTE-ACS addressed that the role of FFR in NSTE-ACS still needs to be defined (Table 28.3).

28.2.2 Non-Culprit Vessels

In NSTE-ACS, it can often be difficult to discriminate IRA if there is no locating electrocardiographic sign, no regional wall motion abnormality on 2D-echocardiogram, or no typical angiographic feature such as haziness, luminal irregularity, and filling defect. Invasive image such as intravascular ultrasound, optical coherence tomography may be helpful to identify plaque rupture or dissection. However, these lesions may not be culprit lesion in terms of FFR if there is no flow limitation. Therefore, FFR has the ability to identify the vessel with physiologically reduced flow and hemodynamic instability. As mentioned above, FFR of non-culprit coronary stenoses during PCI has also reliability as those of STEMI [3, 19].

28.3 Ongoing Clinical Trials of FFR-Guided PCI in Patients with Acute Coronary Syndrome

FFR-guided decision making in patients with STEMI and MVD is now tested in a series of RCTs including COMPARE-ACUTE, COMPLETE, FRAME-AMI, FLOWER-MI, and FULL REVASC (Table 28.4). PRESSUREWire is an international observational registry to compare resting indices with FFR values in patients with ACS and stable coronary artery disease.

28.4 Summary

Despite the theoretical concerns, the use of FFR is well validated in non-culprit vessels both in patients with STEMI and NSTE-ACS (Fig. 28.4). Data regarding the safety and efficacy of FFR in culprit vessels of patients with NSTE-ACS are controversial in recent clinical trials. Measuring FFR in the culprit vessel of patients with STEMI at the time of primary PCI is not recommended, and may be reliable 6 days after STEMI. Current guidelines recommended a great caution for interpreting FFR value in ACS. Therefore, we would summarize several considerations regarding FFR measurement for ACS in Table 28.5.

COMPARE-ACUTE comparison between FFR-guided revascularization versus conventional strategy in acute STEMI patients with MVD, *COMPLETE* complete vs culprit-only revascularization to treat multivessel disease after primary PCI for STEMI, *FRAME-AMI*

Table 28.4 Ongoing clinical trials of fractional flow reserve-guided percutaneous coronary intervention in patients with acute coronary syndrome

Clinical trials	Allocation	Arms	Primary endpoint
COMPARE-ACUTE	RCT ($n = 800$)	Immediate FFR-guided complete revascularization versus staged non-culprit PCI (ischemia-driven) by proven ischemia or recurrent symptoms	Composite of death, non-fatal MI, CVA, or revascularization at 12 months
COMPLETE	RCT ($n = 3900$)	FFR-guided revascularization within 72 h of primary PCI versus optimal medical therapy	Composite of cardiovascular death or MI at 4 years
FRAME-AMI	RCT ($n = 1400$)	Immediate FFR-guided complete revascularization versus immediate angiography-guided PCI of non-culprit lesions	Composite of death or non-fatal MI at 2 years
FLOWER-MI	RCT ($n = 1170$)	FFR-guided PCI versus angiography-guided PCI of non-culprit lesions	Composite of death, non-fatal MI, or repeat revascularization at 12 months
FULL REVASC	RCT ($n = 4052$)	FFR-guided PCI to non-culprit lesions during index admission versus initial conservative management of non-culprit lesions	Composite of death and non-fatal MI at 12 months
PRESSUREWire	Registry ($n = 2000$)	Patient is presenting with STEMI, NSTEMI, unstable angina, or stable coronary artery disease. Patients where FFR has been performed or is planned to be performed for further evaluation of PCI procedures, as per physician clinical practice	Characterize the 12 month clinical outcomes by FFR values and resting indices

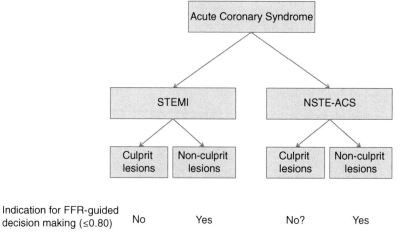

Fig. 28.4 Algorithm for fractional flow reserve (FFR)-guided decision making based on contemporary study for acute coronary syndrome. *STEMI* ST-segment elevation myocardial infarction, *NSTE-ACS* non-ST segment elevation acute coronary syndrome

Table 28.5 Considerations for use of fractional flow reserve in patients with acute coronary syndrome

General consideration
The diagnostic validity of FFR is less certain due to concerns of potential disruption of microcirculation to achieve maximal hyperemia
ST-segment elevation myocardial infarction
Measuring FFR in the culprit vessel is not recommended
Measuring FFR in the culprit vessel more than 6 days after STEMI is relatively reliable
FFR value of non-culprit vessels should be interpreted with caution, particularly in case of hemodynamically unstable or extensive microvascular dysfunction
Measuring FFR in the non-culprit vessel is useful for decision making
Non-ST elevation acute coronary syndrome
Measuring FFR in the culprit vessel appears reliable. However, further studies are required due to conflicting results
Measuring FFR in the non-culprit vessel is useful for decision making
Measuring FFR influence on making a revascularization decision

FFR fractional flow reserve, *STEMI* ST-segment elevation myocardial infarction

FFR versus angiography-guided strategy for management of AMI with multivessel disease, *FLOWER-MI* FLOW evaluation to guide revascularization in multivessel ST-elevation myocardial infarction, *FULL REVASC* Ffr-gUidance for complete non-cuLprit REVAScularization, *PRESSUREWire* practical evaluation of fractional flow reserve and its associated alternative indices during routine clinical procedures, *RCT* randomized controlled trial, *FFR* fractional flow reserve, *PCI* percutaneous coronary intervention, *STEMI* ST-segment elevation myocardial infarction, *NSTEMI* non-ST segment elevation myocardial infarction, *MI* myocardial infarction, *CVA* cerebrovascular accident

FFR in Non-Coronary Disease

The renal artery stenosis (RAS) and peripheral artery disease (PAD) are the two main indications to apply FFR.

28.5 Renal Artery Stenosis

Atherosclerotic RAS is one of the common causes for secondary hypertension [30] and increased hazard of cardiovascular death [31]. The occurrence of atherosclerotic RAS rises with age and is up to about 20% in patients with known coronary artery disease [32, 33]. In spite of a high procedural success rate of renal artery stenting, an improvement in hypertension has been uneven. Large randomized controlled trials comparing percutaneous angioplasty and optimal medical therapy did not demonstrate revascularization benefits [34]. Furthermore, in hypertensive patients the incidental discovery of RAS has become common, even though just a small portion is in charge of renovascular hypertension [33]. Thus, it is essential to detect those patients who might benefit from percutaneous renovascular intervention.

Because a reduction of renal pressure distal to the stenosis and its subsequent release of renin are the important cause of renovascular hypertension, assessment of transstenotic pressure gradient with pressure wires offers the most precise methods of hemodynamic assessment [35]. Furthermore, there have been great interests in using invasive procedures for evaluation of the physiological significance of the RAS. In this regard, numerous studies proved poor correlations comparing diameter stenosis by quantitative renal angiography with renal FFR and hyperemic systolic gradient (HSG), hyperemic mean gradient, and resting systolic gradient (RSG) [35–37]. It was described that renal FFR is a promising tool to detect hypertensive patients with RAS who would probably advantageous from renal artery stenting [38].

28.5.1 Induction of Maximal Hyperemia

Intracoronary papaverine has been used for the physiological evaluation of coronary artery stenosis; but, this was associated with prolongation of the QT interval and generation of polymorphic ventricular tachycardia [39]. Adenosine, meanwhile, is currently universally used instead of papaverine for the physiological evaluation of a coronary artery stenosis. However, intrarenal adenosine can decrease the glomerular filtration rate by constricting afferent arterioles and therefore it is not an appropriate agent to induce hyperemia [40]. On the contrary, numerous studies have revealed that intrarenal papaverine significantly improved renal flow reserve [41]. Nevertheless, no uniform approach has been used to evaluate the physiological significance of the RAS. The guidelines for renal artery revascularization suggest that a hemodynamically significant RAS is defined as a resting systolic translesional gradient ≥20 mm Hg or a resting mean gradient ≥10 mm Hg [42]. One study described that stenting in 17 patients with RAS resulted in a substantial hypertension improvement in patients who had a renal FFR <0.80 compared with those with an FFR >0.80 [33]. Other studies reported that hyperemic systolic gradient of 21 mm Hg and a renal FFR of 0.90 were considered to represent hemodynamically significant stenosis and predict hypertension improvement after stenting [43]. On the other hand, diameter stenosis measured by quantitative renal angiography did not predict blood pressure improvement [43].

28.6 Peripheral Artery Disease

Though angiography is considered the gold standard for detecting PAD, it can overestimate or underestimate disease in the peripheral arteries, and countless interobserver variability occurs [44, 45]. Contemporary researches have revealed that FFR can be successfully measured in peripheral arteries, as it is for coronary vasculatures, during angiography using a standard catheter or pressure guide wire to compare the arterial pressure distal to a lesion with the pressure proximal [46, 47]. Although there is a lack of literature describing the usage of FFR in the management of PAD, FFR has the potential to allow for better identification of which peripheral lesions would most benefit from endovascular therapy.

28.6.1 FFR in Management of Superficial Femoral Artery

Superficial femoral artery (SFA) lesions are distinctive as the SFA moves in multiple planes during leg movement resulting in a higher occurrence of restenosis with a primary patency of 33% at 1 year follow-up. Only few researches have executed FFR measurements for SFA lesions [46, 47]. Especially, in patients who require SFA intervention, physiologic measurements such as FFR and peak systolic velocity (PSV) may aid to recognize patients with a poorer prognostic result and help guide upcoming procedures [46]. FFR < 0.95 might be considered as a predictor for future restenosis. As baseline FFR correlated well with baseline PSV, FFR may be a valuable method for detecting hemodynamically substantial stenosis in the absence of noninvasive PSV examination. FFR measurement may improve predictive value in the treatment of SFA disease and help detect patients who require closer observations for restenosis.

One of the problems concerning the assessing procedure for FFR for SFA lesions was the location of the guide sheath where pressure equalization was accomplished. One study considered the level below the common iliac artery as one component as well as the level below the left main coronary artery for measuring FFR in coronary artery disease (CAD) [47]. One of the drawbacks of measuring FFR in CAD is that the guide catheter can limit coronary blood flow especially if a stenosis is existing at the left main ostium or if a large sized catheter is used [48]. If a big guide sheath is positioned at the iliac artery, the guide sheath may act as a false stenosis proximal to the SFA lesion. Consequently, the severity of the SFA lesion may be undervalued by FFR measurement. The best measuring procedure for FFR for SFA lesions as well as for CAD is that a catheter must not be located in the iliac artery but only a pressure wire. Only a 0.014-inch pressure wire was positioned in the iliac artery, and thus blood flow was not disturbed [47].

28.6.2 Induction of Maximal Hyperemia

Several studies showed that measurement of FFR under hyperemia induced by papaverine is a feasible and harmless procedure in patients with PAD [47]. Papaverine is an endothelium-independent vasodilator that is useful for inducing hyperemia in physiological assessment study involving the aortoiliac arteries [49]. One study also presented that the required papaverine dosage for inducing maximal hyperemia in SFA lesions is between 20 mg and 30 mg [47]. There were no major differences in the required papaverine amount according to sex difference and the amount of infrapopliteal arteries.

28.7 Conclusion and Future Directions

Gradually FFR will play an essential role in the managing of patients with non-coronary diseases beyond RAS and PAD. In particular, result produced is used to determine the need for revascularization. Cumulative gaining of concurrent FFR data and angiography has been shown to rationalize evidence based patient care.

References

1. Puymirat E, Muller O, Sharif F, Dupouy P, Cuisset T, de Bruyne B, et al. Fractional flow reserve: concepts, applications and use in France in 2010. Arch Cardiovasc Dis. 2010;103:615–22.
2. Legalery P, Schiele F, Seronde MF, Meneveau N, Wei H, Didier K, et al. One-year outcome of patients submitted to routine fractional flow reserve assessment to determine the need for angioplasty. Eur Heart J. 2005;26:2623–9.
3. Layland J, Carrick D, McEntegart M, Ahmed N, Payne A, McClure J, et al. Vasodilatory capacity of the coronary microcirculation is preserved inselected patients with non-ST-segment-elevation myocardial infarction. Circ Cardiovasc Interv. 2013;6:231–6.
4. Tamita K, Akasaka T, Takagi T, Yamamuro A, Yamabe K, Katayama M, et al. Effects of microvascular dysfunction on myocardial fractional flow reserveafter percutaneous coronary intervention in patients with acute myocardialinfarction. Catheter Cardiovasc Interv. 2002;57:452–9.
5. De Bruyne B, Pijls NH, Bartunek J, Kulecki K, Bech JW, De Winter H, et al. Fractional flow reserve in patients with prior myocardial infarction. Circulation. 2001;104:157–62.
6. Samady H, Lepper W, Powers ER, Wei K, Ragosta M, Bishop GG, et al. Fractional flow reserve of infarct-related arteries identifies reversible defect-son noninvasive myocardial perfusion imaging early after myocardial infarction. J Am Coll Cardiol. 2006;47:2187–93.
7. Uren NG, Crake T, Lefroy DC, de Silva R, Davies GJ, Maseri A. Reducedcoronary vasodilator function in infarcted and normal myocardium aftermyocardial infarction. N Engl J Med. 1994;331:222–7.
8. Steg PG, James SK, Atar D, Badano LP, Blömstrom-Lundqvist C, Borger MA, et al. ESC guidelines for the management of acute myocardial infarction in patients presenting with ST-segment elevation. Eur Heart J. 2012;33:2569–619.
9. O'Gara PT, Kushner FG, Ascheim DD, Casey DE Jr, Chung MK, de Lemos JA, et al. ACCF/AHA guideline for the management of ST-elevation myocardial infarction: a report of the American College of Cardiology Foundation/American Heart Association Task Force on Practice Guidelines. J Am Coll Cardiol. 2013;2013:e78–140.
10. Muller DW, Topol EJ, Ellis SG, Sigmon KN, Lee K, Califf RM, Thrombolysis and Angioplasty in Myocardial Infarction (TAMI) Study Group. Multivessel coronary artery disease: a key predictor of short-term prognosis after reperfusion therapy for acute myocardial infarction. Am Heart J. 1991;121:1042–9.
11. Jaski BE, Cohen JD, Trausch J, Marsh DG, Bail GR, Overlie PA, et al. Outcome of urgent percutaneous transluminal coronary angioplasty in acute myocardial infarction: comparison of single-vessel versus multivessel coronary artery disease. Am Heart J. 1992;124:1427–33.
12. Windecker S, Kolh P, Alfonso F, Collet JP, Cremer J, Falk V, et al. 2014 ESC/EACTS guidelines on myocardial revascularization: the Task Force on Myocardial Revascularization of the European Society of Cardiology (ESC) and the European Association for Cardio-Thoracic Surgery (EACTS) Developed with the special contribution of the European Association of Percutaneous Cardiovascular Interventions (EAPCI). Eur Heart J. 2014;35:2541–619.
13. Levine GN, Bates ER, Blankenship JC, Bailey SR, Bittl JA, Cercek B, et al. 2015 ACC/AHA/SCAI focused update on primary percutaneous coronary intervention for patients with ST-elevation myocardial infarction: an update of the 2011 ACCF/AHA/SCAI guideline for percutaneous coronary intervention and the 2013 ACCF/AHA guideline for the management of ST-elevation myocardial infarction. J Am Coll Cardiol. 2016;67:1235–50.

14. Toma M, Buller CE, Westerhout CM, Fu T, O'Neill WW, Holmes DR Jr, APEX-AMI Investigators, et al. Non-culprit coronary artery percutaneous coronary intervention during acute ST-segment elevation myocardialinfarction: insights from the APEX-AMI trial. Eur Heart J. 2010;31:1701–7.
15. Kornowski R, Mehran R, Dangas G, Nikolsky E, Assalli A, Claessen BE, HORIZONS-AMI Trial Investigators, et al. Prognostic impact of staged versus "one-time" multivesselpercutaneous intervention in acute myocardial infarction: analysis from the HORIZONS-AMI (harmonizing outcomes with revascularization and stents in acute myocardial infarction) trial. J Am Coll Cardiol. 2011;58:704–11.
16. Wald DS, Morris JK, Wald NJ, Chase AJ, Edwards RJ, Hughes LO, PRAMI Investigators, et al. Randomizedtrial of preventive angioplasty in myocardial infarction. N Engl J Med. 2013;369:1115–23.
17. Uren NG, Marraccini P, Gistri R, de Silva R, Camici PG. Altered coronary vasodilator reserve and metabolism in myocardiumsubtended by normal arteries in patients with coronary artery disease. J Am Coll Cardiol. 1993;22:650–8.
18. Pizzuto F, Voci P, Mariano E, Puddu PE, Spedicato P, Romeo F. Coronary flow reserve of the angiographically normal left anterior descending coronary artery in patients with remote coronary artery disease. Am J Cardiol. 2004;94:577–82.
19. Ntalianis A, Sels JW, Davidavicius G, Tanaka N, Muller O, Trana C, et al. Fractional flow reserve for the asessment of nonculprit coronary artery stenoses in patients with acute myocardial infarction. JACC Cardiovasc Interv. 2010;3:1274–81.
20. Wood DA, Poulter R, Boone R, Owens C, Starovoytov A, Lim I, et al. Stability of non culprit vessel fractional flow reserve in patients with St-segment elevation myocardial infarction. Can J Cardiol. 2013;29:S291–2.
21. Engstrøm T, Kelbæk H, Helqvist S, Høfsten DE, Kløvgaard L, Holmvang L, DANAMI-3—PRIMULTI Investigators, et al. Complete revascularisation versus treatment of the culprit lesion only in patients with ST-segment elevation myocardial infarction and multivessel disease (DANAMI-3—PRIMULTI): an open-label, randomised controlled trial. Lancet. 2015;386:665–71.
22. Sels JW, Tonino PA, Siebert U, Fearon WF, Van't Veer M, De Bruyne B, et al. Fractional flow reserve in unstable angina and non-ST-segment elevation myocardial infarction experience from the FAME (Fractional flow reserve versus Angiographyfor Multivessel Evaluation) study. JACC Cardiovasc Interv. 2011;4:1183–9.
23. Pim AL, Tonino PA, De Bruyne B, Pijls NH, Siebert U, Ikeno F, et al. Fractional flow reserve versus angiography for guiding percutaneous coronary angiography. N Engl J Med. 2009;360:213–24.
24. Carrick D, Behan M, Foo F, Christie J, Hillis WS, Norrie J, et al. Usefulness offractional flow reserve to improve diagnostic efficiency in patients with non-ST elevation myocardial infarction. Am J Cardiol. 2013;111:45–50.
25. Van Belle E, Rioufol G, Pouillot C, Cuisset T, Bougrini K, Teiger E, et al. Outcome impact of coronary revascularization strategy reclassification with fractional flow reserve at time of diagnostic angiography: insights from a large French multicenter fractional flow reserve registry. Circulation. 2014;129:173–85.
26. Leesar MA, Abdul-Baki T, Akkus NI, Sharma A, Kannan T, Bolli R. Use offractional flow reserve versus stress perfusion scintigraphy after unstableangina. Effect on duration of hospitalization, cost, procedural characteristics, and clinical outcome. J Am Coll Cardiol. 2003;41:1115–21.
27. Layland J, Oldroyd KG, Curzen N, Sood A, Balachandran K, Das R, onbehalf of the FAMOUS-NSTEMI investigators, et al. Fractional flow reserve vs. angiography in guiding management to optimize outcomes in non-STsegmentelevation myocardial infarction: the British Heart Foundation FAMOUS-NSTEMI randomized trial. Eur Heart J. 2014;36(2):100–11. pii: ehu338.
28. Potvin JM, Rodés-Cabau J, Bertrand OF, Gleeton O, Nguyen CN, Barbeau G, et al. Usefulness of fractional flow reserve measurements todefer revascularization in patients with stable or unstable angina pectoris, non-ST-elevation and ST-elevation acute myocardial infarction, or atypicalchest pain. Am J Cardiol. 2006;98:289–97.
29. Hakeem A, Edupuganti MM, Almomani A, Pothineni NV, Payne J, Abualsuod AM, et al. Long-term prognosis of deferred acute coronarysyndrome lesions based on nonischemic fractionalflow reserve. J Am Coll Cardiol. 2016;68:1181–91.
30. Caps MT, Perissinotto C, Zierler RE, et al. Prospective study of atherosclerotic disease progression in the renal artery. Circulation. 1998;98:2866–72.
31. Conlon PJ, Little MA, Pieper K, Mark DB. Severity of renal vascular disease predicts mortality in patients undergoing coronary angiography. Kidney Int. 2001;60:1490–7.
32. Buller CE, Nogareda JG, Ramanathan K, et al. The profile of cardiac patients with renal artery stenosis. J Am Coll Cardiol. 2004;43:1606–13.
33. Weber-Mzell D, Kotanko P, Schumacher M, Klein W, Skrabal F. Coronary anatomy predicts presence or absence of renal artery stenosis. A prospective study in patients undergoing cardiac catheterization for suspected coronary artery disease. Eur Heart J. 2002;23:1684–91.
34. Wheatley K, Ives N, Gray R, et al. Revascularization versus medical therapy for renal-artery stenosis. N Engl J Med. 2009;361:1953–62.
35. Colyer WR Jr, Cooper CJ, Burket MW, Thomas WJ. Utility of a 0.014″ pressure-sensing guidewire to assess renal artery translesional systolic pressure gradients. Catheter Cardiovasc Interv. 2003;59:372–7.
36. Subramanian R, White CJ, Rosenfield K, et al. Renal fractional flow reserve: a hemodynamic evaluation of moderate renal artery stenoses. Catheter Cardiovasc Interv. 2005;64:480–6.

37. Siddiqui TS, Elghoul Z, Reza ST, Leesar MA. Renal hemodynamics: theory and practical tips. Catheter Cardiovasc Interv. 2007;69:894–901.
38. Mitchell JA, Subramanian R, White CJ, et al. Predicting blood pressure improvement in hypertensive patients after renal artery stent placement: renal fractional flow reserve. Catheter Cardiovasc Interv. 2007;69:685–9.
39. Talman CL, Winniford MD, Rossen JD, Simonetti I, Kienzle MG, Marcus ML. Polymorphous ventricular tachycardia: a side effect of intracoronary papaverine. J Am Coll Cardiol. 1990;15:275–8.
40. Vallon V, Muhlbauer B, Osswald H. Adenosine and kidney function. Physiol Rev. 2006;86:901–40.
41. Manoharan G, Pijls NH, Lameire N, et al. Assessment of renal flow and flow reserve in humans. J Am Coll Cardiol. 2006;47:620–5.
42. Hirsch AT, Haskal ZJ, Hertzer NR, et al. ACC/AHA 2005 Practice Guidelines for the management of patients with peripheral arterial disease (lower extremity, renal, mesenteric, and abdominal aortic): a collaborative report from the American Association for Vascular Surgery/Society for Vascular Surgery, Society for Cardiovascular Angiography and Interventions, Society for Vascular Medicine and Biology, Society of Interventional Radiology, and the ACC/AHA Task Force on Practice Guidelines (Writing Committee to Develop Guidelines for the Management of Patients With Peripheral Arterial Disease): endorsed by the American Association of Cardiovascular and Pulmonary Rehabilitation; National Heart, Lung, and Blood Institute; Society for Vascular Nursing; Trans Atlantic Inter-Society Consensus; and Vascular Disease Foundation. Circulation. 2006;113:e463–654.
43. Leesar MA, Varma J, Shapira A, et al. Prediction of hypertension improvement after stenting of renal artery stenosis: comparative accuracy of translesional pressure gradients, intravascular ultrasound, and angiography. J Am Coll Cardiol. 2009;53:2363–71.
44. Arvela E, Dick F. Surveillance after distal revascularization for critical limb ischaemia. Scand J Surg. 2012;101:119–24.
45. Fischer JJ, Samady H, McPherson JA, et al. Comparison between visual assessment and quantitative angiography versus fractional flow reserve for native coronary narrowings of moderate severity. Am J Cardiol. 2002;90:210–5.
46. Lotfi AS, Sivalingam SK, Giugliano GR, Ashraf J, Visintainer P. Use of fraction flow reserve to predict changes over time in management of superficial femoral artery. J Interv Cardiol. 2012;25:71–7.
47. Kobayashi N, Hirano K, Nakano M, et al. Measuring procedure and maximal hyperemia in the assessment of fractional flow reserve for superficial femoral artery disease. J Atheroscler Thromb. 2016;23: 56–66.
48. De Bruyne B, Paulus WJ, Pijls NH. Rationale and application of coronary transstenotic pressure gradient measurements. Catheter Cardiovasc Diagn. 1994;33:250–61.
49. Sensier YJ, Thrush AJ, Loftus I, Evans DH, London NJ. A comparison of colour duplex ultrasonography, papaverine testing and common femoral Doppler waveform analysis for assessment of the aortoiliac arteries. Eur J Vasc Endovasc Surg. 2000;20:29–35.

Fractional Flow Reserve in Specific Lesion Subsets

Hyun-Hee Choi and Sang Yeub Lee

29.1 Fractional Flow Reserve (FFR) in Post-percutaneous Coronary Intervention (PCI)

Pre-PCI FFR has been recommended and used to assess of ischemia in the angiographically intermediate lesion and use of percutaneous coronary intervention [1, 2]. In contrast, post-PCI FFR has been less frequently performed and rarely recommended.

Nevertheless, post-PCI FFR can support useful information to perform functional optimization of PCI and be a strong predictor of clinical outcomes. Even though angiographic result seems optimal, post-PCI FFR could give an opportunity for identifying patient with a suboptimal interventional result and higher risk for poor clinical outcome who might advantage from further intervention [3, 4].

There are four mechanisms of low post-PCI FFR (Fig. 29.1). First, unmasked or unappreciated tandem lesions will occasionally increase their gradients after PCI of the primary stenosis. Second, diffuse disease frequently coexists with focal lesions and remains untreated after PCI. Third, pressure drift can cause an artifactual FFR that does not reflect true condition. Fourth, stent implantation itself causes a gradient as demonstrated by longitudinal observations [5].

In previous studies included relatively simple lesions with an overall low coronary artery disease burden, post-PCI FFR > 0.90 has been considered an optimal functional endpoint of PCI and has been associated with favorable clinical outcomes [6]. Shiv et al. reported the post-PCI FFR identified 20% of angiographically satisfactory lesions, which required further intervention, thereby providing an opportunity for functional optimization of PCI results at the time of the index procedure, and further optimization intervention improved the post-PCI FFR by approximately 0.05; furthermore, final FFR ≤ 0.86 had incremental prognostic value over clinical and angiographic variables for major cardiovascular events (MACE) prediction [7].

29.2 FFR in Myocardial Bridge

Myocardial bridging (MB) is a common incidental finding noted on coronary angiography and has been considered a benign condition. However, there are a number of reports that have related MB with myocardial ischemia, acute coronary syndrome, arrhythmia, and sudden cardiac death [8, 9]. So the

H.-H. Choi
Department of internal Medicine, Chunchoen Sacred Heart Hospital, Hallym University, Chunchoen, South Korea

S.Y. Lee (✉)
Division of Cardiology, Chungbuk National University Hospital, Chungbuk National University School of Medicine, Cheongju, South Korea
e-mail: louisahj@gmail.com

Unmasked 2nd lesion

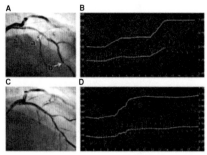

- Tandem or serial lesions
- Post-PCI FFR mandatory
- Largest gains in FFR

Diffuse disease

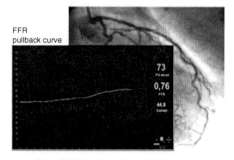

- Pre-PCI selection vital
- High risk post-PCI
- Untreatable with more PCI

Pressure drift

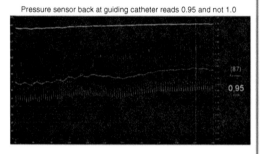

- Technical artifact
- 10% incidence
- Re-equalize wire

Optimization necessary

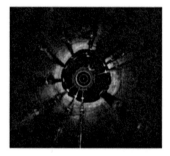

- Stent itself causes gradient
- Larger balloon, higher pressure
- Expect FFR gain of 0.05

Fig. 29.1 The mechanism of a low FFR after PCI

functional assessment of MB is important, and, recently, diastolic FFR was reported to be useful in evaluation of functional significance of MB.

Diefenbach et al. were the first to demonstrate that inotropic stimulation (a beta-agonist) unmasked angiographically silent MB [10], and Escaned et al. reported that the diagnostic value of dobutamine challenges for physiologic assessment of MB in 12 symptomatic patients with positive stress test. Although both FFR and diastolic FFR decreased significantly after dobutamine infusion, diastolic FFR identified hemodynamic significance of MB in five patients (diastolic FFR < 0.76), whereas FFR was <0.75 in only one patient [11].

In contrast to patients with fixed coronary stenosis, FFR measurement after adenosine infusion underestimates the significance of stenosis in patient with MB, but FFR measurement after high-dose dobutamine infusion (40 μg/kg/min) is a promising strategy to unmask the significance of MB [12]. (Fig. 29.2).

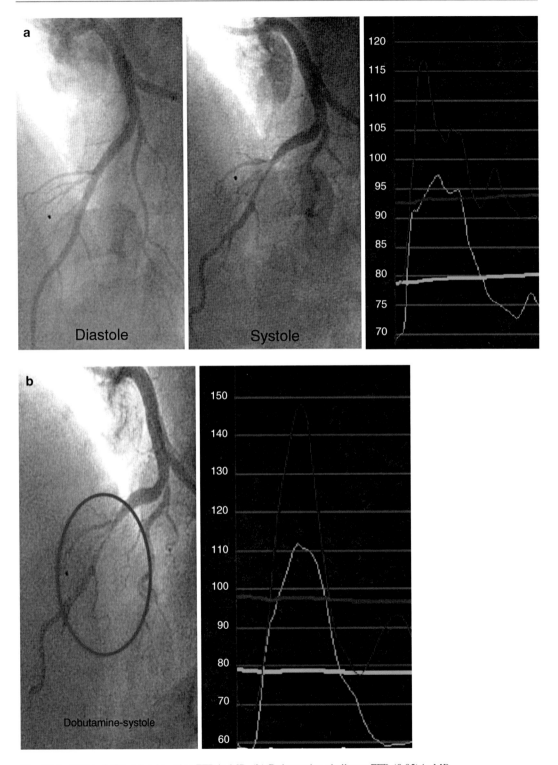

Fig. 29.2 FFR in MB). (**a**) Adenosine FFR in MB. (**b**) Dobutamine challenge FFR (0.85) in MB

29.3 FFR in Non-culprit Vessel

Multivessel (MV) coronary artery disease (CAD) exists in approximately half of acute myocardial infarction (AMI) patients [13, 14]. In these patients, the proper management for non-culprit lesions remains still controversial.

In AMI patients with angiographically significant MV CAD, the incidence of heart failure [15] and recurrent acute coronary syndrome [16] and need for further revascularization [17] have been reported to be significantly higher, and survival has been reported to be significantly lower [18] than single-vessel disease CAD.

Even though the measuring FFR at the time of primary PCI has disadvantages such as higher amount of contrast medium and radiation, the need for additional instruments, and prolonged procedure time, current studies have confirmed the reliability and safety of FFR measurements in setting of MI.

Ntalianis et al. reported the reliability of FFR measurements of non-culprit lesions during the acute phase of MI and a very good reproducibility of FFR. In this study, FFR measurements of 112 non-culprit lesions were done immediately after PCI of the culprit vessel and were repeated within 35 days. The FFR value of the non-culprit lesion did not change between the acute and follow-up. In only two patients, the FFR value changed from >0.8 during primary PCI to <0.75 at follow-up [19].

Functional assessment with FFR of non-culprit lesions in AMI patients could be a valuable guide of decision about the additional revascularization and might contribute to a better risk stratification. (Fig. 29.3).

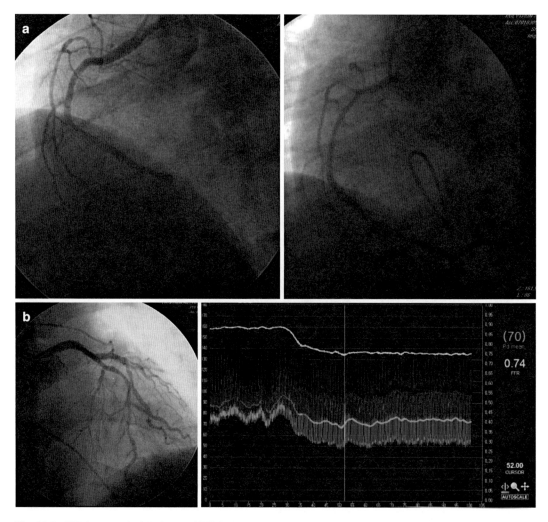

Fig. 29.3 FFR in non-culprit lesion (**a**) STEMI (inf. Wall MI). (**b**) Non-culprit lesion of LAD FFR (0.74)

29.4 FFR in In-Stent Restenosis

FFR has been validated as a useful index to determine the functional significance of coronary stenosis and the performance of percutaneous coronary intervention [20]. In-stent restenosis (ISR) after stent implantation cannot be easily measured with conventional angiography because metallic component of stent makes it difficult to estimate the severity of stenosis [21].

In patients with restenosis after bare-metal stent, a poor correlation between angiographic quantification and FFR of moderate ISR was found. Conservative management of moderate 40–70% in-stent restenotic lesions with FFR value ≥0.75 is safe avoiding unnecessary revascularizations based solely on the angiography (Fig. 29.4) [22]. In patients with restenosis after drug-eluting stent (DES), a discrepancy was found between functional ischemia measured by the FFR and the angiographic % diameter stenosis, in moderate- or diffuse-type restenotic lesions after DES implantation. The incidence of adverse events during the 12 months of follow-up after FFR-guided treatment was 18.0% (23.3% in the FFR < 0.80 group and 10.0% in FFR > 0.80 group). The outcome of FFR-guided deferral in patients with DES in-stent restenosis seems favorable [23]. Another study in patients with moderate angiographic restenosis after paclitaxel-eluting stent (PES) implantation revealed that FFR was also preserved, and the functional severity of restenosis is often limited. Although percent diameter stenosis was not significantly different between the two groups (PES group, 40.6 ± 11.2%; de novo group, 40.6 ± 9.0%, P=0.981), the functional severity of stenosis was significantly less in the PES group than in the de novo group (FFR: PES group, 0.86 ± 0.07; de novo group, 0.79 ± 0.10, P=0.002). Revascularization should be performed with caution for patients with moderate angiographic restenosis after drug-eluting stent deployment [24]. In summary, a favorable prognosis was found in patients who had angiographic restenosis and preserved FFR regardless of stent types (both BMS and DES).

29.5 FFR in Left Ventricular Hypertrophy

Left ventricular hypertrophy (LVH) has been well known as a marker of hypertension-related target organ damage and an important independent predictor of adverse cardiovascular (CV) events [25]. The precise mechanisms underlying the adverse CV events in patients with LVH have not been identified. Coronary microvascular dysfunction (CMD) with structural abnormalities has been accepted as a potential pathophysiology of adverse

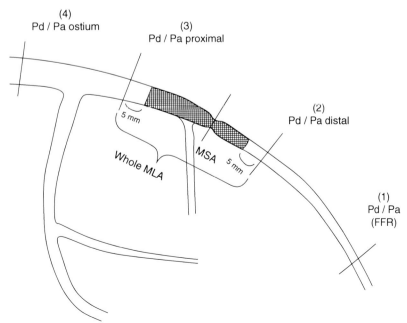

Fig. 29.4 Illustration of pressure wire and intravascular ultrasound (IVUS) examination. The pressure wire measurement at far distal Pd/Pa: fractional flow reserve (FFR) (2), Pd/Pa at stent distal (3), Pd/Pa at stent proximal (4), and Pd/Pa at ostium (5). The IVUS measurement was performed at stent and both stent edge segments. MSA minimum stent area, whole MLA minimum lumen area at whole analyzed region

events in patients with hypertrophic cardiomyopathy (HCM) and essential hypertension [26]. However, there are no definite diagnostic tools to visualize directly the coronary microcirculation in humans. Functional assessments, such as myocardial blood flow and coronary flow reserve which is an integrated measure of flow through both the large epicardial coronary arteries and the microcirculation, have been studied in patients with LVH [26]. Structural abnormalities have been presumed pathologically responsible for CMD in patients with LVH. Morphologic changes are characterized by an adverse remodeling of intramural coronary arterioles consisting of vessel wall thickening, mainly due to hypertrophy of smooth muscle cells and increased collagen deposition in the tunica media, with variable degrees of intimal thickening [26]. In the absence of epicardial obstruction, therefore, the abnormal coronary circulation appeared to be mainly based on CMD in patients with pathologic LVH (Fig. 29.5) [27]. Previous studies showed the growth of vascular structure in patients with left ventricular hypertrophy is not proportional to increase of muscular mass [28]. Therefore, it was believed that the coronary flow reserve in the myocardial vascular bed would be reduced as left ventricular hypertrophy develops and the cutoff value of 0.75 would probably not to be valid anymore. It seemed to become higher with more severely hypertrophied. But recent study revealed that FFR of coronary lesions in patients with high LVMI is no different than FFR of angiographically matched lesions in patients with normal LVMI, suggesting that high LV mass should not limit the utility of FFR as an index of coronary lesion severity [29]. Since there is controversy in this issue, it is recommended to be careful in the interpretation of FFR in patients with left ventricular hypertrophy.

29.6　FFR in Post-heart Transplantation

Cardiac allograft vasculopathy (CAV) after organ transplantation remains a major cause of morbidity and mortality in cardiac transplant recipients [30]. It is limited to detect CAV with noninvasive imaging or coronary angiography [31]. Intravascular ultrasound (IVUS) could easily identify the anatomic evidence of transplant arteriopathy involving the epicardial arterial system and the progression of CAV over time [32]. The functional evaluation of the coronary vasculature with fractional flow reserve (FFR) and with the index of microcirculatory resistance (IMR) for epicardial and microvascular structure predicts clinical outcomes in patients with ischemic heart diseases [33]. In cardiac transplant recipients, changes in FFR have been shown to correlate with IVUS parameters, whereas IMR is a predictor of development of CAV and poor cardiac function in this population [34]. Invasive measures of coronary physiology (fractional flow reserve and IMR) determined early after heart transplantation are significant predictors of late death or retransplantation (Fig. 29.6).

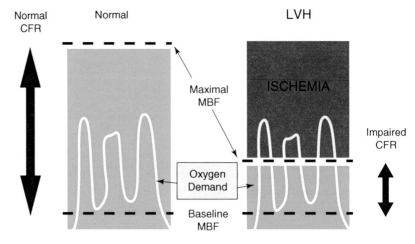

Fig. 29.5 In patients with left ventricular hypertrophy, coronary flow reserve is impaired due to different mechanisms and exposes the myocardium to recurrent microvascular ischemia when increased oxygen demand cannot be adequately met

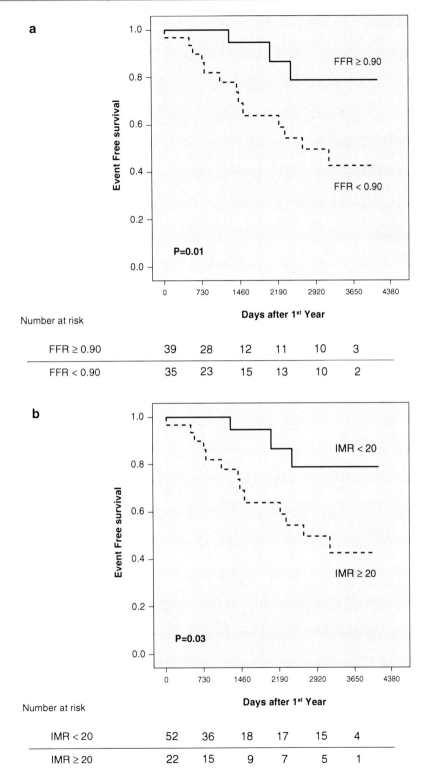

Fig. 29.6 (**a**) Kaplan–Meier analysis demonstrating a lower survival free of death or retransplantation in patients with fractional flow reserve (FFR) < 0.90 at baseline. (**b**) Kaplan–Meier analysis demonstrating a lower survival free of death or retransplantation in patients with an index of microcirculatory resistance (IMR) ≥ 20 measured at 1 year after transplantation

Moreover, patients with an improvement in microvascular function as assessed by a decrease in IMR from baseline to 1 year had better survival compared with those with worsening microvascular function [35].

References

Fractional Flow Reserve (FFR) in Post-Percutaneous Coronary Intervention (PCI)

1. Kolh P, Windecker S, Alfonso F, Collet J-P, Cremer J, Falk V, et al. 2014 ESC/EACTS guidelines on myocardial revascularization the task force on myocardial revascularization of the European Society of Cardiology (ESC) and the European Association for Cardio-Thoracic Surgery (EACTS) developed with the special contribution of the European Association of Percutaneous Cardiovascular Interventions (EAPCI). Eur J Cardiothorac Surg. 2014;46(4):517–92.
2. Levine GN, Bates ER, Blankenship JC, Bailey SR, Bittl JA, Cercek B, et al. 2011 ACCF/AHA/SCAI guideline for percutaneous coronary intervention. A report of the American College of Cardiology Foundation/American Heart Association Task Force on Practice Guidelines and the Society for Cardiovascular Angiography and Interventions. J Am Coll Cardiol. 2011;58(24):e44–122.
3. Leesar MA, Satran A, Yalamanchili V, Helmy T, Abdul-Waheed M, Wongpraparut N. The impact of fractional flow reserve measurement on clinical outcomes after transradial coronary stenting. EuroIntervention. 2011;7(8):917–23.
4. Pijls NHJ. Coronary pressure measurement after stenting predicts adverse events at follow-up: a multicenter registry. Circulation. 2002;105(25):2950–4.
5. Tonino PA, Johnson NP. Why is fractional flow reserve after percutaneous coronary intervention not always 1.0? JACC Cardiovasc Interv. 2016;9(10):1032–5.
6. Leesar MA. Baseline fractional flow reserve and stent diameter predict event rates after stenting: a further step, but still much to learn. JACC Cardiovasc Interv. 2009;2(4):364–5.
7. Agarwal SK, Kasula S, Hacioglu Y, Ahmed Z, Uretsky BF, Hakeem A. Utilizing post-intervention fractional flow reserve to optimize acute results and the relationship to long-term outcomes. JACC Cardiovasc Interv. 2016;9(10):1022–31.

FFR in Myocardial Bridge

8. Bourassa MG, Butnaru A, Lespérance J, Tardif J-C. Symptomatic myocardial bridges: overview of ischemic mechanisms and current diagnostic and treatment strategies. J Am Coll Cardiol. 2003;41(3):351–9.
9. Mookadam F, Green J, Holmes D, Moustafa SE, Rihal C. Clinical relevance of myocardial bridging severity: single center experience. Eur J Clin Investig. 2009;39(2):110–5.
10. Diefenbach C, Erbel R, Treese N, Bollenbach E, Meyer J. Incidence of myocardial bridges after adrenergic stimulation and decreasing afterload in patients with angina pectoris, but normal coronary arteries. Z Kardiol. 1994;83(11):809–15.
11. Escaned J, Cortés J, Flores A, Goicolea J, Alfonso F, Hernández R, et al. Importance of diastolic fractional flow reserve and dobutamine challenge in physiologic assessment of myocardial bridging. J Am Coll Cardiol. 2003;42(2):226–33.
12. Yoshino S, Cassar A, Matsuo Y, Herrmann J, Gulati R, Prasad A, et al. Fractional flow reserve with dobutamine challenge and coronary microvascular endothelial dysfunction in symptomatic myocardial bridging. Circ J. 2014;78(3):685–92.

FFR in Non-culprit Vessel

13. Grines CL, Cox DA, Stone GW, Garcia E, Mattos LA, Giambartolomei A, et al. Coronary angioplasty with or without stent implantation for acute myocardial infarction. N Engl J Med. 1999;341(26):1949–56.
14. Sorajja P, Gersh BJ, Cox DA, McLaughlin MG, Zimetbaum P, Costantini C, et al. Impact of multivessel disease on reperfusion success and clinical outcomes in patients undergoing primary percutaneous coronary intervention for acute myocardial infarction. Eur Heart J. 2007;28(14):1709–16.
15. Kim DH, Burton JR, Fu Y, Lindholm L, Van de Werf F, Armstrong PW, et al. What is the frequency and functional and clinical significance of complex lesions in non-infarct-related arteries after fibrinolysis for acute ST-elevation myocardial infarction? Am Heart J. 2006;151(3):668–73.
16. Goldstein JA, Demetriou D, Grines CL, Pica M, Shoukfeh M, O'Neill WW. Multiple complex coronary plaques in patients with acute myocardial infarction. N Engl J Med. 2000;343(13):915–22.
17. Lee SG, Lee CW, Hong MK, Kim JJ, Park SW, Park SJ. Change of multiple complex coronary plaques in patients with acute myocardial infarction: a study with coronary angiography. Am Heart J. 2004;147(2):281–6.
18. Parodi G, Memisha G, Valenti R, Trapani M, Migliorini A, Santoro GM, et al. Five year outcome after primary coronary intervention for acute ST elevation myocardial infarction: results from a single centre experience. Heart. 2005;91(12):1541–4.
19. Nam CW, Hur SH, Cho YK, Park HS, Yoon HJ, Kim H, et al. Relation of fractional flow reserve after drug-eluting stent implantation to one-year outcomes. Am J Cardiol. 2011;107(12):1763–7.

FFR in In-Stent Restenosis

20. Tanaka N, Takazawa K, Shindo N, Kobayashi H, Teramoto T, Yamashita J, et al. Decrease of fractional flow reserve shortly after percutaneous coronary intervention. Circ J. 2006;70:1327–31.
21. Pijls NH, De Bruyne B, Peels K, Van Der Voort PH, Bonnier HJ, Bartunek JKJJ, Koolen JJ. Measurement of fractional flow reserve to assess the functional severity of coronary-artery stenoses. N Engl J Med. 1996;334:1703–8.
22. Lopez-Palop R, Pinar E, Lozano I, Saura D, Pico F, Valdes M. Utility of the fractional flow reserve in the evaluation of angiographically moderate in-stent restenosis. Eur Heart J. 2004;25:2040–7.
23. Nam CW, Rha SW, Koo BK. Usefulness of coronary pressure measurement for functional evaluation of drug-eluting stent restenosis. Am J Cardiol. 2011;107:1783–6.
24. Yamashita J, Tanaka N, Fujita H. Usefulness of functional assessment in the treatment of patients with moderate angiographic paclitaxel-eluting stent restenosis. Circ J. 2013;77:1180–5.

FFR in Left Ventricular Hypertrophy

25. Ruilope LM, Schmieder RE. Left ventricular hypertrophy and clinical outcomes in hypertensive patients. Am J Hypertens. 2008;21:500–8.
26. Camici PG, Olivotto I, Rimoldi OE. The coronary circulation and blood flow in left ventricular hypertrophy. J Mol Cell Cardiol. 2012;52:857–64.
27. Lanza GA, Crea F. Primary coronary microvascular dysfunction: clinical presentation, pathophysiology, and management. Circulation. 2010;121:2317–25.
28. Marcus ML, Mueller TM, Gascho JA, Kerber RE. Effects of cardiac hypertrophy secondary to hypertension on coronary circulation. Am J Cardiol. 1979;44:1023–31.
29. Kenichi Tsujita K, Yamanaga K, Komura N. Impact of left ventricular hypertrophy on impaired coronary microvascular dysfunction. Int J Cardiol. 2015;187:411–3.

FFR in Post-heart Transplantation

30. Hosenpud JD, Bennett LE, Keck BM, Boucek MM, Novick RJ. The registry of the International Society for Heart and Lung Transplantation: eighteenth official report—2001. J Heart Lung Transplant. 2001;20:805–15.
31. Fang JC, Rocco T, Jarcho J, Ganz P, Mudge GH. Noninvasive assessment of transplant-associated arteriosclerosis. Am Heart J. 1998;135:980–7.
32. Mehra MR, Ventura HO, Stapleton DD, et al. Presence of severe intimal thickening by intravascular ultrasonography predicts cardiac events in cardiac allograft vasculopathy. J Heart Lung Transplant. 1995;14:632–9.
33. Johnson NP, Tóth GG, Lai D, Zhu H, Açar G, Agostoni P, Appelman Y, Arslan F, Barbato E, Chen SL, Di Serafino L, Domínguez-Franco AJ, Dupouy P, Esen AM, Esen OB, Hamilos M, Iwasaki K, Jensen LO, Jiménez-Navarro MF, Katritsis DG, Kocaman SA, Koo BK, López-Palop R, Lorin JD, Miller LH, Muller O, Nam CW, Oud N, Puymirat E, Rieber J, Rioufol G, Rodés-Cabau J, Sedlis SP, Takeishi Y, Tonino PA, Van Belle E, Verna E, Werner GS, Fearon WF, Pijls NH, De Bruyne B, Gould KL. Prognostic value of fractional flow reserve: linking physiologic severity to clinical outcomes. J Am Coll Cardiol. 2014;64:1641–54.
34. Fearon WF, Nakamura M, Lee DP, Rezaee M, Vagelos RH, Hunt SA, Fitzgerald PJ, Yock PG, Yeung AC. Simultaneous assessment of fractional and coronary flow reserves in cardiac transplant recipients: physiologic investigation for transplant arteriopathy (PITA study). Circulation. 2003;108:1605–10.
35. Yang HM, Khush K, Luikart H, Okada K, Lim HS, Kobayashi Y, Honda Y, Yeung AC, Valantine H, Fearon WF. Invasive assessment of coronary physiology predicts late mortality after heart transplantation. Circulation. 2016;133:1945–50.

Invasive Assessment for Microcirculation

Kyungil Park and Myeong-Ho Yoon

Clinical trials show that microvascular coronary disease is an independent predictor of poor prognosis in patients with or without significant epicardial coronary disease [1–3]. However, currently, it is not possible to allow the direct visualization of coronary microvasculature in human. Most parameters for evaluation of microvascular function rely on the quantification of coronary blood flow. Many studies using invasive techniques for the assessment of coronary physiology have produced a large wealth of data leading to a better understanding of coronary microvascular dysfunction.

30.1 Thrombolysis in Myocardial Infarction Myocardial Perfusion Grade (TMPG)

The TMPG is a widely used method for the assessment of coronary artery flow in coronary artery disease. Following contrast injection into the coronary arteries, there is late filling of the distal capillaries, which appears as a blushing of contrast in the myocardium between the epicardial coronary vessels. In order to visualize myocardial blush, it is important to remain on the cine pedal for an extended period longer than is customary for routine coronary angiography. Reduced or absent blush was correlated with persistent ST-segment elevation, larger infarcts, and higher mortality. Flow in coronary arteries is classified as presented in Table 30.1.

Although the TMPG is widely used to assess angiographic outcomes, it is limited by poor reproducibility and its semiquantitative and subjective nature. Furthermore, in addition to its subjective nature, the conventional flow-grading system is categorical, and no continuous angiographic index of coronary flow currently exists. In order to overcome these problems, the Thrombolysis in Myocardial Infarction (TIMI) frame count was developed as a more quantitative index of coronary artery flow [4] (Figs. 30.1, 30.2, 30.3, 30.4, 30.5, 30.6, and 30.7).

Table 30.1 TIMI myocardial perfusion grading

Blush 0	No appearance of blush or opacification of the myocardium
Blush 1	Presence of blush but no clearance of contrast (stain is present on the next injection)
Blush 2	Blush clears slowly—clears minimally or not at all during three cardiac cycles
Blush 3	Blush begins to washout and is only minimally persistent after three cardiac cycles

K. Park
Department of Internal Medicine, Dong-A University Hospital, Dong-A University College of Medicine, Busan, South Korea

M.-H. Yoon (✉)
Department of Cardiology, Ajou University School of Medicine, Suwon, South Korea
e-mail: yoonmh65@hanmail.net

Fig. 30.1 Definitions of the first frame used for TIMI frame counting. *TIMI* thrombolysis in myocardial infarction

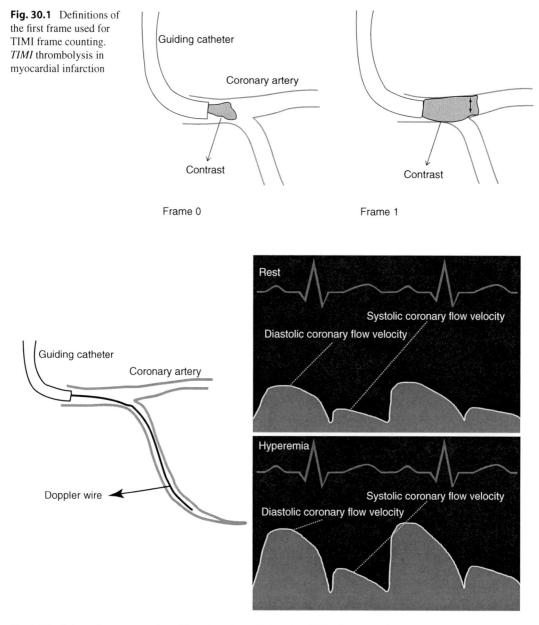

Fig. 30.2 Schematic representation of Doppler wire technique for CFR. The ratio of hyperemic average peak velocity to resting average peak velocity was calculated as the CFR. *CFR* coronary flow reserve

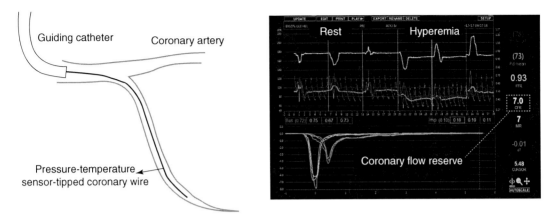

Fig. 30.3 Schematic representation of thermodilution technique for coronary flow reserve (CFR). The mean transit time at rest was 0.72 s (*blue*), and the mean transit time during hyperemia was 0.10 s (*yellow*). The CFR was 7.0

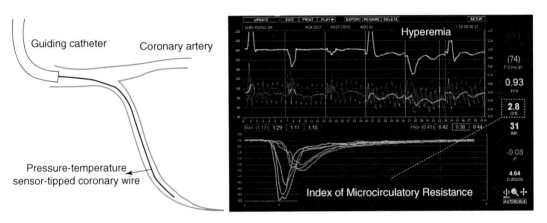

Fig. 30.4 Schematic representation of thermodilution technique for IMR. The mean transit time during hyperemia was 0.41 s (*yellow*). Distal pressure was 74 mmHg (*green*). The IMR was 31(= 0.41 × 74). In a simplified form, assuming coronary flow and myocardial flow are equal and that the contribution of collateral flow is negligible. *IMR* index of microcirculatory resistance

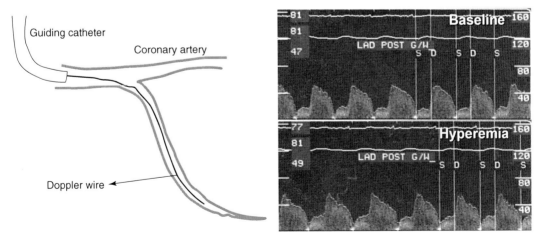

Fig. 30.5 A case of forty-eight year old female with anterior AMI. Although CFR was one point three seven relatively low, there are showed high baseline the average peak velocity (*APV*) and hyperemic APV and decreased hyperemic MVRI and have a favorable diastolic decelertion time (*DDT*) patterns with longer than six hundred miliseconds. Baseline APV 30 cm/s, systolic APV 17 cm/s, and DDT 712 msec; hyperemic APV 41 cm/s, systolic APV 18 cm/s, and DDT 764 msec

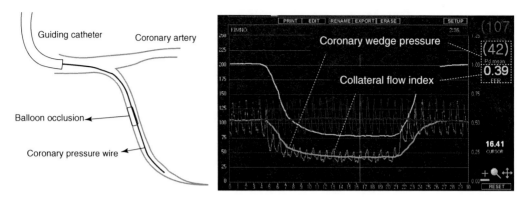

Fig. 30.6 Coronary wedge pressure (Pcw): distal coronary pressure during balloon occlusion. Pressure derived collateral flow index (CFI): (Pcw − Pv)/(Pa − Pv), simplified by the ratio of Pcw and Pa. *Pa* aortic pressure

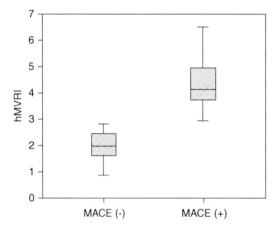

Fig. 30.7 Comparison of HMVRI in patients with and without MACEs. *HMVRI* hyperemic microvascular resistance index, *MACE* major adverse cardiovascular event (From Jin et al. [5])

30.2 Thrombolysis in Myocardial Infarction (TIMI) Frame Count (TFC)

The TFC is a simple clinical tool for microcirculation assessment and was suggested by C. Michael Gibson [4]. It is defined as the number of cineframes required for contrast to reach a standardized distal coronary landmark in the culprit vessel. The first frame used for TFC is the first frame in which dye fully enters the artery (Fig. 30.1) [4]. The last frame counted is that in which contrast enters a distal landmark. Full opacification of the branch is not required. The number is expressed based upon a cinefilming rate of 30 frames/s. Therefore, a frame count of 30 would mean that 1 s was required for dye to traverse the artery. The TIMI frame count is counted using an electronic frame counter.

These landmarks are as follows: the distal bifurcation in the left anterior descending artery, the distal branch of the lateral left ventricular wall artery furthest from the coronary ostium in the circumflex system, and the first branch of the posterolateral artery in the right coronary artery [4]. In general, the TFC for the left anterior descending (LAD) and the circumflex arteries is assessed in a right anterior oblique projection with caudal angulation (RAO caudal view) and the TFC for the right coronary artery in a left anterior oblique projection with cranial angulation (LAO cranial view).

The TIMI frame count was not significantly affected by increasing the dye injection rate or by changing catheter size. However, the use of intracoronary nitrates significantly increased the TIMI frame count in normal and diseased coronary arteries. In addition, TIMI frame count varies with body size, systemic arterial pressure, age, or gender [6]. The impact of the mechanical force of injection may also impact on TIMI frame count although it was small and insignificant.

30.2.1 Corrected TIMI Frame Count (CTFC)

The frame count number after adjustment for vessel length is given the term "corrected TIMI frame count." In the corrected TIMI frame count, a correction factor is needed to compensate for the longer length of the left anterior descending artery (LAD) compared with the circumflex and right coronary arteries. A 1.7 correction factor is used to correct the TIMI frame counts for the average greater length of the LAD. The CTFC is a simple, more objective continuous variable index of coronary blood flow that can be broadly and inexpensively applied. This technique is highly correlated with coronary flow reserve measurements obtained using the Doppler guidewire [7]. It is also correlated with volumetric flow and resting distal average peak velocity [8].

30.2.2 Converted TIMI Frame Count

A conversion factor of 2.4, 2, and 1.2 can be used to convert the frame rate values when filmed at 12.5, 15, and 25 frames/s, respectively, to adjust for the 30 frames/s acquisition speed used in the original cineangiographic studies.

30.3 Coronary Flow Reserve (CFR)

CFR is the magnitude of the increase in coronary flow that can be measured as the ratio of coronary flow during maximal microvascular dilation and basal coronary flow. CFR can be measured by intracoronary Doppler wire or thermodilution technique. Using intracoronary Doppler wire, CFR can be calculated as the ratio of hyperemic average peak velocity (hAPV) during maximal hyperemia induced by adenosine or others and baseline average peak velocity (bAPV) (hAPV/bAPV) (Fig. 30.2).

Besides, by thermodilution technique, it is possible to measure pressure and to estimate coronary artery flow simultaneously with a single pressure-temperature sensor-tipped coronary wire (PressureWire Certus, St. Jude Medical, MN, USA) [9, 10]. For measurement of CFR, coronary flow under basal conditions was determined by intracoronary administration of 3 ml of room-temperature saline three times in succession manually (3 ml/s). Maximal hyperemia was then induced, and three additional room temperature saline boluses of 3 ml were administered intracoronarily to determine peak coronary flow presented as peak mean transit time. In this method, the mean transit time (Tmn) of room-temperature saline injected down a coronary artery can be determined and has been shown to correlate inversely with absolute flow [11]. CFR was calculated based on the ratio of the mean transit times during hyperemia and at baseline (Fig. 30.3). A thermodilution-based CFR can be derived that has been shown to correlate well with Doppler velocity wire-derived CFR both in their experimental model and in humans [9, 10]. A CFR less than 2.5 was considered to be abnormal [12, 13].

CFR interrogates the entire coronary system, including the epicardial artery and microcirculation. A normal CFR indicates that epicardial and minimally achievable microvascular bed resistances are low and normal. However, CFR is unable to differentiate which component is affected when it is abnormal. Furthermore, CFR was largely affected by the baseline coronary flow velocity (CFV) associated with heart rate, preload, contractility, and after percutaneous coronary intervention (PCI) [14, 15]. Therefore, using a CFR to evaluate the microcirculation has a few limitations.

30.4 Index of Microcirculatory Resistance (IMR)

With recent technological advances, coronary microcirculation can be measured simultaneously with the same pressure wire by calculating the IMR using this thermodilution technique. By using this wire and modified software, it is able to calculate the mean transit time (Tmn) of room-temperature saline injected down a coronary artery. The inverse of the hyperemic Tmn has been shown to correlate with absolute flow. For measure of IMR, a 0.014 coronary temperature and pressure-sensing guidewire (PressureWire Certus, St. Jude Medical, MN, USA) was calibrated for the pressure recording. It was then equalized with aortic pressure (Pa) in the guiding catheter. The tip pressure sensor was advanced across the stented segment and beyond the mid-to-distal portion of the culprit vessel. In a simplified form, assuming coronary flow and myocardial flow are equal and that the contribution of collateral flow is negligible, then IMR was calculated as mean distal coronary pressure multiplied by the thermodilution-derived hyperemic Tmn (Fig. 30.4) [11].

IMR is a well-validated tool of representing microvascular function both experimentally and clinically. Compared to CFR, it is less influenced by the presence of epicardial disease and the variation of hemodynamics [15]. The IMR correlated significantly with 3-month echocardiographic wall motion score (WMS) ($r = 0.59$, $p = 0.002$). The WMS at 3-month follow-up was significantly worse in the group with an IMR >32 U compared with ≤32 U [16]. IMR correlated significantly with regional myocardial Fludeoxyglucose (FDG) uptake ($r = -0.738$, $p < 0.001$), and it demonstrated significant correlation with percent change in anterior WMS ($r = -0.464$, $p = 0.003$) [17]. Patients with an IMR > 40 had a higher rate of the primary endpoint at 1 year than patients with an IMR ≤ 40 (17.1% versus 6.6%; $P = 0.027$). An IMR > 40 was associated with an increased risk of death or rehospitalization for heart failure (hazard ratio [HR], 2.1; $P = 0.034$) and of death alone (HR, 3.95; $P = 0.028$) [18].

In the presence of severe epicardial stenosis, myocardial flow is the composite of both the coronary and collateral flows [19]. This will produce an overestimation of IMR if not corrected for collateral flow. As a result, to calculate the true IMR in this setting, a more complex formula has been developed, which includes measurement of the coronary wedge pressure as a measure of collateral pressure [19]. Therefore, the coronary wedge pressure (Pcw) and venous pressure (Pv) should be used to estimate IMR when IMR is measured in an obstructed coronary artery, according to the following equation:

Corrected IMR = [(Pa − Pv) × Tmn] × [(Pd − Pcw)/(Pa − Pcw)]

When wedge and venous pressure are not available, IMR may be estimated using this equation [19]:

IMR = Pa × Tmn × FFRcor
where, FFRcor = 1.34 × FFR − 0.32

30.5 Phasic coronary flow velocity pattern (CFV Pattern)

Microvascular injury can be assessed more quantitatively from CFV patterns. A 0.014 in. intracoronary Doppler flow wire was used to measure coronary flow velocity. Frequency analysis of the Doppler signals was carried out in real time by fast Fourier transform using a velocimeter. ECG and blood pressure were monitored continuously. Once baseline flow velocity data had been obtained, 12–18 μg intracoronary adenosine was given to obtain data during hyperemia. Phasic patterns of coronary flow by intracoronary Doppler wire are also related with myocardial damage because microvascular pool is decreased when the microvasculature of the myocardium is damaged, and it can be reflected to CFV patterns.

In patients with severe larger transmural myocardial infarction, the microvasculature shows loss of its anatomic integrity [20, 21] which may markedly affect the CFV pattern. During PCI in AMI, insufficient coronary microcirculation may be developed by many mechanisms such as oxygen-free radical injury calcium overload, microvascular spasm, neutrophil plugging of the microvessels, and tissue edema or embolization of the coronary microvasculature attributable to dissemination of thrombi and the contents of the disintegrated atheroma in the lesion to distal parts of the coronary vasculature [20, 22–24]. These phenomena result in decreased vascular pool of infarct-related artery and microvasculature of infarct-related myocardium. Therefore, systolic coronary flow velocity can be decreased or early systolic reversal flow can be occurred and rapid decreased diastolic flow velocity can be developed. This method was limited by substantial interpretation variability.

30.6 Coronary Collateral Flow Index or Coronary Wedge Pressure

Coronary collateral flow index (CFI) was determined by simultaneous measure of mean aortic pressure, the distal mean coronary pressure during balloon occlusion, and the mean central venous pressure (Fig. 30.6) [25]. The pressure-derived fractional coronary collateral flow index takes the venous pressure into account and can be calculated according to the following equation:

$CFI = (Pw - Pv)/(Pa - Pv)$, where Pv is venous pressure ideally measured from the right atrium and Pa is the aortic pressure measured from the guide catheter.

Coronary wedge pressure, which is a distal coronary pressure during balloon occlusion, can be measured by pressure wire. It can be measured also during PCI and it can be elevated with microvascular dysfunction or damage.

Coronary collateral flow index or coronary wedge pressure were well correlated with left ventricular wall motion changes and final wall motion score index [26, 27]. It can be affected also by collateral flows and preload, and there was no clinical outcome data.

30.7 Hyperemic Microvascular Resistance Index (HMVRI)

Using a pressure wire, HMVRI can be measured as the ratio of distal coronary pressure and distal coronary low velocity (averaged peak velocity) during maximal hyperemia. Be measured simultaneously by ComboWire which has a dual sensor with pressure sensor and temperature sensor. The tip of the wire was placed just distal to the site of the culprit lesion to assess the microvascular function of the entire region at risk. HMVRI was able to distinguish between nontransmural and transmural myocardial infarction immediately after primary PCI at the cardiac catheterization laboratory [28]. And HMVRI measured immediately after primary PCI in ST-segment elevation myocardial infarction (STEMI) was correlated with LV wall motion at 6-month follow-up and FDG uptake rate in cardiac PET [27]. In recent study, HMVRI measured by Doppler wire immediately after primary PCI in patients with STEMI was higher in patients with cardiovascular events compared to that of patients without cardiovascular events (Fig. 30.7) [5].

References

1. Fearon WF, Balsam LB, Farouque HM, Caffarelli AD, Robbins RC, Fitzgerald PJ, Yock PG, Yeung AC. Novel index for invasively assessing the coronary microcirculation. Circulation. 2003;107:3129–32.
2. Rosenblum WI, El-Sabban F. Platelet aggregation in the cerebral microcirculation: effect of aspirin and other agents. Circ Res. 1977;40:320–8.
3. Wu Z, Ye F, You W, Zhang J, Xie D, Chen S. Microcirculatory significance of periprocedural myocardial necrosis after percutaneous coronary intervention assessed by the index of microcirculatory resistance. Int J Cardiovasc Imaging. 2014;30:995–1002.
4. Gibson CM, Cannon CP, Daley WL, Dodge JT, Alexander B, Marble SJ, McCabe CH, Raymond L, Fortin T, Poole WK, Braunwald E. TIMI frame count: a quantitative method of assessing coronary artery flow. Circulation. 1996;93:879–88.
5. Jin X, Yoon MH, Seo KW, Tahk SJ, Lim HS, Yang HM, Choi BJ, Choi SY, Hwang GS, Shin JH, Park JS. Usefulness of hyperemic microvascular resistance index as a predictor of clinical outcomes in patients with ST-segment elevation myocardial infarction. Korean Circ J. 2015;45:194–201.
6. Faile B, Guzzo J, Tate D, et al. Effect of sex, hemodynamics, body size, and other clinical variables on the corrected thrombolysis in myocardial infarction frame count used as an assessment of coronary blood flow. Am Heart J. 2000;140:308–14.
7. Manginas A, Gatzov P, Chasikidis C, Voudris V, Pavlides G, Cokkinos DV. Estimation of coronary flow reserve using the thrombolysis in myocardial infarction (TIMI) frame count method. Am J Cardiol. 2000;83:1562–5.
8. Stankovic G, Manginas A, Voudris V, Pavlides G, Athanassopoulos G, Ostojic M, Cokkinos DV. Prediction of restenosis after coronary angioplasty by use of a new index: TIMI frame count/minimal luminal diameter ratio. Circulation. 2000;101:962–8.
9. De Bruyne B, Pijls NH, Smith L, Wievegg M, Heyndrickx GR. Coronary thermodilution to assess flow reserve: experimental validation. Circulation. 2001;104:2003–6.
10. Pijls NH, De Bruyne B, Smith L, Aarnoudse W, Barbato E, Bartunek J, Bech GJ, Van De Vosse

F. Coronary thermodilution to assess flow reserve: validation in humans. Circulation. 2002;105:2482–6.
11. Fearon WF, Balsam LB, Farouque HM, Caffarelli AD, Robbins RC, Fitzgerald PJ, Yock PG, Yeung AC. Novel index for invasively assessing the coronary microcirculation. Circulation. 2003;107:3129–32.
12. Reis SE, Holubkov R, Conrad Smith AJ, Kelsey SF, Sharaf BL, Reichek N, Rogers WJ, Merz CN, Sopko G, Pepine CJ, WISE Investigators. Coronary microvascular dysfunction is highly prevalent in women with chest pain in the absence of coronary artery disease: results from the NHLBI WISE study. Am Heart J. 2001;141:735–41.
13. Serruys PW, di Mario C, Piek J, Schroeder E, Vrints C, Probst P, de Bruyne B, Hanet C, Fleck E, Haude M, Verna E, Voudris V, Geschwind H, Emanuelsson H, Mühlberger V, Danzi G, Peels HO, Ford AJ Jr, Boersma E. Prognostic value of intracoronary flow velocity and diameter stenosis in assessing the short- and long-term outcomes of coronary balloon angioplasty: the DEBATE study (Doppler endpoints balloon angioplasty trial Europe). Circulation. 1997;96:3369–77.
14. Leung DY, Leung M. Non-invasive/invasive imaging: significance and assessment of coronary microvascular dysfunction. Heart. 2011;97:587–95.
15. Ng MK, Yeung AC, Fearon WF. Invasive assessment of the coronary microcirculation: superior reproducibility and less hemodynamic dependence of index of microcirculatory resistance compared with coronary flow reserve. Circulation. 2006;113:2054–61.
16. Fearon WF, Shah M, Ng M, Brinton T, Wilson A, Tremmel JA, Schnittger I, Lee DP, Vagelos RH, Fitzgerald PJ, Yock PG, Yeung AC. Predictive value of the index of microcirculatory resistance in patients with ST-segment elevation myocardial infarction. J Am Coll Cardiol. 2008;51:560–5.
17. Lim HS, Yoon MH, Tahk SJ, Yang HM, Choi BJ, Choi SY, Sheen SS, Hwang GS, Kang SJ, Shin JH. Usefulness of the index of microcirculatory resistance for invasively assessing myocardial viability immediately after primary angioplasty for anterior myocardial infarction. Eur Heart J. 2009;30:2854–60.
18. Fearon WF, Low AF, Yong AS, et al. Prognostic value of the index of microcirculatory resistance measured after primary percutaneous coronary intervention. Circulation. 2013;127:2436–41.
19. Yong AS, Layland J, Fearon WF, Ho M, Shah MG, Daniels D, Whitbourn R, Macisaac A, Kritharides L, Wilson A, Ng MK. Calculation of the index of microcirculatory resistance without coronary wedge pressure measurement in the presence of epicardial stenosis. JACC Cardiovasc Interv. 2013;6:53–8.
20. Kloner RA, Ganote CE, Jennings RB. The "no-reflow" phenomenon after temporary coronary occlusion in the dog. J Clin Invest. 1974;54:1496–508.
21. Van't Hof AWJ, Liem A, Suryapranata H, Hoorntje JC, de Boer MJ, Zijlstra F. Angiographic assessment of myocardial reperfusion in patients treated with primary angioplasty for acute myocardial infarction: myocardial blush grade. Circulation. 1998;97:2302–6.
22. Antoniucci D, Valenti R, Migliorini A, Moschi G, Bolognese L, Cerisano G, Buonamici P, Santoro GM. Direct infarct artery stenting without predilation and no-reflow in patients with acute myocardial infarction. Am Heart J. 2001;142:684–90.
23. Grech ED, Dodd NJF, Jackson MJ, Morrison WL, Faragher EB, Ramsdale DR. Evidence for free radical generation after primary percutaneous transluminal coronary angioplasty recanalization in acute myocardial infarction. Am J Cardiol. 1996;77:122–7.
24. Tanaka A, Kawarabayashi T, Nishibori Y, Sano T, Nishida Y, Fukuda D, Shimada K, Yoshikawa J. No-reflow phenomenon and lesion morphology in patients with acute myocardial infarction. Circulation. 2002;105:2148–52.
25. Seiler C, Fleisch M, Garachemani A, Meier B. Coronary collateral quantitation in patients with coronary artery disease using intravascular flow velocity or pressure measurements. J Am Coll Cardiol. 1998;32:1272–9.
26. Yamamoto K, Ito H, Iwakura K, et al. Pressure-derived collateral flow index as a parameter of microvascular dysfunction in acute myocardial infarction. J Am Coll Cardiol. 2001;38:1383–9.
27. Yoon MH, Tahk SJ, Yang HM, et al. Comparison of accuracy in the prediction of left ventricular wall motion changes between invasively assessed microvascular integrity indexes and fluorine-18 fluorodeoxyglucose positron emission tomography in patients with ST-elevation myocardial infarction. Am J Cardiol. 2008;102:129–34.
28. Kitabata H, Imanishi T, Kubo T, et al. Coronary microvascular resistance index immediately after primary percutaneous coronary intervention as a predictor of the transmural extent of infarction in patients with ST-segment elevation anterior acute myocardial infarction. JACC Cardiovasc Imaging. 2009;2:263–72.

Non-invasive Assessment of Myocardial Ischemia

Jin-Ho Choi, Ki-Hyun Jeon, and Hyung-Yoon Kim

31.1 The Need of Non-invasive Physiological Assessment

Despite highly advanced technologies and devices in invasive physiological modalities, there is still enormous clinical need of non-invasive physiological assessment as follows. First, indirect assessment of physiological parameters is based on the profound understanding of coronary pathophysiology and exact modeling of coronary physiology. For example, computational modeling of fractional flow reserve highly depends on the exact modeling of coronary circulation. Second, non-invasive assessment enables a large-scale or population-scale study of coronary physiology which might be limited by invasive physiological assessment. Third, non-invasive physiological assessment before sending the patients to catheterization procedure might find out patients who would not benefit from invasive angiography or physiological study and reduce unnecessary procedure. Finally, replacing invasive physiological assessment with non-invasive technology might greatly reduce the burden of medical cost.

Coronary computed tomography angiography (CCTA) is the best non-invasive modality that depicts anatomy of coronary artery. However, anatomical stenosis is a poor predictor of physiological severity and frequently underestimates or overestimates physiological severity of stenosis. Fractional flow reserve (FFR) < 0.80, a widely accepted gold standard of vessel-specific physiologically significant stenosis which may evoke myocardial ischemia, is identified in less than a half of vessel with significant stenosis defined by diameter stenosis (DS) ≥ 50%, and the discordance between anatomical stenosis and physiological severity is found as high as 40% [1, 2]. The key role of coronary artery is supplying sufficient blood flow that contains vital materials such as oxygen or glucose required by myocardium. Therefore, the insufficiency of blood supply can be defined by decreased myocardial perfusion, decreased pressure gradient or arterial flow across stenosis, or relative ratio of minimal luminal area which represents the maximal blood supply to the subtended myocardial mass. These concepts constitute the principles of non-invasive assessment of myocardial ischemia (Fig. 31.1).

J.-H. Choi (✉)
Departments of Emergency Medicine and Internal Medicine, Samsung Medical Center, Sungkyunkwan University School of Medicine, Seoul, South Korea
e-mail: jhchoimd@gmail.com

K.-H. Jeon
Department of Medicine, Sejong General Hospital, Bucheon, South Korea

H.-Y. Kim
Department of Medicine, Jeju National University Hospital, Jeju National University College of Medicine, Jeju, South Korea

© Springer Nature Singapore Pte Ltd. 2018
M.-K. Hong (ed.), *Coronary Imaging and Physiology*,
https://doi.org/10.1007/978-981-10-2787-1_31

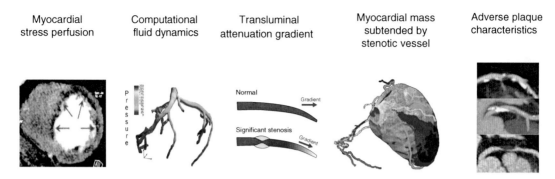

Fig. 31.1 Concepts for non-invasive assessment of myocardial ischemia

31.2 Perfusion CT

The strength of perfusion imaging is visualizing the myocardial blood flow on which myocardial metabolism depends. Perfusion MR uses similar concept used in nuclear perfusion imaging or perfusion cardiac magnetic resonance imaging (CMR). From the myocardial and left ventricular cavity arterial input function or time-attenuation curves, the extent of regional myocardial perfusion is calculated or compared with the other regional myocardial perfusion. Perfusion is imaged in a complete cardiac cycle (dynamic perfusion imaging) or as a snapshot (static perfusion imaging). Scanners equipped with dual energy source can be used for perfusion imaging and mostly used for static perfusion imaging (Fig. 31.2). The performance of perfusion CT for predicting functionally significant stenosis is considered to be similar to nuclear perfusion imaging, stress CMR, or stress echocardiography, and is being validated against FFR [3–5]. Standard coronary angiography can be done along with perfusion imaging, which enables simultaneous anatomic evaluation of coronary arteries with functional evaluation of heart. Therefore, perfusion CT combined with coronary CT angiography can be a one-stop shop modality that assesses both anatomical and functional stenosis within a single session [6].

31.2.1 Technical Aspect of Perfusion CT Imaging

Hyperemia is induced by pharmacological stress agents. Intravenous adenosine is widely used in a continuous dose of 140 μg/kg/min for 2 or 3 min. Regadenoson has longer plasma half time than adenosine and is administered in a single agent. Also it is a selective adenosine 2A receptor agonist and can be used in patients with asthma or airway disease. Dobutamine, a myocardial beta-1 agonist, or dipyridamole, adenosine receptor blocker, is not commonly used (Table 31.1).

Static or snap-shot perfusion CT assesses myocardial contrast distribution in a single time and doable in most CT scanners with lesser radiation exposure to dynamic perfusion CT. With sophisticated mathematical modeling, dynamic perfusion CT enables direct quantification of myocardial blood flow (MBF), myocardial blood volume, and myocardial flow reserve (Table 31.2). Regarding the diagnostic performance, static perfusion CT showed sensitivity = 0.85 (95% confidence interval = 0.70–0.93), specificity = 0.81 (0.59–0.93), area under curve = 0.90 (0.87–0.92) [7]. A recent dynamic perfusion CT showed comparable performance compared to CMR (Table 31.3) [8–20]. Also perfusion CT is better suited for quantification of myocardial blood flow than perfusion MR. Based on the nuclear perfusion studies, the nominal value of resting myocardial blood flow is

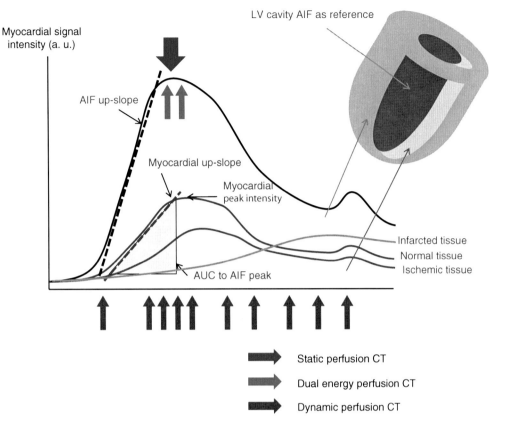

Fig. 31.2 Principle of myocardial perfusion CT. The difference between myocardial blood flow correlates with the myocardial up-slope normalized by arterial input function (AIF) up-slope, area under curve (AUC) of myocardial signal intensity up to AIF peak, or myocardial peak signal intensity. The difference between normal tissue and ischemic tissue is imaged as perfusion defect (line with *red arrows*)

Table 31.1 Stress agents for perfusion imaging

	Advantage	Disadvantage
Exercise	Most physiological	Motion artifact → not practical for CT or MR
	Least expensive	Effort-dependent
Adenosine	Current de facto standard	Potential bronchospasm (not good for chronic obstructive lung disease, asthma, caffeine user)
		Tachycardia, AV block
Dipyridamole	Inexpensive	Prolonged effect
		Tachycardia, AV block
		May require aminophylline
Regadenoson, binodenoson	Bolus injection	Expensive
	Fewer side effects in COPD/asthma	Prolonged effect
		Tachycardia
Dobutamine	Physiological	Lower sensitivity/specificity
		Tachycardia
		Can provoke ischemia

Table 31.2 Techniques for myocardial perfusion CT

	Strength	Weak points
Static perfusion CT	• Doable in most CT scanner	• Highly affected by image acquisition timing
	• Doable with standard CCTA	
	• Minimal radiation (1–3 mSv)	• May need multiphase reconstruction to reduce artifacts (beam hardening, motion, partial scan artifacts)
	• TPR (trans-myocardial perfusion ratio)	
Dynamic perfusion CT	• Less susceptible to artifact	• Need specific scanner (256 or 320-slice, or 128-slice with shuttle mode)
	• Quantitative blood flow analysis (myocardial blood flow or flow reserve)	• High radiation (>10 mSv)
		• Axial coverage might be insufficient
		• Need separated CCTA scanning
		• Limited clinical data
Dual energy perfusion CT	• Iodine distribution map → better tissue discrimination	• Affected by image acquisition timing
	• Quantitation of myocardial blood pool	• Needs standardization of iodine map interpretation
		• Mostly static perfusion CT

known to be 0.9 ml/μg/min. The cut-off value of hemodynamically significant stenosis in perfusion CT was reportedly 0.75–0.78 ml/μg/min [16].

31.3 Computational Simulation of Fractional Flow Reserve

Increase of myocardial blood flow by 2 to 3-fold is required to match the increased need of cardiac output in most activities. Coronary microvessel accounts for most resistance or pressure drop in coronary circulation. The increase of myocardial blood flow is mainly controlled by decrease in microvascular resistance. Therefore functionally significant epicardial coronary artery stenosis can be defined by failure to increase blood flow during hyperemia which induces maximal dilatation of resistance vessel. Fractional flow reserve (FFR) is defined by the ratio of hyperemic coronary flow through stenotic vessel to the hypothetical normal vessel. Because flow is proportional to pressure in fixed stenosis, FFR can be measured by average pressure gradient. Pressure drop of more than 20% or FFR ≤ 0.80 is widely advocated as a gold standard of vessel-specific physiologically significant stenosis which may evoke myocardial ischemia.

FFR is measured during invasive cardiac catheterization and requires insertion of a pressure wire inserted through the stenosis. There may be and instability of measurement and signal shift. Placement of a pressure wire near the stenosis or pressure recovery zone may lead to overestimation of FFR. A non-invasive simulation of FFR would be very valuable to avoid these procedural shortcomings and the expense of pressure wire and invasive cardiac catheterization.

31.3.1 Computation of Simulated FFR

Like the other fluid systems, blood flow in the cardiovascular system is ruled by the physical laws of mass conservation, momentum conservation, and energy conservation. Therefore it can be calculated by mathematical models. For patient-specific coronary circulation, 3-dimensional numerical models based on computational flow dynamics which can compute complex flow patterns are preferred to zero dimensional models or lumped parameter model which is employed in large systemic vessels. Computational FFR is derived based on the regional physical geometry, the boundary condition which is the behavior and

Table 31.3 Diagnostic performance of perfusion CT

Study and publication	Techniques (Scanner)	N	Reference	Stenosis (%)	Sensitivity	Specificity	PPV	NPV	Basis of analysis
Rocha-Filho et al. [8]	Static (64-DSCT)	35	ICA	50	91	91	86	93	Vessel
				70	91	78	53	97	
Feuchtner et al. [9]	Static (128-DSCT)	25	ICA	70	100	74	97	100	Vessel
Nasis et al. [10]	Static (320-MDCT)	20	SPECT/ICA	50	94	98	94	98	Vessel/territory
				70	79	91	73	93	
Carrascosa et al. [11]	Static, rapid kV switching	25	SPECT	–	73	95	–	–	Vessel
Magalhaes et al. [12]	Static (320-MDCT)	381	ICA + SPECT,MR	50	98	96	98	96	Territory
					58	86	87	55	Vessel
Huber et al. [13]	Static (256-MDCT)	32	ICA	50	76	100	90	100	Vessel
Rossi et al. [14]	Dynamic (128-DSCT)	80	ICA	50	78	75	91	51	Territory
					88	90	98	77	Vessel
Bamberg et al. [15]	Dynamic (128-DSCT)	33	FFR	50	93	87	75	97	Vessel
Bamberg et al. [16]	Dynamic (128-DSCT)	31	Cardiac MR	–	100	75	100	92	Vessel
Ko et al. [17]	Dynamic (64-DSCT)	45	ICA	50	93	86	88	91	Vessel
Wang et al. [18]	Dynamic (128-DSCT)	30	ICA/SPECT	50	90	81	58	97	Vessel/territory
Kim et al. [19]	Dynamic (128-DSCT)	50	Cardiac MR	–	77	94	53	98	Vessel/territory
Wichmann et al. [20]	Dynamic (128-DSCT)	71	Visual assessment	50	100	88	100	43	Territory

properties at the boundaries of the region, and the physical laws of fluid in the region.

FFR can be described as a pressure gradient across stenotic segment during maximal hyperemia. Anatomical stenosis, myocardial mass, and microvascular resistance constitute FFR value and can be calculated from patient-specific sophisticated coronary arterial anatomical model, vessel-specific myocardial mass, and microvascular resistance which determine the outlet boundary condition [21, 22]. CT images provide patient-specific anatomy model of local geometry, individual coronary artery morphology, volume, and myocardial mass. From these data, cardiac output and baseline coronary blood flow can be calculated by using allometric scaling laws [23–25]. This computational approach was derived from a general model that describes the transport of essential materials through space-filling fractal branched networks, and is based on a form-function relationship [26]. The diameter-flow rate relation is determined according to Murray's law [27] and Poiseuille's equation, which considers shear stress on the endothelial surface and remodeling to maintain homeostasis [28]. Morphometry laws are also adapted to obtain the physiological resistance to flow aroused by coronary artery branches [29].

Microvascular resistance at baseline and during maximal hyperemia, which is fundamental for FFR measurement, can be approximated using population-based data on the effect of adenosine on coronary flow [30] (Fig. 31.3).

31.3.2 Clinical Results of Computational FFR

Landmark trials including DISCOVER-FLOW [31], DeFACTO [32], and NXT [33] showed that FFR-CT, a proprietary computational FFR, showed high diagnostic performance in discriminating ischemia in patients who had intermediate coronary artery stenosis. The NXT trial reported sensitivity and negative predictive value of FFR-CT in diagnosis of ischemia (defined as invasive FFR < 0.80) in patients with intermediate stenosis severity were 80% and 92%, respectively [33]. In a recent meta-analysis of FFR-CT based on 833 patients and 1377 vessels, FFR-CT showed a moderate diagnostic performance for identification of ischemic vessel with pooled sensitivity = 84% and specificity = 76% at a per-vessel basis [34] (Table 31.4). The PLATFORM study showed that a decision-making strategy

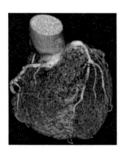
Coronary CT raw data

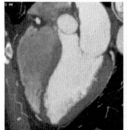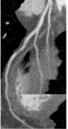
Segmentation of patient-specific myocardium and coronary arterial tree

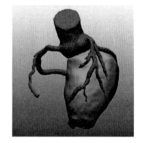

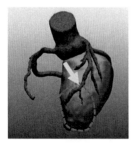
Calculate vessel-specific myocardial mass

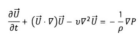
$$\frac{\partial \vec{U}}{\partial t} + (\vec{U} \cdot \nabla)\vec{U} - \nu \nabla^2 \vec{U} = -\frac{1}{\rho}\nabla P$$
Apply virtual hyperemia and do computational flow dynamics

Computational FFR

Fig. 31.3 Concept of computational FFR

Table 31.4 Results of non-invasive computational FFR technologies

Study	Technology	N of vessels	Sensitivity (%)	Specificity (%)	PPV (%)	NPV (%)	Accuracy (%)	Area under curve, per vessel	Correlation coefficient compared with FFR	Computation time
FFR-CT, DISCOVER-FLOW, Koo et al. [31]	Heartflow ver 1.0	159	88	82	74	92	84	0.90	0.72	Hours, off-site
FFR-CT, DeFACTO, Min et al. [32]	Heartflow ver 1.2	408	80	63	56	84	69	0.81	0.63	Hours, off-site
FFR-CT, NXT, Norgaard et al. [33]	Heartflow ver 1.4	484	84	87	62	95	86	0.93	0.82	Hours, off-site
FFR-CT, Kim et al. [19]	Heartflow ver 1.2	48	85	57	83	62	77	–	0.60	Hours, off-site
cFFR	Siemens cFFR ver 1.4	67	85	85	71	93	85	0.92	0.66	<1 h, standalone
cFFR	Siemens cFFR ver 1.4	189	88	65	65	88	75	0.83	0.59	<1 h, standalone
cFFR	Siemens cFFR ver 1.7	23	83	76	56	93	78	–	0.77	<1 h, standalone
Pooled analysis	–	1330	84	76	63	91	79	0.86	–	–

Per-vessel data is shown. *PPV* positive predictive value, *NPV* negative predictive value

using CCTA with FFR-CT was associated with clinical outcomes comparable to using invasive FFR and a 33% cost reduction [35]. Therefore, FFR-CT can effectively rule out intermediate lesions that cause ischemia and could also reduce the unnecessary ICA and invasive FFR.

31.4 Intracoronary Transluminal Attenuation Gradient Analysis

31.4.1 Transluminal Attenuation Gradient (TAG) and Corrected Contrast Opacification (CCO)

Standard coronary CT image is a snapshot of dynamic transit of intravascular contrast driven by blood flow. Therefore, coronary CT is not only a simple static anatomical imaging but also contains information of coronary hemodynamics. Intracoronary contrast filling is governed by arterial input function from coronary ostium and the flow or velocity of intracoronary flow. Based on this intuitive concept, transluminal attenuation gradient (TAG) was defined as the difference of intracoronary attenuation along vessel axis that reflects contrast kinetics and is readily available from conventional CCTA image without additional radiation or off-site long time computation [36]. TAG theoretically depends on the temporal uniformity of Z-axis coverage and adequate contrast enhancement curve (Fig. 31.4). TAG has been tested in both animal and human studies and showed consistently poor correlation with anatomical and functional stenosis [37–42]. Adjustment with descending aortic opacification (corrected contrast opacification, CCO) or exclusion of nonlinear values caused by stented or calcified segment has been proposed but with mixed results [38, 42, 43]. Because coronary CT image is a snapshot of convection of intracoronary time-varying contrast bolus, TAG represents the spatial dispersion of contrast concentration along vessel axis. Therefore, the discordance among TAG and anatomical or functional stenosis is no wonder considering the well-known discordance among anatomical stenosis, fractional flow reserve (FFR), and coronary flow reserve (CFR).

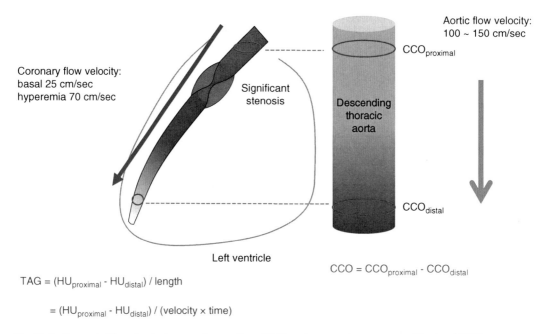

Fig. 31.4 Concept of transluminal attenuation gradient (TAG) and corrected contrast opacification (CCO)

31.4.2 Transluminal Attenuation Flow Encoding (TAFE)

The principle of myocardial blood flow assessment in perfusion scan based on the comparison of enhancement dynamics between left ventricular cavity and myocardium can be applied with modification to standard CCTA data [44, 45]. This concept enables calculation of CBF from the time-dependent change of contrast density proximal to coronary artery as input function of contrast cohort, arterial volume to be filled by the contrast cohort, and the gradient of intraluminal contrast density which reflects blood flow velocity. All these input parameters are readily and rapidly available from current conventional CT suite [46]. Based on this concept, Lardo et al. reported an elegant engineering solution named transluminal attenuation flow encoding (TAFE) (Fig. 31.5) [47]. Coronary CT image is a snapshot of convection of intracoronary time-varying contrast bolus. Therefore TAG represents the spatial dispersion of contrast concentration along vessel axis. With additional temporal data from arterial input function, TAFE formula decodes the spatial dispersion of TAG into temporal dispersion of vessel-specific CBF. TAFE showed excellent correlation with myocardial blood flow (MBF) in animal microsphere model and warrants validation in human study.

31.5 Coronary Artery Stenosis and Subtended Myocardial Mass

FFR is a mean pressure gradient across stenosis with maximal myocardial blood flow. Anatomical stenosis, myocardial mass, and microvascular resistance are major constituents of FFR value [21]. The major unknowns in anatomical measurement are myocardial mass and microvascular resistance. Therefore the anatomic-physiological discordance can be reduced by addition of downstream myocardial mass to anatomical stenosis of supplying artery (Fig. 31.6). Based on the fluid continuity principle, functional severity of stenosis was shown to increase proportionally to the ratio of flow demand represented by subtended myocardial mass to flow supply represented by luminal area or diameter of supplying vessel [48, 49]. Principle of efficiency or minimum energy loss concept is considered in the structure of human vascular tree and myocardial territory based on the fact that energy-efficient provision of materials such as oxygen in hierarchical fractal-like network of branching tubes plays a key role in the mechanism of living organism [50].

Two mathematical principles that have been used extensively in life science can be applied to calculate the relationship between vessel dimension and subtended myocardial mass (Table 31.5).

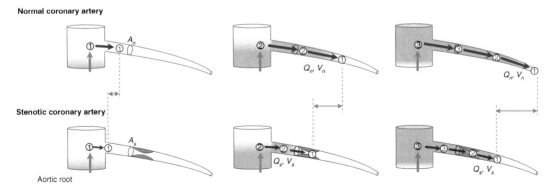

Fig. 31.5 Concept of transluminal arterial flow encoding (TAFE). Addition of the arterial input function (AIF, *yellow arrow*) adds time domain to TAG (=ΔHU/arterial length) and enables calculation of vessel-specific coronary blood flow. Compared to normal artery, the flow of stenotic artery is slower and has lower flow rate. Numbers in *circle* represent time points

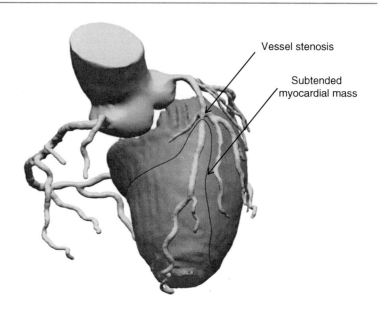

Fig. 31.6 Concept of myocardial mass subtended by the vessel

Table 31.5 Mathematical principles for the relationship between vessel dimension and subtended myocardial mass

Methods	Principle	Mathematics
Voronoi tessellation	Geometry and mathematics	$R_k = \{x \in X / d(x, P_k) \leq d(x, P_j)$ for all $j \neq k\}$
Allometric scaling law	Hypothesis of logarithmic correlations in life science	$Y = K\, X_b^a$ or $\log(Y) = a\, \log(X) + \log K$

Voronoi tessellation is based on the geometrical characteristics of vessel course and myocardial geometry. Allometric scaling law is a simple and universally observed logarithmic relationship among size, function, and energy expenditure in life science [26]. Stem-and-crown models describing scaling power between structures and functions were developed theoretically and validated experimentally in both animal and human studies [51, 52]. In clinical study, both Voronoi- and allometric scaling law-based study showed similar findings for the relation between vessel size and subtended myocardial mass (Table 31.6) [53, 54].

The concept of myocardial mass subtended by specific coronary artery can be extended beyond vessel-specific ischemia and may lead to better diagnostic and therapeutic decision in cardiovascular medicine including the following clinical issues. It might be used for adjudicating myocardial infarction caused by supply and demand mismatch (type 2) [55]. It also may clarify the appropriateness and optimal threshold of revascularization. Direct assessment of the amount of ischemic myocardium as well as myocardium to be revascularized has been estimated semi-quantitatively by angiographic scoring systems. As the FFR could

Table 31.6 Results of vessel dimension to subtended myocardial mass for detection of functionally significant stenosis

Study	Technology	N of vessels	Index parameter	Reference	Sensitivity (%)	Specificity (%)	PPV (%)	NPV (%)	Accuracy (%)	Area under curve, per vessel	Correlation coefficient compared with FFR
Kim et al. [53]	Allometric scaling law	724	Myocardial mass, angiographic minimal luminal diameter	FFR ≤ 0.80	78	72	75	75	75	0.84	0.61
Kang et al. [54]	Voronoi tessellation	103	Myocardial mass, IVUS minimal luminal area	FFR ≤ 0.80	88	90	86	92	90	0.94	0.78

Per-vessel data is shown. *PPV* positive predictive value, *NPV* negative predictive value

reclassify the need of revascularization based on the presence of ischemia, myocardial mass subtended by specific vessel might reclassify the strategy of revascularization based on the amount of ischemic myocardium to be saved [56–59]. The concept of vessel-specific myocardial mass explains the limited clinical benefit of bifurcation side branch and chronic total occlusion (CTO) revascularization [60], because both side branch of bifurcation and CTO vessel supply smaller or infarcted myocardial mass [61–64].

31.6 Limitations

The most important limitation of non-invasive physiological assessment is radiation exposure required by CT image, especially in perfusion CT imaging. A combined rest and stress myocardial perfusion CT may reach radiation dose of >15 mSv. Although the radiation exposure of CT is regarded as lower than those with nuclear imaging, appropriate radiation reducing strategy should be applied as reasonable as possible (Fig. 31.7).

Insufficient spatial and temporal resolution is the major cause of inadequate results. Typical isotropic spatial resolution of CT image is 0.5 mm at best. Therefore even single voxel difference in 3.0 mm sized vessel results in 17% difference in diameter. Such vessel with 50% diameter stenosis would have just 7–9 voxels in the lumen. Addition or deletion of single voxel causes 33% difference in minimal luminal diameter or 11% difference in minimal luminal area (Fig. 31.8). Mathematical correction by subvoxel resolution technique and avoidance of partial volume effect is being developed.

Mismatch of perfusion defect and stenotic or non-stenotic coronary artery may occur as cardiac positron emission tomography (PET) and coronary CT. Concept of vessel-specific myocardial territory rather than traditional 17-segment model may reduce misregistration error [53, 54, 65].

Boundary conditions in computational flow dynamics are critical in the result of computational FFR but include several assumed parameters which cannot be determined from conventional coronary CT. The individual variation of blood pressure, heart rate, coronary flow reserve, extent of collateral flow may explain the discrepancy between computational FFR and invasively acquired FFR (correlation coefficient $r = 0.72$ in DISCOVER-FLOW study) [31]. The time to calculation and

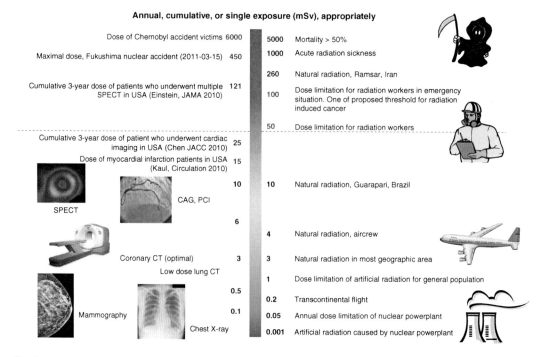

Fig. 31.7 Radiation exposure

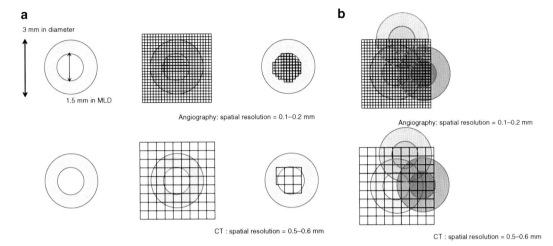

Fig. 31.8 Limitations in spatial resolution and motion artifact. (**a**) Less than 10 voxels consists lumen of typical coronary artery disease with 3.0 mm diameter and 50% stenosis. Omission or addition of single voxel affects significantly the result of computational FFR. (**b**) The limitation of spatial resolution may be worsen by the motion artifact

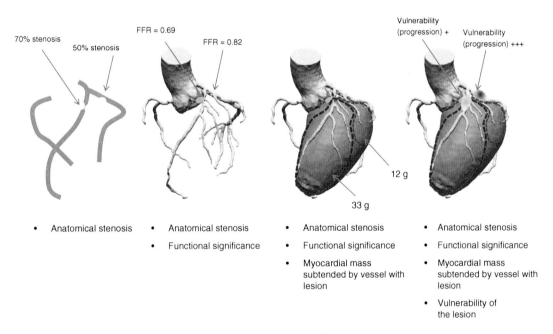

Fig. 31.9 Multifactorial non-invasive evaluation of coronary artery disease

heavy computational resource is another limitation of computational FFR but may be overcome by big-data based machine learning [66].

Single measurement or modality may represent but cannot show every aspect of coronary artery disease and is not sufficient for decision of treatment strategy. Revascularization by percutaneous coronary intervention or bypass surgery relieves symptom but does not improve clinical outcome of all patients [67]. Non-invasive physiological assessment may vastly improve the predictive value of coronary artery disease evaluation and be additive to the current decision-making strategy (Fig. 31.9).

References

1. Blankstein R, Di Carli MF. Integration of coronary anatomy and myocardial perfusion imaging. Nat Rev Cardiol. 2010;7:226–36
2. Toth G, Hamilos M, Pyxaras S, et al. Evolving concepts of angiogram: fractional flow reserve discordances in 4000 coronary stenoses. Eur Heart J. 2014;35:2831–8
3. Yang DH, Kim YH, Roh JH, et al. Stress myocardial perfusion CT in patients suspected of having coronary artery disease: visual and quantitative analysis-validation by using fractional flow reserve. Radiology. 2015;276:715
4. Greenwood JP, Maredia N, Younger JF, et al. Cardiovascular magnetic resonance and single-photon emission computed tomography for diagnosis of coronary heart disease (CE-MARC): a prospective trial. Lancet. 2012;379:453–60
5. Takx RA, Blomberg BA, El Aidi H, et al. Diagnostic accuracy of stress myocardial perfusion imaging compared to invasive coronary angiography with fractional flow reserve meta-analysis. Circ Cardiovasc Imaging. 2015;8:e002666
6. Williams MC, Newby DE. CT myocardial perfusion imaging: current status and future directions. Clin Radiol. 2016;71:739–49
7. Sorgaard MH, Kofoed KF, Linde JJ, et al. Diagnostic accuracy of static CT perfusion for the detection of myocardial ischemia. A systematic review and meta-analysis. J Cardiovasc Comput Tomogr. 2016;10:450
8. Rocha-Filho JA, Blankstein R, Shturman LD, et al. Incremental value of adenosine-induced stress myocardial perfusion imaging with dual-source CT at cardiac CT angiography. Radiology. 2010;254:410–9
9. Feuchtner G, Goetti R, Plass A, et al. Adenosine stress high-pitch 128-slice dual-source myocardial computed tomography perfusion for imaging of reversible myocardial ischemia: comparison with magnetic resonance imaging. Circ Cardiovasc Imaging. 2011;4:540–9
10. Nasis A, Ko BS, Leung MC, et al. Diagnostic accuracy of combined coronary angiography and adenosine stress myocardial perfusion imaging using 320-detector computed tomography: pilot study. Eur Radiol. 2013;23:1812–21
11. Carrascosa PM, Deviggiano A, Capunay C, et al. Incremental value of myocardial perfusion over coronary angiography by spectral computed tomography in patients with intermediate to high likelihood of coronary artery disease. Eur J Radiol. 2015;84:637–42
12. Magalhaes TA, Kishi S, George RT, et al. Combined coronary angiography and myocardial perfusion by computed tomography in the identification of flow-limiting stenosis - the CORE320 study: an integrated analysis of CT coronary angiography and myocardial perfusion. J Cardiovasc Comput Tomogr. 2015;9:438–45
13. Huber AM, Leber V, Gramer BM, et al. Myocardium: dynamic versus single-shot CT perfusion imaging. Radiology. 2013;269:378–86
14. Rossi A, Uitterdijk A, Dijkshoorn M, et al. Quantification of myocardial blood flow by adenosine-stress CT perfusion imaging in pigs during various degrees of stenosis correlates well with coronary artery blood flow and fractional flow reserve. Eur Heart J Cardiovasc Imaging. 2013;14:331–8
15. Bamberg F, Becker A, Schwarz F, et al. Detection of hemodynamically significant coronary artery stenosis: incremental diagnostic value of dynamic CT-based myocardial perfusion imaging. Radiology. 2011;260:689–98
16. Bamberg F, Marcus RP, Becker A, et al. Dynamic myocardial CT perfusion imaging for evaluation of myocardial ischemia as determined by MR imaging. JACC Cardiovasc Imaging. 2014;7:267–77
17. Ko BS, Cameron JD, Leung M, et al. Combined CT coronary angiography and stress myocardial perfusion imaging for hemodynamically significant stenoses in patients with suspected coronary artery disease: a comparison with fractional flow reserve. JACC Cardiovasc Imaging. 2012;5:1097–111
18. Wang Y, Qin L, Shi X, et al. Adenosine-stress dynamic myocardial perfusion imaging with second-generation dual-source CT: comparison with conventional catheter coronary angiography and SPECT nuclear myocardial perfusion imaging. AJR Am J Roentgenol. 2012;198:521–9
19. Kim SM, Chang SA, Shin W, Choe YH. Dual-energy CT perfusion during pharmacologic stress for the assessment of myocardial perfusion defects using a second-generation dual-source CT: a comparison with cardiac magnetic resonance imaging. J Comput Assist Tomogr. 2014;38:44–52
20. Wichmann JL, Meinel FG, Schoepf UJ, et al. Semiautomated global quantification of left ventricular myocardial perfusion at stress dynamic CT: diagnostic accuracy for detection of territorial myocardial perfusion deficits compared to visual assessment. Acad Radiol. 2016;23:429–37
21. Gould KL, Johnson NP, Bateman TM, et al. Anatomic versus physiologic assessment of coronary artery disease. Role of coronary flow reserve, fractional flow reserve, and positron emission tomography imaging in revascularization decision-making. J Am Coll Cardiol. 2013;62:1639–53
22. Min JK, Taylor CA, Achenbach S, et al. Noninvasive fractional flow reserve derived from coronary CT angiography: clinical data and scientific principles. JACC Cardiovasc Imaging. 2015;8:1209–22
23. Choy JS, Kassab GS. Scaling of myocardial mass to flow and morphometry of coronary arteries. J Appl Physiol. 2008;104:1281–6
24. Kassab GS. Scaling laws of vascular trees: of form and function. Am J Physiol Heart Circ Phys. 2006;290:H894–903
25. Lindstedt SL, Schaeffer PJ. Use of allometry in predicting anatomical and physiological parameters of mammals. Lab Anim. 2002;36:1–19
26. West GB, Brown JH, Enquist BJ. A general model for the origin of allometric scaling laws in biology. Science. 1997;276:122–6

27. Murray CD. The physiological principle of minimum work: I. The vascular system and the cost of blood volume. Proc Natl Acad Sci U S A. 1926;12:207–14
28. Kamiya A, Togawa T. Adaptive regulation of wall shear stress to flow change in the canine carotid artery. Am J Phys. 1980;239:H14–21
29. Wischgoll T, Choy JS, Kassab GS. Extraction of morphometry and branching angles of porcine coronary arterial tree from CT images. Am J Physiol Heart Circ Physiol. 2009;297:H1949–55
30. Wilson RF, Wyche K, Christensen BV, Zimmer S, Laxson DD. Effects of adenosine on human coronary arterial circulation. Circulation. 1990;82:1595–606
31. Koo BK, Erglis A, Doh JH, et al. Diagnosis of ischemia-causing coronary stenoses by noninvasive fractional flow reserve computed from coronary computed tomographic angiograms. Results from the prospective multicenter DISCOVER-FLOW (diagnosis of ischemia-causing Stenoses obtained via noninvasive fractional flow reserve) study. J Am Coll Cardiol. 2011;58:1989–97
32. Min JK, Leipsic J, Pencina MJ, et al. Diagnostic accuracy of fractional flow reserve from anatomic CT angiography. JAMA. 2012;308:1237–45
33. Norgaard BL, Leipsic J, Gaur S, et al. Diagnostic performance of noninvasive fractional flow reserve derived from coronary computed tomography angiography in suspected coronary artery disease: the NXT trial (analysis of coronary blood flow using CT angiography: next steps). J Am Coll Cardiol. 2014;63:1145–55
34. Wu W, Pan DR, Foin N, et al. Noninvasive fractional flow reserve derived from coronary computed tomography angiography for identification of ischemic lesions: a systematic review and meta-analysis. Sci Rep. 2016;6:29409
35. Douglas PS, De Bruyne B, Pontone G, et al. 1-year outcomes of FFRCT-guided care in patients with suspected coronary disease: the PLATFORM study. J Am Coll Cardiol. 2016;68:435–45
36. Choi JH, Min JK, Labounty TM, et al. Intracoronary transluminal attenuation gradient in coronary CT angiography for determining coronary artery stenosis. JACC Cardiovasc Imaging. 2011;4:1149–57
37. Choi JH, Kim EK, Kim SM, et al. Noninvasive evaluation of coronary collateral arterial flow by coronary computed tomographic angiography. Circ Cardiovasc Imaging. 2014;7:482–90
38. Stuijfzand WJ, Danad I, Raijmakers PG, et al. Additional value of transluminal attenuation gradient in CT angiography to predict hemodynamic significance of coronary artery stenosis. JACC Cardiovasc Imaging. 2014;7:374–86
39. Wong DT, Ko BS, Cameron JD, et al. Comparison of diagnostic accuracy of combined assessment using adenosine stress computed tomography perfusion + computed tomography angiography with transluminal attenuation gradient + computed tomography angiography against invasive fractional flow reserve. J Am Coll Cardiol. 2014;63:1904–12
40. Steigner ML, Mitsouras D, Whitmore AG, et al. Iodinated contrast opacification gradients in normal coronary arteries imaged with prospectively ECG-gated single heart beat 320-detector row computed tomography. Circ Cardiovasc Imaging. 2010;3:179–86
41. Yoon YE, Choi JH, Kim JH, et al. Noninvasive diagnosis of ischemia-causing coronary stenosis using CT angiography: diagnostic value of transluminal attenuation gradient and fractional flow reserve computed from coronary CT angiography compared to invasively measured fractional flow reserve. JACC Cardiovasc Imaging. 2012;5:1088–96
42. Choi JH, Koo BK, Yoon YE, et al. Diagnostic performance of intracoronary gradient-based methods by coronary computed tomography angiography for the evaluation of physiologically significant coronary artery stenoses: a validation study with fractional flow reserve. Eur Heart J Cardiovasc Imaging. 2012;13:1001–7
43. Chow BJ, Kass M, Gagne O, et al. Can differences in corrected coronary opacification measured with computed tomography predict resting coronary artery flow? J Am Coll Cardiol. 2011;57:1280–8
44. Gewirtz H, Dilsizian V. Integration of quantitative positron emission tomography absolute myocardial blood flow measurements in the clinical management of coronary artery disease. Circulation. 2016;133:2180–96
45. Bovenschulte H, Krug B, Schneider T, et al. CT coronary angiography: coronary CT-flow quantification supplements morphological stenosis analysis. Eur J Radiol. 2013;82:608–16
46. Ko BS, Wong DT, Norgaard BL, et al. Diagnostic performance of transluminal attenuation gradient and noninvasive fractional flow reserve derived from 320-detector row CT angiography to diagnose he modynamically significant coronary stenosis: an NXT substudy. Radiology. 2016;279:75–83
47. Lardo AC, Rahsepar AA, Seo JH, et al. Estimating coronary blood flow using CT transluminal attenuation flow encoding: formulation, preclinical validation, and clinical feasibility. J Cardiovasc Comput Tomogr. 2015;9:559–66. e1
48. Anderson HV, Stokes MJ, Leon M, Abu-Halawa SA, Stuart Y, Kirkeeide RL. Coronary artery flow velocity is related to lumen area and regional left ventricular mass. Circulation. 2000;102:48–54
49. Leone AM, De Caterina AR, Basile E, et al. Influence of the amount of myocardium subtended by a stenosis on fractional flow reserve. Circ Cardiovasc Interv. 2013;6:29–36
50. Dewey FE, Rosenthal D, Murphy DJ Jr, Froelicher VF, Ashley EA. Does size matter? Clinical applications of scaling cardiac size and function for body size. Circulation. 2008;117:2279–87
51. Seiler C, Kirkeeide RL, Gould KL. Measurement from arteriograms of regional myocardial bed size distal to any point in the coronary vascular tree for assessing anatomic area at risk. J Am Coll Cardiol. 1993;21:783–97

52. Huo Y, Kassab GS. Intraspecific scaling laws of vascular trees. J R Soc Interface. 2012;9:190–200
53. Kim HY, Lim HS, Doh JH, et al. Physiological severity of coronary artery stenosis depends on the amount of myocardial mass subtended by the coronary artery. JACC Cardiovasc Interv. 2016;9:1548–60
54. Kang SJ, Kweon J, Yang DH, et al. Mathematically derived criteria for detecting functionally significant stenoses using coronary computed tomographic angiography-based myocardial segmentation and intravascular ultrasound-measured minimal lumen area. Am J Cardiol. 2016;118:170–6
55. Sandoval Y, Smith SW, Thordsen SE, Apple FS. Supply/demand type 2 myocardial infarction: should we be paying more attention? J Am Coll Cardiol. 2014;63:2079–87
56. Gibbons RJ, Miller TD. Should extensive myocardial ischaemia prompt revascularization to improve outcomes in chronic coronary artery disease? Eur Heart J. 2015;36:2281–7
57. De Bruyne B, Fearon WF, Pijls NH, et al. Fractional flow reserve-guided PCI for stable coronary artery disease. N Engl J Med. 2014;371:1208–17
58. Johnson NP, Toth GG, Lai D, et al. Prognostic value of fractional flow reserve: linking physiologic severity to clinical outcomes. J Am Coll Cardiol. 2014;64:1641–54
59. Van Belle E, Rioufol G, Pouillot C, et al. Outcome impact of coronary revascularization strategy reclassification with fractional flow reserve at time of diagnostic angiography: insights from a large French multicenter fractional flow reserve registry. Circulation. 2014;129:173–85
60. Christakopoulos GE, Christopoulos G, Carlino M, et al. Meta-analysis of clinical outcomes of patients who underwent percutaneous coronary interventions for chronic total occlusions. Am J Cardiol. 2015;115:1367–75
61. Hachamovitch R, Hayes SW, Friedman JD, Cohen I, Berman DS. Comparison of the short-term survival benefit associated with revascularization compared with medical therapy in patients with no prior coronary artery disease undergoing stress myocardial perfusion single photon emission computed tomography. Circulation. 2003;107:2900–7
62. Chen SL, Ye F, Zhang JJ, et al. Randomized comparison of FFR-guided and angiography-guided provisional stenting of true coronary bifurcation lesions: the DKCRUSH-VI trial (double kissing crush versus provisional stenting technique for treatment of coronary bifurcation lesions VI). JACC Cardiovasc Interv. 2015;8:536–46
63. Ladwiniec A, Cunnington MS, Rossington J, et al. Collateral donor artery physiology and the influence of a chronic total occlusion on fractional flow reserve. Circ Cardiovasc Interv. 2015;8:e002219
64. Choi JH, Chang SA, Choi JO, et al. Frequency of myocardial infarction and its relationship to angiographic collateral flow in territories supplied by chronically occluded coronary arteries. Circulation. 2013;127:703–9
65. Ortiz-Perez JT, Rodriguez J, Meyers SN, Lee DC, Davidson C, Wu E. Correspondence between the 17-segment model and coronary arterial anatomy using contrast-enhanced cardiac magnetic resonance imaging. JACC Cardiovasc Imaging. 2008;1:282–93
66. Itu L, Rapaka S, Passerini T, et al. A machine-learning approach for computation of fractional flow reserve from coronary computed tomography. J Appl Physiol. 2016;121:42–52
67. Fokkema ML, James SK, Albertsson P, et al. Population trends in percutaneous coronary intervention: 20-year results from the SCAAR (Swedish coronary angiography and angioplasty registry). J Am Coll Cardiol. 2013;61:1222–30

Index

A
Accordion effect, 237, 238
Acoustic shadowing, 29
ACS. *See* Acute coronary syndrome (ACS)
Acute coronary syndrome (ACS), 29, 146, 151, 152, 279, 280, 283–286
Ambiguous lesions, 145–147, 260, 263
Angiographic aneurysms, 32, 33
Artifacts
    air bubble, 14, 15
    blood speckles, 13, 14
    ghost, 12, 14
    NURD, 9, 12
    post-acoustic shadowing, 9, 10
    reverberations, 11, 13
    ring-down artifacts, 9, 11
    sawtooth, 15
    side lobes, 10, 12
Asymmetry index (AI), 63, 64
Attenuated plaque, 31, 54

B
Bifurcated lesions, 41, 53, 139, 140, 269
Bioresorbable vascular scaffold (BVS), 177–185

C
Calcified nodule, 32
Calcified plaque, 28, 29
Calcium measurements, 20, 22, 23
Cardiac allograft vasculopathy (CAV), 298
Catheter-induced pressure damping, 215
Chemogram, 88
Comet Pressure Guidewire, 213
Coronary anatomy
    and physiologic stenosis, 250
    prognosis, 251
    revascularization, 251
Coronary arterial circulation
    functional anatomy, 203, 204
    physiologic characteristics, 203–206
    regulation and resistance, 203, 204

Coronary autoregulation, 206–208
Coronary collateral flow index (CFI), 309
Coronary flow reserve (CFR), 207, 208, 307
    Doppler wire technique, 304
    epicardial stenosis, 241, 242
    thermodilution technique, 305
Coronary flow velocity (CFV) patterns, 307, 308
Coronary plaque rupture, 193
Coronary spasm, 237
Coronary wedge pressure, 306, 309
Corrected contrast opacification (CCO), 318
Corrected TIMI frame count (CTFC), 307

D
Damping pressure, 236
Diffuse long lesion, 41–42

E
Eccentricity index (EI), 63, 64
Echolucent plaque. *See* Soft plaque
External elastic membrane (EEM), 115, 259
    diameter, 116
    measurements, 20–22

F
Fibrous plaque, 28, 125, 128
Fractional flow reserve (FFR), 116, 208–211, 260–266, 279, 288, 289, 314, 316–318
    accordion effect, 237, 238
    ACS (*see* Acute coronary syndrome (ACS))
    animal validation, 223
    basic setting, 233, 234
    bifurcation study, 230
    biological variability, 227
    calibrating pressure system, 214
    catheter and, 234
    coronary spasm, 237
    cutoff value, 224–227
    damping pressure during pullback, 235, 236
    definition, 213

Fractional flow reserve (FFR) (cont.)
    equalization of pressure wire, 235
    human validation, 223, 224
    hyperemia, 238, 239
    in-stent restenosis, 297
    intermediate lesion
        ACCF/AHA/SCAI guidelines, 266
        anatomical-functional mismatch, 263
        angiographic morphology and, 260, 261
        arterial narrowing, 260, 262
        discrepancies, 260, 262
        ESC/EACTS guidelines, 266
        significance, 261, 262
    intracoronary adenosine, 217, 218
    intracoronary nicorandil, 218
    intravenous adenosine, 216, 217
    maximal hyperemia, 287, 288
    multivessel disease, 275
    myocardial bridge, 293–296
    myocardial ischemia
        clinical results, 316–318
        computational approach, 316
        definiton, 314
        Morphometry laws, 316
        zero-dimensional/lumped parameter
            model, 314
    PAD
        maximal hyperemia, 289
        SFA lesions, 288
    post-heart transplantation, 298–300
    post-PCI, 293, 294
    pressure drift, 236
    pressure equalization, 215, 216
    pressure sensor positioning, 216
    pullback tracing, 220
    SB lesions, 269, 270
    serial lesions, 273–275
    signal drift, 221
    value determination, 219, 220
Frequency-domain optical coherence tomography
    (FD-OCT), 110–112
Functional SYNTAX score (FSS), 275

G
Ghost artifact, 12, 14

H
Half-moon phenomenon, 33
Hyperemic average peak velocity (hAPV), 307
Hyperemic microvascular resistance index (HMVRI),
    306, 309
Hyperemic stenosis resistance (HSR), 242, 243

I
Incomplete stent apposition (ISA), 66
Index of microcirculatory resistance (IMR), 242, 283, 298

    coronary wedge pressure, 308
    hyperemic mean transit time, 307
    microvascular function, 308
    thermodilution technique, 305, 307
    venous pressure, 308
Instantaneous wave-free ratio (iFR), 243–245, 255
In-stent neoatherosclerosis, 79, 81
In-stent restenosis (ISR), 42, 297
Intermediate coronary lesion, 259–266
    definition, 259
    FFR
        ACCF/AHA/SCAI guidelines, 266
        anatomical-functional mismatch, 260, 263
        angiographic morphology and, 260, 261
        arterial narrowing, 260, 262
        discrepancies, 260, 262
        ESC/EACTS guidelines, 266
        significance, 261, 262
Internal elastic membrane (IEM), 115
Intra-stent dissection, 157, 159, 160
Invasive coronary physiology, 203–206
    coronary arterial circulation
        functional anatomy, 203, 204
        physiologic characteristics, 203, 205, 206
        regulation and resistance, 203, 204
    coronary autoregulation, 206, 207
    coronary reserve, 206–208
    maximal perfusion, 208, 210
    stenosis pressure-flow relationship, 208, 209
IVUS-guided percutaneous coronary intervention (PCI),
    37–40, 42–45
    bifurcation lesion, 41
    chronic total occlusion, 41
    clinical benefit, 37
    diffuse long lesion, 41–42
    in-stent restenosis, 42
    left main lesion, 40
    patient with CKD, 42
    stent optimization, 45
        after post-stent adjuvant ballooning, 43, 44
        DES implantation predicting angiographic
            restenosis/MACE, 42, 43
        edge dissection and residual plaque, 44, 45
        underexpansion, 43

J
Jailed side branch, 230

K
Kissing balloon dilation, 230

L
Late stent malapposition
    clinical impact, 79
    definition, 77
    prognostic impact, 79

Late stent thrombosis, 158
Left main (LM) coronary artery stenosis, 269, 273
Left ventricular hypertrophy (LVH), 297, 298
Lesional assessment
    computational flow dynamics model, 134
    fibrocalcific plaques, 135
    fibrotic plaques, 135
    fractional flow reserve simulation, 134
    lipid-rich plaques, 135
    minimal lumen area, 134
    post-interventional myocardial infarction, 135, 136
Lipid core burden index (LCBI), 88, 89
Long-term complications, 75, 77–80
    late stent malapposition
        clinical impact, 79
        definition, 77
        prognostic impact, 79
    stent underexpansion, 75, 76
Lumen CSA, 116
Lumen diameter, 116
Lumen eccentricity, 116
Lumen measurements, 19, 20

M

Maximal coronary flow, 259
Maximal hyperemia, 216–218
maxLCBI$_{4mm}$, 88, 89
Mean minimum lumen diameter (MLD), 116
Micro-OCT (μOCT), 104
Microvasculature stunning, 280
Minimal lumen area (MLA), 252
Minimal stent area (MSA), 64, 65
Mixed plaque, 28, 29
Myocardial blood flow, 259, 260
Myocardial bridging (MB), 32, 33, 293, 295
Myocardial ischemia, 312–314, 316–318
    computational FFR
        clinical results, 316–318
        computational approach, 316
        definition, 314
        Morphometry laws, 316
        zero-dimensional/lumped parameter model, 314
Myocardial perfusion CT
    diagnostic performance, 312, 315
    principles, 312–314
    techniques, 312, 314
Myocardial perfusion scintigraphy, 262

N

Napkin-ring stenosis, 260
Near-infrared autofluorescence (NIRAF), 104
Near-infrared fluorescence (NIRF), 104, 193–195
Near-infrared spectroscopy (NIRS), 56, 89–92
    chemogram, 88
    clinical studies
        endothelial dysfunction, 92
        outcomes, 91, 92
        PCI guidance, 90
        periprocedural myocardial infarction, 89, 90
        plaque progression, 92
    intravascular imaging modalities, 85
    LCBI, 88, 89
    limitation, 93
    maxLCBI$_{4mm}$, 88, 89
    measurement, 88
    mechanism, 86, 87
    validation, 87
Neoatherosclerosis
    characteristics, 171
    definition, 170
    incidence, 170
    in-stent neoatherosclerosis, 171, 172
    prevalence, 172
Neointimal characteristics, 168–171
Neovascularization, 128

O

Optical coherence tomography (OCT)
    IVUS *vs.* FD-OCT, 107
    lesion severity assessment, 147–148
    3D
        ambiguous angiographic lesions, 189, 190
        coronary bifurcation lesions, 189–192
        coronary stent fracture, 192
        jailed side branch evaluation, 190, 191
        ultrahigh-speed, 192, 193
    time and frequency domain, 98, 99
    tissue microstructures, 97

P

Papaverine, 288, 289
Physical principles, 3, 4
    attenuation, 5
    beam geometry, 3, 4
    blood stasis, 5
    contrast resolution, 4
    image quality, 3
    piezoelectric crystal material, 3
    scattering, 5
    shorter and longer wavelengths, 3
    spatial resolution
        axial, 3
        lateral, 4
    transducer frequency, 3, 4
    ultrasound and body tissue interaction, 5
Plaque, 27–32
    characteristics, 70
    dissections, 126, 129
    erosion, 128, 130
    measurements, 20, 22
    morphology, 255, 256
        calcified, 28, 29

Plaque (cont.)
   fibrous, 28
   mixed, 28, 29
   soft, 28
   ulceration/rupture, 126, 130
   vulnerable
      attenuated, 31
      calcified nodule, 32
      rupture, 29, 30
      thrombus, 30, 31
Poiseuille's law, 241
Polarization-sensitive OCT (PS-OCT), 104, 196, 198

Q
Qualitative assessment, 125–129, 131
   atherosclerosis
      calcifications within plaques, 125, 126
      components, 125, 126
      fibrous plaques, 125, 128
      inflammatory cells, 129
      intracoronary thrombi, 126
      necrotic lipid pools, 125
      neovascularization/angioneogenesis, 128, 131
      red thrombi, 126, 129
      white thrombi, 126, 128, 129
   plaque
      composition and histology, 125, 127
      dissections, 126, 129
      erosion, 128, 130
      hemorrhage, 129
      ulceration/rupture, 126, 130
Quantitative coronary angiography (QCA), 259
Quantitative measurement
   stent measurements
      area, 119
      bioabsorbable vascular scaffolds, 119
      biological and clinical significance, 119
      malapposition, 122
      neointima area, 122
      post-percutaneous coronary intervention OCT, 122
      pre-percutaneous coronary intervention OCT, 122
      stent apposition, 122
      strut coverage thickness, 122

R
Resting coronary flow, 259, 260
Resting indices, 243
   iFR, 243–245
   resting Pd/Pa, 245
Resting Pd/Pa, 245
Reverse gradient, 234, 235

S
Serial contour plot analyses, 157, 159
Serial lesions, 273–275

Side branch (SB) ostial stenosis, 269, 270
Spectroscopic OCT (S-OCT), 194–196
Spontaneous coronary artery dissection, 33, 34
Stenosis pressure-flow relationship, 208, 209
Stent edge dissection, 66, 67
Stent fracture
   classification, 77
   complete and partial, 77
   definition, 75
Stent optimization
   DES implantation predicting angiographic restenosis/MACE, 42, 43
   edge dissection and residual plaque, 44, 45
   after post-stent adjuvant ballooning, 43, 44
   underexpansion, 43
Stent strut coverage
   application, 167
   contour plot images, 168
   follow-up strut coverage, 167
   malapposed and uncovered struts, 167, 168
   neointimal formation, 165
   uncovered stent struts, 166
   within incomplete endothelialization, 165, 166
Strut coverage thickness, 122
Superficial microcalcifications, 125
Symmetry index (SI), 63, 64

T
Thermodilution technique, 307
Three-dimensional optical coherence tomography (3D-OCT)
   ambiguous angiographic lesions, 189, 190
   coronary bifurcation lesions, 189–192
   coronary stent fracture, 192
   jailed side branch evaluation, 190, 191
   ultrahigh-speed, 192, 193
Thrombolysis In Myocardial Infarction (TIMI), 279
Thrombolysis in myocardial infarction (TIMI) frame count (TFC), 304, 306
Thrombolysis in myocardial infarction myocardial perfusion grade (TMPG), 303–306
Thrombus, 30, 31
Time gain compensation (TGC), 16
Tissue prolapse, 159, 162
Tissue protrusion (TP), 67–69
Transluminal attenuation flow encoding (TAFE), 319
Transluminal attenuation gradient (TAG), 318

U
Ultrahigh-speed OCT (UHS OCT), 192, 193

V
Vascular remodeling, 22, 24
Ventricularization, 215, 220
Virtual histology and intravascular ultrasound (VH-IVUS), 54

Vulnerable plaque
    attenuated, 31, 54
    calcified nodule, 32, 55
    informative color-coded tissue characterization
        technology, 55
    NIRS, 56
    no-reflow phenomenon/CK-MB elevation, 53
    rupture, 29, 30
    thrombus aspiration/distal protection device
        deployment, 55

W

Whipping artifact, 237, 238